Harmonic Analysis and Boundary Value Problems in the Complex Domain

Mkhitar M. Djrbashian

Springer Basel AG

Author

Mkhitar M. Djrbashian
Academy of Sciences of Armenia
Institute of Mathematics
Marshal Bagramian Ave. 24-B
Yerevan, 375019
Armenia

A CIP catalogue record for this book is available from the Library of Congress, Washington D.C., USA

Deutsche Bibliothek Cataloging-in-Publication Data
Djrbashian, Mkhitar M.:
Harmonic analysis and boundary value problems in the complex
domain / Mkhitar M. Djrbashian. - Basel ; Boston ; Berlin
Birkhäuser, 1993
  (Operator theory ; Vol. 65)
  ISBN 978-3-0348-9674-0   978-3-0348-8549-2 (eBook)
  DOI 10.1007/978-3-0348-8549-2
NE: GT

© 1993 Springer Basel AG
Originally published by Birkhäuser Verlag in 1993
Softcover reprint of the hardcover 1st edition 1993
Printed on acid-free paper produced from chlorine-free pulp
Cover design: Heinz Hiltbrunner, Basel

ISBN 978-3-0348-9674-0

9 8 7 6 5 4 3 2 1

*Dedicated
to the memory of my parents*

Table of Contents

Preface

As is well known, the first decades of this century were a period of elaboration of new methods in complex analysis. This elaboration had, in particular, one characteristic feature, consisting in the interfusion of some concepts and methods of harmonic and complex analyses. That interfusion turned out to have great advantages and gave rise to a vast number of significant results, of which we want to mention especially the classical results on the theory of Fourier series in $L_2(-\pi, \pi)$ and their continual analog – Plancherel's theorem on the Fourier transform in $L_2(-\infty, +\infty)$. We want to note also two important Wiener and Paley theorems on parametric integral representations of a subclass of entire functions of exponential type in the Hardy space H^2 over a half-plane.

Being under the strong influence of these results, the author began in the fifties a series of investigations in the theory of integral representations of analytic and entire functions as well as in the theory of harmonic analysis in the complex domain. These investigations were based on the remarkable properties of the asymptotics of the entire function

$$E_\rho(z; \mu) = \sum_{k=0}^{\infty} \frac{z^k}{\Gamma(\mu + k/\rho)} \qquad (\rho, \mu > 0),$$

which was introduced into mathematical analysis by Mittag-Leffler for the case $\mu = 1$. In the process of investigation, the scope of some classical results was essentially enlarged, and the results themselves were evaluated. Thus the author established far-reaching generalizations of Planchrel's theorem for the case of an arbitrary finite system of rays starting from the point $z = 0$ of the complex plane. These generalizations were further used to establish some essentially new Wiener-Paley type theorems for various classes of entire functions of finite order and for weighted H^2 classes of analytic functions over certain corner domains.

These and other results of the author's research were summarized in his monograph published in 1966. Here we present results in the theory of discrete generalized harmonic analysis and in the closely related theory of boundary value problems in the complex domain. Except for a small number of results obtained in publications preceding this monograph, and including solutions of certain particular problems, the results given here are essentially new in their totality and published for the first time.

The contents of the book may be described briefly as follows. Chapters 1 and 2 present some preliminaries without proofs from the above mentioned monograph of 1966. In Chapter 3 we establish a series of auxiliary estimates for certain weighted classes of entire functions of exponential type.

Further chapters of the book may be conditionally diveded into three stages.

At the first stage, in Chapters 4, 6 and 8, using the asymptotic properties of the function $E_\rho(z; \mu)$, we establish interpolation expansions in some Banach spaces of entire functions of orders $\rho = s - 1/2$ and $\rho = s$, where s is any natural number.

The second stage (Chapters 5, 7 and 9) is based mainly on Wiener-Paley type theorems for some Hilbert spaces of entire functions established in the monograph of 1966. The application of these theorems enables us to pass from the interpolation expansions mentioned above to an explicit construction of systems of vector functions (which may have both even or odd numbers of components of $E_\rho(z; \mu)$ type) forming biorthogonal Riesz bases.

At the third and concluding stage (Chapters 10, 11 and 12) of our investigation, it is established that the constructed basic biorthogonal systems of vector functions have very deep roots. Namely, they may be interpreted as systems of functions on a collection of intervals, arbitrary in number (either odd or even), which come out of the point $z = 0$ of the Riemann surface G^∞ of function $\mathrm{Ln}z$. In turn, these systems (like the classical Fourier systems) represent the systems of eigenfunctions and associated functions for certain non-ordinary boundary value problems formulated in terms of differential operators of fractional order on the suitable sets of intervals of the surface G^∞. The solution of these boundary value problems is conducted up to its logical completion, i.e. to the theorems on expansions in terms of the mentioned eigenfunctions and associated functions and also to the theorems on basis properties of these systems in the Riesz sense. Finally, a passage is made from the surface G^∞ to the complex plane $\mathbb{C}$, and, as a result, we obtain systems of entire functions similar to the Fourier systems and forming Riesz bases in the weighted classes L_2 over the suitable sets of intervals in the complex plane $\mathbb{C}$.

Notes to each of the chapters of the book contain necessary references. The results which are proved but have no references were obtained by the author directly in the process of writing the present book.

Certain specific features of this monograph should be noted. First, the harmonic relation between the contents of all its chapters. Therefore we strongly recommend that the potential reader not skip over any significant fragments of the book. Further, the monograph may be hardly considered as a textbook. It is not designed for easy reading and will require a certain mathematical background. Finally, one should stress the deep analytic character of the applied research methods.

I am very grateful to my pupil S. G. Raphaelian (Yerevan State University) for his contributions in the initial stage of the present investigations.

I should like to thank also V. M. Martirosian and A. O. Karapetian — my pupils and colleagues in the Institute of Mathematics, who read and checked the whole manuscript and improved it with their helpful suggestions.

The laborious work of the translation of the manuscript into English was carried out by my daughter H.M. Jerbashian and my son, A. M. Jerbashian, who was also greatly helpful for his valuable remarks.

I am very glad to express my gratitude to Prof. Israel Gohberg (University of Tel-Aviv), whose moral support and encouragement helped me to write this book.

Yerevan, July, 1992

MKHITAR M. DJRBASHIAN
Armenian Academy of Sciences
Institute of Mathematics

1 Preliminary results.
Integral transforms in the complex domain

1.1 Introduction

This chapter contains the most important properties and some of the fundamental applications of the classical Mittag-Leffler type function used in the construction of the theory of integral transforms in the complex domain. As it is well known, this function is defined as the sum of the power series

$$E_\rho(z;\mu) = \sum_{k=0}^{\infty} \frac{z^k}{\Gamma(\mu + k/\rho)} \qquad (\rho > 0).$$

It is an entire function of order ρ and of type 1, while the parameter μ is assumed to be any complex number in general. The theorems and lemmas of this chapter, the proofs of most of them are omitted, will mainly be used as a base for later chapters of the book. The reader may find these proofs in M.M.Djrbashian's monograph [5]. Other necessary references will be made in suitable places.

1.2 Some identities

(a) This group of formulas follows at once from the definition of the function $E_\rho(z;\mu)$:

$$E_{1/2}(z;1) = \cosh\sqrt{z}, \qquad E_{1/2}(z;2) = \frac{\sinh\sqrt{z}}{\sqrt{z}} \tag{1}$$

$$E_1(z;1) = e^z, \qquad E_1(z;2) = \frac{e^z - 1}{z}, \tag{2}$$

$$E_\rho(z;\mu) = \frac{1}{\Gamma(\mu)} + zE_\rho(z;\mu + 1/\rho), \tag{3}$$

$$E_\rho(z;\mu) = \mu E_\rho(z;\mu + 1) + \frac{z}{\rho}E'_\rho(z;\mu + 1). \tag{4}$$

Further, the termwise integration of the power expansion of the function $E_\rho(z;\mu)$ along the interval $(0, z)$ leads to the integral relation

$$\int_0^z E_\rho\left(\lambda t^{1/\rho};\mu\right) t^{\mu-1}\, dt = z^\mu E_\rho\left(\lambda z^{1/\rho};1 + \mu\right) \qquad (\mu > 0) \tag{5}$$

and its generalization

$$\frac{1}{\Gamma(\alpha)} \int_0^z (z - t)^{\alpha-1} E_\rho\left(\lambda t^{1/\rho};\mu\right) t^{\mu-1}\, dt$$

$$= z^{\mu+\alpha-1} E_\rho\left(\lambda z^{1/\rho};\mu + \alpha\right) \qquad (\mu > 0,\ \alpha > 0). \tag{6}$$

The following identities arise particularly from (6), when $\mu > 0$:

$$\frac{1}{\Gamma(\mu)} \int_0^z (z-t)^{\mu-1} \cosh \sqrt{\lambda} t \, dt = z^\mu E_{1/2}\left(\lambda z^2; 1+\mu\right), \tag{7}$$

$$\frac{1}{\Gamma(\mu)} \int_0^z (z-t)^{\mu-1} \frac{\sinh \sqrt{\lambda} t}{\sqrt{\lambda} t} \, dt = z^{1+\mu} E_{1/2}\left(\lambda z^2; 2+\mu\right), \tag{8}$$

$$\frac{1}{\Gamma(\mu)} \int_0^z (z-t)^{\mu-1} e^{t\lambda} \, dt = z^\mu E_1(\lambda z; 1+\mu). \tag{9}$$

In the same way, termwise integration leads to the equality

$$\int_0^\sigma E_\rho\left(zt^{1/\rho}; \alpha\right) t^{\alpha-1} E_\rho\left(\lambda(\sigma-t)^{1/\rho}; \beta\right) (\sigma-t)^{\beta-1} \, dt$$

$$= \frac{E_\rho\left(\sigma^{1/\rho} z; \alpha+\beta-1/\rho\right) - E_\rho\left(\sigma^{1/\rho}\lambda; \alpha+\beta-1/\rho\right)}{z-\lambda} \sigma^{\alpha+\beta-1/\rho-1} \tag{10}$$

$$= \frac{z E_\rho\left(\sigma^{1/\rho} z; \alpha+\beta\right) - \lambda E_\rho\left(\sigma^{1/\rho}\lambda; \alpha+\beta\right)}{z-\lambda} \sigma^{\alpha+\beta-1} \qquad (\alpha, \beta > 0),$$

which is true for any $\sigma > 0$ and, in general, for any complex values of the parameters z and λ. This formula remains true also when $\alpha, \beta \geq 0$.

(b) Four identities for linear combinations of Mittag-Leffler type functions will also be used in the later chapters. Suppose that $s \geq 0$ is an integer, and denote

$$\alpha_s = \exp\left[i\frac{2\pi}{2s+1}\right] \qquad (s \geq 0), \qquad \beta_s = \exp\left[i\frac{\pi}{s}\right] \qquad (s \geq 1). \tag{11}$$

Further, we shall write $k \equiv l(\mod n)$, where k, l and n are integers, if the number $k - l$ is a multiple of n. Then the following relations can easily be verified

$$\sum_{h=-s}^{s} \alpha_s^{kh} = \begin{cases} 2s+1 & \text{when } k \equiv 0(\mod 2s+1) \\ 0 & \text{when } k \not\equiv 0(\mod 2s+1) \end{cases} \qquad (s \geq 0), \tag{12}$$

$$\sum_{h=0}^{2s-1} \beta_s^{kh} = \begin{cases} 2s & \text{when } k \equiv 0(\mod 2s) \\ 0 & \text{when } k \not\equiv 0(\mod 2s) \end{cases} \qquad (s \geq 1). \tag{13}$$

Now, by the use of the power expansion of the function $E_\rho(z; \mu)$ and (12), the following identities can be obtained for any $s \geq 0$:

$$z^{s+j} E_\rho\left(z^{2s+1}; \mu + \frac{s+j}{(2s+1)\rho}\right)$$

$$= \frac{1}{2s+1} \sum_{h=-s}^{s} \alpha_s^{-(s+j)h} E_{(2s+1)\rho}(\alpha_s^{hz}; \mu) \qquad (-s \leq j \leq s), \tag{14}$$

$$E_{(2s+1)\rho}(z; \mu) = \sum_{j=-s}^{s} z^{s+j} E_\rho\left(z^{2s+1}; \mu + \frac{s+j}{(2s+1)\rho}\right).$$

Similarly, (13) leads to the following identities, which hold for any natural $s \geq 1$:

$$z^j E_\rho \left(z^{2s}; \mu + \frac{j}{2s\rho} \right) = \frac{1}{2s} \sum_{h=0}^{2s-1} \beta_s^{-jh} E_{2s\rho} \left(\beta_s^h z; \mu \right), \qquad 0 \leq j \leq 2s - 1,$$

$$E_{2s\rho}(z; \mu) = \sum_{j=0}^{2s-1} z^j E_\rho \left(z^{2s}; \mu + \frac{j}{2s\rho} \right). \tag{15}$$

(c) The application of identity (5) makes it possible to find the explicit solution of a special Volterra type integral equation.

Theorem 1.2. *If $f(x) \in L_1(0, l)$, $(0 < l < +\infty)$, then the integral equation*

$$u(x) = f(x) + \frac{\lambda}{\Gamma(1/\rho)} \int_0^x (x - t)^{1/\rho - 1} u(t)\, dt, \qquad x \in (0, l), \tag{16}$$

where $\rho > 0$ and λ is any complex parameter, has as its only solution

$$u(x) = f(x) + \lambda \int_0^x (x - t)^{1/\rho - 1} E_\rho \left(\lambda(x - t)^{1/\rho}; 1/\rho \right) f(t) dt, \qquad x \in (0, l) \tag{17}$$

in the class $L_1(0, l)$.

1.3 Integral representations and asymptotic formulas

(a) First we shall state some results relating to the function $E_{1/2}(z; \mu)$.

Theorem 1.3-1. *Let the parameter μ satisfy the condition*

$$0 \leq \mu < 3. \tag{1}$$

Then the function $E_{1/2}(z; \mu)$ may be represented as follows:
1°. If $|\arg z| < \pi$, then

$$E_{1/2}(z; \mu) = \frac{1}{2} z^{(1-\mu)/2} e^{z^{1/2}}$$

$$+ \frac{1}{2\pi} \int_0^{+\infty} \frac{\sin \left(\sqrt{t} + \frac{\pi}{2}(1 - \mu) \right)}{t + z} t^{(1-\mu)/2}\, dt \qquad (\mu \neq 0). \tag{2}$$

2°. If $0 \leq \arg z \leq \pi$ or $-\pi \leq \arg z \leq 0$, then, correspondingly,

$$E_{1/2}(z; \mu) = \frac{1}{2} z^{(1-\mu)/2} \left\{ e^{z^{1/2}} + e^{\mp i\pi(1-\mu)} e^{-z^{1/2}} \right\}$$

$$+ e^{\pm i\pi(1+\mu)/4} \frac{\sin \pi\mu}{2\pi} \int_0^{+\infty} \frac{\exp\left\{ -\frac{1 \mp i}{\sqrt{2}} t^{1/2} \right\}}{\pm it + z} t^{(1-\mu)/2}\, dt. \tag{3}$$

From this theorem follows

Theorem 1.3-2. *If condition (1) is satisfied, then the following asymptotic relations are true for any natural $n \geq 1$:*
1°. If $0 \leq \arg z \leq \pi$, or $-\pi \leq \arg z \leq 0$, then correspondingly

$$
\begin{aligned}
E_{1/2}(z;\mu) =& \frac{1}{2} z^{(1-\mu)/2} \left\{ e^{z^{1/2}} + e^{\mp i\pi(1-\mu)} e^{-z^{1/2}} \right\} \\
& - \sum_{k=1}^{n} \frac{z^{-k}}{\Gamma(\mu - 2k)} + O(|z|^{-n-1}), \qquad \text{as } z \to \infty \qquad (\mu \neq 0).
\end{aligned}
\tag{4}
$$

2°. If $x \in (0, +\infty)$, then

$$
\begin{aligned}
E_{1/2}(-x;\mu) =& x^{(1-\mu)/2} \cos\left(\sqrt{x} + \frac{\pi}{2}(1-\mu) \right) \\
& - \sum_{k=1}^{n} (-1)^k \frac{x^{-k}}{\Gamma(\mu - 2k)} + O(x^{-n-1}), \text{ as } x \to +\infty \qquad (\mu \neq 0).
\end{aligned}
\tag{5}
$$

Note that formulas (4) and (5) remain true when $\mu = 0$, but in this case all terms of the right-hand sides of these formulas, except the first terms, vanish.

(b) Now we shall pass to the similar results for the function $E_\rho(z;\mu)$, where $\rho > 1/2$. To this end it is necessary to introduce some notation. Let the contour $\Gamma(\alpha)$ $(0 < \alpha \leq \pi)$ be the sum of the rays $\arg \zeta = \pm\alpha$ running in the direction in which $\arg \zeta$ does not decrease. Obviously, $\Gamma(\alpha)$ is the common boundary of two mutually complementary corner domains

$$
\Delta_\alpha = \{\zeta : |\operatorname{Arg}\zeta| < \alpha\} \qquad \text{and} \qquad \Delta_\alpha^* = \{\zeta : |\pi - \operatorname{Arg}\zeta| < \pi - \alpha\} \tag{6}
$$

with corresponding openings 2α and $2(\pi - \alpha)$.

Theorem 1.3-3. *Let parameters ρ and μ satisfy the conditions*

$$
\rho > 1/2, \qquad -\infty < \mu < 1 + 1/\rho \tag{7}
$$

and let

$$
\pi/2\rho < \alpha < \min\{\pi; \pi/\rho\}. \tag{8}
$$

Then the following integral representations are true:

$$
E_\rho(z;\mu) = \frac{\rho}{2\pi i} \int_{\Gamma(\alpha)} \frac{e^{\zeta^\rho} \zeta^{\rho(1-\mu)}}{\zeta - z} \, d\zeta, \qquad z \in \Delta_\alpha^*, \tag{9}
$$

$$
E_\rho(z;\mu) = \rho z^{\rho(1-\mu)} e^{z^\rho} + \frac{\rho}{2\pi i} \int_{\Gamma(\alpha)} \frac{e^{\zeta^\rho} \zeta^{\rho(1-\mu)}}{\zeta - z} \, d\zeta, \qquad z \in \Delta_\alpha. \tag{10}
$$

Theorem 1.3-4. *If conditions (7) and (8) are satisfied, then the following asymptotic relations are true for any natural $n \geq 1$:*
1°. If $z \in \overline{\Delta_\alpha}$, then

$$E_\rho(z;\mu) = \rho z^{\rho(1-\mu)} e^{z^\rho} - \sum_{k=1}^{n} \frac{z^{-k}}{\Gamma(\mu - k/\rho)} + O(|\,z\,|^{-n-1}) \qquad as \qquad z \to \infty. \quad (11)$$

2°. If $z \in \overline{\Delta_\alpha^*}$, then

$$E_\rho(z;\mu) = -\sum_{k=1}^{n} \frac{z^{-k}}{\Gamma(\mu - k/\rho)} + O(|\,z\,|^{-n-1}) \qquad as \qquad z \to \infty. \qquad (12)$$

(c) Along with representations of Theorems 1.3-1 and 1.3-3, some other integral representations of Mittag-Leffler type functions are true.

Theorem 1.3-5. *If*

$$\rho > 1 \qquad and \qquad 0 < \mu < 1 + 1/\rho, \tag{13}$$

then the following formula is true:

$$\begin{aligned}
& E_\rho\left(-x^{1/\rho};\mu\right) x^{\mu-1} \\
& = \frac{1}{\pi} \int_0^{+\infty} \frac{\sin\pi(\mu - 1/\rho) + \tau^{1/\rho} \sin\pi\mu}{1 + 2\tau^{1/\rho}\cos\frac{\pi}{\rho} + \tau^{2/\rho}} \tau^{1/\rho-\mu} e^{-x\tau}\, d\tau, \qquad x \in (0, +\infty).
\end{aligned} \tag{14}$$

Theorem 1.3-6. *If*

$$1 < \rho < +\infty \qquad and \qquad 1/\rho \leq \mu < +\infty, \tag{15}$$

then the function $E_\rho(z;\mu)$ is representable in the whole z-plane as a Laplace integral:

$$E_\rho(-z;\mu) = \int_0^{+\infty} e^{-z\tau} \Phi_{\rho,\mu}(\tau)\, d\tau, \qquad |\,z\,| < +\infty. \tag{16}$$

Here

$$\Phi_{\rho,\mu}(s) = -\frac{1}{\pi} \sum_{k=0}^{\infty} (-1)^k \frac{\Gamma\left(1 - \mu + \frac{k+1}{\rho}\right)}{\Gamma(1 + k)} \sin\pi\left(\frac{k+1}{\rho} - \mu\right) s^k, \qquad |\,s\,| < +\infty \tag{17}$$

is an entire function of order $\lambda = \rho/(\rho-1) > 1$ and of type $\sigma = (1-\rho^{-1})\cdot\rho^{-1/(\rho-1)}$.
In addition,

$$\Phi_{\rho,\mu}(\tau) \geq 0, \qquad \tau \in [0, +\infty); \qquad \limsup_{\tau \to +\infty} \tau^{-\lambda} \log|\,\Phi_{\rho,\mu}(\tau)\,| \leq -\sigma\cos\frac{\pi}{2\rho}. \tag{18}$$

A more general result is contained in the next theorem.

Theorem 1.3-7. *Let ρ_j and μ_j $(j = 1, 2)$ be arbitrary parameters which satisfy the conditions*

$$0 < \rho_1 < \rho_2 < +\infty \qquad \text{and} \qquad 0 < \mu_j < +\infty \qquad (j = 1, 2). \tag{19}$$

Then the integral identity

$$E_{\rho_2}(z; \mu_2) = \int_0^{+\infty} E_{\rho_1}\left(z\tau^{1/\rho_1}; \mu_1\right) \tau^{\mu_1 - 1} \Phi_{\rho, \mu}(\tau)\, d\tau, \qquad |z| < +\infty \tag{20}$$

is true provided

$$\rho = \rho_2/\rho_1, \qquad \mu = \mu_2 + (1 - \mu_1)\rho_1/\rho_2 \tag{21}$$

in definition (17) of the function $\Phi_{\rho, \mu}(s)$. If, in addition,

$$\mu_2 \geq \frac{\rho_1}{\rho_2}\mu_1, \tag{22}$$

then the inequalities (18) are true again.

(d) As is well known, a function $\Phi(x)$, which is infinitely differentiable in $[0, +\infty)$ (or in $(0, +\infty)$), is called completely monotonic if

$$(-1)^n \Phi^{(n)}(x) \geq 0, \qquad \text{when} \qquad x \in [0, +\infty) \qquad (\text{or } x \in (0, +\infty)) \tag{23}$$

for every integer $n \geq 0$. The following assertion is an immediate consequence of Theorems 1.3-5 and 1.3-6.

Theorem 1.3-8. *If $\rho \geq 1$ and $1/\rho \leq \mu < +\infty$ (or $\rho \geq 1$ and $1/\rho \leq \mu \leq 1$), then the function*

$$E_\rho(-x; \mu) \qquad \left(\text{or } E_\rho\left(-x^{1/\rho}; \mu\right) x^{1/\rho - 1}\right) \tag{24}$$

is completely monotonic in $[0, +\infty)$ (or in $(0, +\infty)$).

1.4 Distribution of zeros

(a) Obviously the entire function $E_{1/2}(z; \mu)$ of order $\rho = 1/2$ has an infinite set of zeros for any μ. Let $\{\gamma_k\}_1^\infty$ be this set enumerated in the order of non-decreasing modulus and according to multiplicities. Observe now that all zeros of the functions 1.2(1) are simple and are situated on the semi-axis $(-\infty, 0]$. As to the distribution of zeros of the function $E_{1/2}(z; \mu)$, the following assertion is established using Theorem 1.3-2.

Lemma 1.4-1. *If $0 < \mu < 3$, then:*
1°. All zeros γ_k of the function $E_{1/2}(z; \mu)$, which have sufficiently large modulus, are simple and belong to the semi-axis $(-\infty, 0]$.
2°. The asymptotic formula

$$\gamma_k = -(\pi k)^2 \left(1 + O(k^{-1})\right) \qquad \text{as } k \to +\infty \tag{1}$$

is true.

As to the distribution of zeros of the function $E_\rho(z;\mu)$ when $\rho > 1/2$, it appears that $E_\rho(z;\mu)$, in this case, always has an infinite set of zeros, except when $\rho = \mu = 1$ and $E_1(z;1) = e^z$.

It easily follows from the asymptotic formulas 1.3(11) and 1.3(12) of Theorem 1.3-1, that, if condition 1.3(7) is satisfied, then all zeros of the function $E_\rho(z;\mu)$, having sufficiently large modulus, must be situated inside the corner domains

$$D_\delta^\pm = \left\{ z : |\arg z \mp \frac{\pi}{2\rho}| < \delta \right\},$$

where $\delta > 0$ is any number from the interval $(0, \min\{\pi/2\rho; \pi - \pi/2\rho\})$. Therefore, $E_\rho(z;\mu)$ may have only a finite number of zeros on the real axis. The non-real zeros of the function $E_\rho(z;\mu)$, which are situated in the half-planes $G_\pm = \{z : \pm Imz > 0\}$, can be enumerated separately according to their multiplicities and in the order of increasing modulus. Let $\{\gamma_k^{(\pm)}\}_1^\infty \subset G_\pm$ be these sequences. Then, obviously, $\gamma_k^{(-)} = \overline{\gamma_k^{(+)}}$ $(1 \le k < +\infty)$ since $Im\mu = 0$, and the following assertion is also true.

Lemma 1.4-2. *If $\rho > 1/2$, $\rho \neq 1$ and $0 < \mu < 3$, then:*
1°. All zeros of the function $E_\rho(z;\mu)$, having sufficiently large modulus, are simple.
2°. The following asymptotic formula is true:

$$\gamma_k^{(\pm)} = e^{\pm i\pi/2\rho}(2\pi k)^{1/\rho} \left(1 + O(k^{-1}\log k)\right) \qquad \text{as } k \to +\infty \qquad (2)$$

(b) The use of one of G. Polya's elegant results allows us to state a result concerning the zeros of the function $E_{1/2}(z;\mu)$, which says considerably more than Lemma 1.4-1. The mentioned result can be formulated as follows.

Theorem 1.4-1. *If the function $f(t) \in L_1(0,1)$ is positive and increasing, then:*
1°. The zeros of the entire functions of exponential type

$$U(z) = \int_0^1 f(t)\cos(zt)dt, \ V(z) = \int_0^1 f(t)\sin(zt)dt \qquad (3)$$

are real and simple.
2°. $U(z)$ is an even function having no zeros in $[0, \pi/2)$, and its positive zeros are situated in the intervals $(\pi k - \pi/2, \pi k + \pi/2)$ $(1 \le k < +\infty)$, one in each. The odd function $V(z)$ has only one zero $z = 0$ in $[0, \pi)$, and its positive zeros are situated in the intervals $(\pi k, \pi(k+1))$ $(1 \le k < +\infty)$, one in each.

The application of this result to entire functions of exponential type $\sigma(0 < \sigma < +\infty)$

$$\mathcal{E}_\sigma(z;\nu) = E_{1/2}\left(-\sigma^2 z; 1 + \nu\right) \qquad (4)$$

leads to the following assertion.

Theorem 1.4-2. *1°. If*

$$0 \le \nu < 2, \tag{5}$$

then the zeros of $\mathcal{E}_\sigma\left(z^2; \nu\right)$ are simple and real and they are symmetric with respect to the point $z = 0$.

2°. If $0 < \nu < 1$, then all zeros of the function $\mathcal{E}_\sigma\left(z^2; \nu\right)$ are situated in the intervals

$$\Delta_k^{(1)} = \left(\frac{\pi}{\sigma}k - \frac{\pi}{2\sigma}, \frac{\pi}{\sigma}k + \frac{\pi}{2\sigma}\right), \quad k = \pm 1, \pm 2, \dots, \tag{6}$$

one in each, and if $\nu = 0$, then the zeros of this function are the endpoints of these intervals.

If $1 < \nu < 2$, then the zeros of $\mathcal{E}_\sigma\left(z^2; \nu\right)$ are situated in the intervals

$$\Delta_k^{(2)} = \begin{cases} \left(\frac{\pi}{\sigma}k, \frac{\pi}{\sigma}(k+1)\right), & 1 \le k < +\infty \\ \left(\frac{\pi}{\sigma}(k-1), \frac{\pi}{\sigma}k\right), & -\infty < k \le -1, \end{cases} \tag{7}$$

one in each, and if $\nu = 1$, then they are the endpoints of these intervals.

Proof. According to formulas 1.2(1) and (4),

$$\mathcal{E}_\sigma\left(z^2; 0\right) = \cos(\sigma z), \qquad \mathcal{E}_\sigma\left(z^2; 1\right) = \frac{\sin(\sigma z)}{\sigma z}. \tag{8}$$

Hence the desired statements follow for the cases $\nu = 0$ and $\nu = 1$. Note now that formula 1.2(6) may be written for $\rho = 1/2$ in the form

$$E_{1/2}\left(-\sigma^2 z^2; \mu + \alpha\right) = \frac{1}{\Gamma(\alpha)} \int_0^1 (1-t)^{\alpha-1} E_{1/2}\left(-\sigma^2 z^2 t^2; \mu\right) t^{\mu-1}\, dt \quad (\mu, \alpha > 0).$$

Here we first put $\mu = 1$, $\alpha = \nu$ $(0 < \nu < 1)$, and next $\mu = 2$, $\alpha = \nu - 1$ $(1 < \nu < 2)$, then, correspondingly, we obtain from definition (4) of the function $\mathcal{E}_\sigma(z; \nu)$ and formulas (8), that

$$\mathcal{E}_\sigma\left(z^2; \nu\right) = \frac{1}{\Gamma(\nu)} \int_0^1 (1-t)^{\nu-1} \cos(\sigma z t)\, dt \quad (0 < \nu < 1),$$
$$\sigma z \mathcal{E}_\sigma\left(z^2; \nu\right) = \frac{1}{\Gamma(\nu-1)} \int_0^1 (1-t)^{\nu-2} \sin(\sigma z t)\, dt \quad (1 < \nu < 2). \tag{9}$$

But both the functions $(1-t)^{\nu-1}(0 < \nu < 1)$ and $(1-t)^{\nu-2}(1 < \nu < 2)$ are increasing when $0 < t < 1$ and belong to $L_1(0,1)$. Therefore, all the desired assertions follow from the integral representations (9) and Theorem 1.4-1.

Finally, note that the first assertion of Theorem 1.4-2 is not true when $\nu = -1$ and $\nu = 2$. Indeed, it easily follows from the corresponding power expansions that all zeros of the functions

$$\mathcal{E}_\sigma\left(z^2; -1\right) = -\sigma z \sin(\sigma z), \qquad \mathcal{E}_\sigma\left(z^2; 2\right) = 2\frac{\sin^2 \frac{\sigma z}{2}}{\sigma^2 z^2}$$

are real, but are not simple. In connection with this and with the first assertion of Lemma $1.4 - 1, a$ question arises: whether a similar theorem is true when $\nu \in (-1, 0)$? This question remains as yet open.

(c) The following theorem is a consequence of Theorem 1.4-2.

Theorem 1.4-3. *1°. All zeros of the function $\mathcal{E}_\sigma(z;\nu)$ are simple and positive when $0 \le \nu < 2$.*
2°. The sequence of zeros $\{\lambda_k \equiv \lambda_k(\sigma,\nu)\}_1^\infty$ $(0 < \lambda_k < \lambda_{k+1},\ 1 \le k < +\infty)$ of the function $\mathcal{E}_\sigma(z;\nu)$ has the following distribution:
When $\nu = 0$ or $0 < \nu < 1$, then, correspondingly,

$$\lambda_k = \left(\frac{\pi}{\sigma}k - \frac{\pi}{2\sigma}\right)^2 \text{ or } \lambda_k \in \left(\left(\frac{\pi}{\sigma}k - \frac{\pi}{2\sigma}\right)^2, \left(\frac{\pi}{\sigma}k + \frac{\pi}{2\sigma}\right)^2\right) \quad (k \ge 1). \tag{10}$$

When $\nu = 1$ or $1 < \nu < 2$, then, correspondingly,

$$\lambda_k = \left(\frac{\pi}{\sigma}k\right)^2 \text{ or } \lambda_k \in \left(\left(\frac{\pi}{\sigma}k\right)^2, \left(\frac{\pi}{\sigma}k + \frac{\pi}{\sigma}\right)^2\right) \quad (k \ge 1). \tag{11}$$

3°. For arbitrary $\nu \in [0,2)$

$$\lambda_k \asymp (1 + k)^2 (1 \le k < +\infty). \tag{12}$$

Here and later on, the symbol $\asymp$ means that the quotient of the quantities, which are on the left and on the right-hand sides of it, varies between two positive constants.

We prove also the following theorem.

Theorem 1.4-4. *1°. If $0 < \nu < 1$ or $1 < \nu < 2$, then*

$$\sqrt{\lambda_k} = \frac{\pi}{\sigma}k + \frac{\pi}{2\sigma}(\nu - 1) + O(k^{\nu-2}), \qquad k \to +\infty, \tag{13}$$

and if $\nu = 0$ or $\nu = 1$, then, correspondingly,

$$\sqrt{\lambda_k} = \frac{\pi}{\sigma}k - \frac{\pi}{2\sigma} \quad \text{or} \quad \sqrt{\lambda_k} = \frac{\pi}{\sigma}k \qquad (1 \le k < +\infty). \tag{14}$$

Proof. The equalities (14) are already proved in Theorem 1.4-3. To prove the asymptotic formula (13), we put $x = \sigma^2\lambda_k$, $\mu = 1 + \nu$ $(0 \le \nu < 2)$ in 1.3(5) and arrive at the equalities

$$\mathcal{E}_\sigma(\lambda_k;\nu) = E_{1/2}(-\sigma^2\lambda_k; 1 + \nu)$$
$$= (\sigma^2\lambda_k)^{-\nu/2}\cos\left(\sigma\sqrt{\lambda_k} - \frac{\pi}{2}\nu\right) + O(\lambda_k^{-1}) = 0 \ (1 \le k < +\infty),$$

which, with two-sided estimates (12), imply

$$\cos\left(\sigma\sqrt{\lambda_k} - \frac{\pi}{2}\nu\right) = O(k^{\nu-2}) \quad \text{as} \quad k \to +\infty.$$

Hence the asymptotic formula (13) follows, if the distribution of zeros given by the relations (10) and (11) and condition $\nu - 2 < 0$ are taken into account.

1.5 Identities between some Mellin Transforms

(a) We state first two well-known theorems on Mellin transforms in $L_2(0, +\infty)$ and their inversions.

Theorem 1.5-1. Let $f(x) \in L_2(0, +\infty)$. Then:
1°. The functions

$$\mathcal{F}(s; a) = \int_{1/a}^{a} f(x) x^{s-1}\, dx \quad (s = 1/2 + it, -\infty < t < +\infty) \tag{1}$$

converge in mean on the line $s = 1/2 + it$ $(-\infty < t < +\infty)$ as $a \to +\infty$, i.e., there exists a function $\mathcal{F}(s) \in L_2(1/2 - i\infty, 1/2 + i\infty)$, such that

$$\lim_{a \to +\infty} \int_{1/2-i\infty}^{1/2+i\infty} |\mathcal{F}(s) - \mathcal{F}(s; a)|^2\, |ds| = 0. \tag{2}$$

2°. Conversely, the functions

$$f(x; a) = \frac{1}{2\pi i} \int_{1/2-ia}^{1/2+ia} \mathcal{F}(s) x^{-s}\, ds, \qquad x \in (0, +\infty) \tag{3}$$

converge in mean on the semi-axis $(0, +\infty)$ to the function $f(x)$, as $a \to +\infty$, i.e.,

$$\lim_{a \to +\infty} \int_{0}^{+\infty} |f(x) - f(x; a)|^2\, dx = 0. \tag{4}$$

In addition,

$$f(x) = \frac{1}{2\pi i} \frac{d}{dx} \int_{1/2-i\infty}^{1/2+i\infty} \mathcal{F}(s) \frac{x^{1-s}}{1-s}\, ds \tag{5}$$

almost everwhere in $(0, +\infty)$.
3°. The following equality is true:

$$\int_{0}^{+\infty} |f(x)|^2\, dx = \frac{1}{2\pi} \int_{-\infty}^{+\infty} \left| \mathcal{F}\left(\frac{1}{2} + it \right) \right|^2 dt. \tag{6}$$

Theorem 1.5-2. 1°. The functions $f(x; a)$, constructed by means of any $\mathcal{F}(s) \in L_2(1/2 - i\infty, 1/2 + i\infty)$, converge in the sense of (4) to a function $f(x) \in L_2(0, +\infty)$ which is representable in the form of (5). Further, the functions $\mathcal{F}(s; a)$ defined by (1) converge, in the sense of (2), to the function $\mathcal{F}(s)$, and equality (6) is true again.
2°. If the functions $f(x)$ and $g(x)$ belong to $L_2(0, +\infty)$, and $\mathcal{F}(s), G(s)$ are their Mellin transforms, then

$$\int_{0}^{+\infty} f(x) g(x)\, dx = \frac{1}{2\pi i} \int_{1/2-i\infty}^{1/2+i\infty} \mathcal{F}(s) G(1-s)\, ds. \tag{7}$$

(b) Along with the well-known formula of Mellin transform

$$\int_0^{+\infty} \frac{e^{\pm ix} - 1}{\pm ix} x^{s-1}\, dx = \frac{H^{(\pm)}(s)}{1-s}, \qquad 0 < Res < 1, \tag{8}$$

where

$$H^{(\pm)}(s) = e^{\pm i\frac{\pi}{2} s}\Gamma(s), \tag{9}$$

we note the following important assertion.

Lemma 1.5-1. *If the parameters ρ, μ and α satisfy the conditions*

$$\rho \geq \frac{1}{2}, \quad \frac{1}{2} < \mu < \frac{1}{2} + \frac{1}{\rho} \quad \text{and} \quad \frac{\pi}{2\rho} \leq \alpha \leq 2\pi - \frac{\pi}{2\rho}, \tag{10}$$

then:
1°. The following formula is true:

$$\int_0^{+\infty} \left\{ E_\rho\left(e^{i\alpha}x^{1/\rho}; 1 + \mu\right) x^{\mu - 1} \right\} x^{s-1}\, dx = \frac{K_\rho(s; \alpha; \mu)}{1-s}, \quad Res = \frac{1}{2}, \tag{11}$$

where the left-hand side integral is convergent in the ordinary sense and

$$K_\rho(s; \alpha; \mu) = \frac{\pi\rho}{\Gamma(1-s)} \frac{\exp\{i\rho(\pi - \alpha)(s + \mu - 1)\}}{\sin \pi\rho(s + \mu - 1)}. \tag{12}$$

2°.

$$\sup_{-\infty < t < +\infty} \left| K_\rho\left(\frac{1}{2} + it; \alpha; \mu\right) \right| < M_\mu < +\infty. \tag{13}$$

Formulas (8)–(9) and (11)–(12) allow us to establish the following lemmas, which contain three important identities between the functions $H^{(\pm)}(s)$ and $K_\rho(s; \alpha; \mu)$.

Lemma 1.5-2. *If $\rho \geq 1/2$ and $1/2 < \mu < 1/2 + 1/\rho$, then the following identities are true on the line $s = 1/2 + it$ $(-\infty < t < +\infty)$:*

$$e^{-i\pi(1-\mu)/2} K_\rho\left(s; \frac{\pi}{2\rho}; \mu\right) H^{(-)}(1-s)$$

$$+ e^{i\pi(1-\mu)/2} K_\rho\left(s; 2\pi - \frac{\pi}{2\rho}; \mu\right) H^{(+)}(1-s) \equiv 2\pi\rho, \tag{14}$$

$$e^{-i\pi(1-\mu)/2} K_\rho\left(s, \frac{\pi}{2\rho}; \mu\right) H^{(+)}(1-s)$$

$$+ e^{i\pi(1-\mu)/2} K_\rho\left(s; 2\pi - \frac{\pi}{2\rho}; \mu\right) H^{(-)}(1-s)$$

$$\equiv 2\pi\rho \frac{\sin\{\pi[(1-\rho)s + \rho(1-\mu)]\}}{\sin \pi\rho(s + \mu - 1)} \equiv \Phi_{\rho,\mu}(s). \tag{15}$$

Besides, the following estimates are true on the same line for the function $\Phi_{\rho,\mu}(s)$:

$$\sup_{-\infty<t<+\infty} e^{\pi|t|}\left|\Phi_{\rho,\mu}\left(\frac{1}{2}+it\right)\right| \le 2\pi\rho P_{\rho,\mu}, \quad \text{if} \quad \rho \ge 1,$$

$$\sup_{-\infty<t<+\infty} e^{\pi(2\rho-1)|t|}\left|\Phi_{\rho,\mu}\left(\frac{1}{2}+it\right)\right| \le 2\pi\rho P_{\rho,\mu}, \quad \text{if} \quad \frac{1}{2} \le \rho \le 1, \tag{16}$$

where the constant $P_{\rho,\mu} > 0$ *depends only on* ρ *and* μ.

Lemma 1.5-3. *If* $\rho \ge 1$, $1/2 < \mu < 1/2 + 1/\rho$ *and*

$$0 \le \alpha \le 2\pi(1 - 1/\rho), \tag{17}$$

then the following identity is true on the line $s = 1/2 + it$ $(-\infty < t < +\infty)$:

$$e^{i\pi(1-\mu)/2} K_\rho\left(s; \frac{\pi}{2\rho} + \alpha; \mu\right) H^{(+)}(1-s)$$

$$+ e^{-i\pi(1-\mu)/2} K_\rho\left(s; \frac{\pi}{2\rho} + \frac{\pi}{\rho} + \alpha; \mu\right) H^{(-)}(1-s) \equiv 0. \tag{18}$$

(c) Finally, note that the behavior of the function $\Phi_{\rho,\mu}(s)$ on the line $s = 1/2 + it$ $(-\infty < t < +\infty)$ is such that the inverse Mellin transform

$$\Psi_{\rho,\mu}(x) = \frac{1}{2\pi i} \int_{1/2-i\infty}^{1/2+i\infty} \frac{\Phi_{\rho,\mu}(s)}{s} x^{-s}\, ds, \quad x \in (0, +\infty) \tag{19}$$

also converges in the ordinary sense and may be calculated in an explicit form.

Lemma 1.5-4. 1°. *If* $\rho = 1/2$ *and* $1/2 < \mu < 5/2$, *then*

$$\Psi_{1/2,\mu}(x) = \begin{cases} -\pi\cos(\pi\mu) + 2\sin(\pi\mu)\left(\text{v.p.} \int_x^\infty \frac{t^{\mu-2}}{1-t^2}\, dt\right), & x \in (0,1), \\ 2\sin(\pi\mu)\int_x^\infty \frac{t^{\mu-2}}{1-t^2}\, dt, & x \in (1,+\infty). \end{cases} \tag{20}$$

2°. *If* $\rho > 1/2$ *and* $1/2 < \mu < 1/2 + 1/\rho$, *then*

$$\Psi_{\rho,\mu}(x) = 2\int_x^{+\infty} \frac{\sin\pi\mu + [\sin\pi(1/\rho - \mu)]t^{1/\rho}}{1 - 2t^{1/\rho}\cos\pi/\rho + t^{2/\rho}} t^{\mu-2} dt, \quad x \in (0,+\infty). \tag{21}$$

1.6 Fourier type transforms with Mittag-Leffler kernels

The identities of Lemmas 1.5-1, 1.5-2 and 1.5-3 make it possible to prove some general results in this section. The proofs are based on the classical Theorems 1.5-1 and 1.5-2 on direct and inverse Mellin transforms and, particularly, on the equality 1.5(7).

(a) As in Section 1.5, we shall assume everywhere that

$$\rho \geq \frac{1}{2} \quad \text{and} \quad \frac{1}{2} < \mu < \frac{1}{2} + \frac{1}{\rho}. \tag{1}$$

Further, denoting by $L_{2,\mu}(0, +\infty)$ the class of those complex valued functions $g(y)$ for which $g(y)y^{\mu-1} \in L_2(0, +\infty)$, we shall agree that the notation

$$g(y) = \mathop{\mathrm{l.\,i.\,m.}}\limits_{\sigma \to +\infty}^{(\mu)} g_\sigma(y) \tag{2}$$

means: $g(y) \in L_{2,\mu}(0, +\infty)$, and the family of functions $\{g_\sigma(y)\} \in L_{2,\mu}(0, +\infty)$ depending on parameter σ is such that

$$\lim_{\sigma \to +\infty} \int_0^{+\infty} |g(y) - g_\sigma(y)|^2 y^{2(\mu-1)} dy = 0. \tag{2'}$$

If $\mu = 1$, then, obviously, $L_{2,1}(0, +\infty) = L_2(0, +\infty)$, and convergence (2) is a convergence in the metric of $L_2(0, +\infty)$. In this case, instead of (2) we simply write $g(y) = \mathrm{l.\,i.\,m.}_{\sigma \to +\infty}\, g_\sigma(y)$.

(b) Theorem 1.6-1. *Let condition (1) be satisfied, also let* $g(y) \in L_{2,\mu}(0, +\infty)$ *be an arbitrary function, and let*

$$f^{(\pm)}(x; \sigma) = \frac{1}{\sqrt{2\pi\rho}} \int_0^\sigma e^{\pm ixy} g(y) y^{\mu-1} dy \qquad (\sigma > 0). \tag{3}$$

Then:
1°. There exist functions $f^{(+)}(x)$ *and* $f^{(-)}(x)$ *in* $L_2(0, +\infty)$*, such that*

$$f^{(\pm)}(x) = \mathop{\mathrm{l.\,i.\,m.}}\limits_{\sigma \to +\infty} f^{(\pm)}(x; \sigma). \tag{4}$$

Conversely, if we denote

$$g(y; \sigma) = \frac{1}{\sqrt{2\pi\rho}} \left\{ e^{-i\frac{\pi}{2}(1-\mu)} \int_0^\sigma E_\rho\left(e^{i\frac{\pi}{2\rho}} y^{\frac{1}{\rho}} x^{\frac{1}{\rho}}; \mu\right) x^{\mu-1} f^{(-)}(x) dx \right.$$

$$\left. + e^{i\frac{\pi}{2}(1-\mu)} \int_0^\sigma E_\rho\left(e^{-i\frac{\pi}{2\rho}} y^{\frac{1}{\rho}} x^{\frac{1}{\rho}}; \mu\right) x^{\mu-1} f^{(+)}(x) dx \right\} \qquad (\sigma > 0), \tag{5}$$

then

$$g(y) = \mathop{\mathrm{l.\,i.\,m.}}\limits_{\sigma \to +\infty}^{(\mu)} g(y; \sigma), \tag{6}$$

and the following Parseval type equality is true:

$$\int_0^{+\infty} |g(y)|^2 y^{2(\mu-1)} dy = \rho \left\{ \int_0^{+\infty} |f^{(+)}(x)|^2 dx + \int_0^{+\infty} |f^{(-)}(x)|^2 dx \right\}. \tag{7}$$

$2°$. The functions $g(y)$ and $f^{(\pm)}(x)$ are connected by the equalities

$$f^{(\pm)}(x) = \frac{1}{\sqrt{2\pi\rho}} \frac{d}{dx} \int_0^{+\infty} \frac{e^{\pm ixy} - 1}{\pm iy} g(y) y^{\mu-1} dy, \tag{8}$$

$$g(y) = \frac{y^{1-\mu}}{\sqrt{2\pi\rho}} \left\{ e^{-i\frac{\pi}{2}(1-\mu)} \frac{d}{dy} \left[y^\mu \int_0^{+\infty} E_\rho \left(e^{i\frac{\pi}{2\rho}} y^{\frac{1}{\rho}} x^{\frac{1}{\rho}}; \mu + 1 \right) x^{\mu-1} f^{(-)}(x)dx \right] \right.$$
$$\left. + e^{i\frac{\pi}{2}(1-\mu)} \frac{d}{dy} \left[y^\mu \int_0^{+\infty} E_\rho \left(e^{-i\frac{\pi}{2\rho}} y^{\frac{1}{\rho}} x^{\frac{1}{\rho}}; \mu + 1 \right) x^{\mu-1} f^{(+)}(x)dx \right] \right\} \tag{9}$$

which are true almost everywhere in $(0, +\infty)$.

Theorem 1.6-2. If $\rho \geq 1$ and instead of $g(y; \sigma)$ the function

$$g(y; \varphi; \sigma) = \frac{1}{\sqrt{2\pi\rho}} \left\{ e^{-i\frac{\pi}{2}(1-\mu)} \int_0^\sigma E_\rho \left(e^{i\left(\frac{\pi}{2\rho}+\varphi\right)} y^{\frac{1}{\rho}} x^{\frac{1}{\rho}}; \mu \right) x^{\mu-1} f^{(-)}(x)dx \right.$$
$$\left. + e^{i\frac{\pi}{2}(1-\mu)} \int_0^\sigma E_\rho \left(e^{-i\left(\frac{\pi}{2\rho}-\varphi\right)} y^{\frac{1}{\rho}} x^{\frac{1}{\rho}}; \mu \right) x^{\mu-1} f^{(+)}(x)dx \right\} \quad (\sigma > 0) \tag{10}$$

is considered, then

$$\operatorname*{l.\,i.\,m.}_{\sigma\to+\infty}{}^{(\mu)}\ g(y; \varphi; \sigma) = \begin{cases} g(y) & \text{when } \varphi = 0, \\ 0 & \text{when } \frac{\pi}{\rho} \leq \rho \leq 2\pi - \frac{\pi}{\rho}, \end{cases} \begin{matrix} (11) \\ (12) \end{matrix}$$

and almost everywhere on the semi-axis $(0, +\infty)$

$$g(y; \varphi) \equiv \frac{y^{1-\mu}}{\sqrt{2\pi\rho}} \left\{ e^{-i\frac{\pi}{2}(1-\mu)} \frac{d}{dy} \left[y^\mu \int_0^{+\infty} E_\rho \left(e^{i\left(\frac{\pi}{2\rho}+\varphi\right)} y^{\frac{1}{\rho}} x^{\frac{1}{\rho}}; \mu + 1 \right) x^{\mu-1} f^{(-)}(x)dx \right] \right.$$
$$\left. + e^{i\frac{\pi}{2}(1-\mu)} \frac{d}{dy} \left[y^\mu \int_0^{+\infty} E_\rho \left(e^{-i\left(\frac{\pi}{2\rho}-\varphi\right)} y^{\frac{1}{\rho}} x^{\frac{1}{\rho}}; \mu + 1 \right) x^{\mu-1} f^{(+)}(x)dx \right] \right\}$$
$$= \begin{cases} g(y) & \text{when } \varphi = 0, \\ 0 & \text{when } \frac{\pi}{\rho} \leq \varphi \leq 2\pi - \frac{\pi}{\rho}. \end{cases} \tag{13}$$

(c)   The following assertion is the converse to Theorem 1.6-1.

Theorem 1.6-3. $1°$. If $f(x) \in L_2(0, +\infty)$, then the limit in mean

$$g^{(\pm)}(y) = \operatorname*{l.\,i.\,m.}_{\sigma\to+\infty}{}^{(\mu)}\ \frac{1}{\sqrt{2\pi\rho}} \int_0^\sigma E_\rho \left(e^{\pm i\pi/2\rho} y^{1/\rho} x^{1/\rho}; \mu \right) x^{\mu-1} f(x)dx \tag{14}$$

exists on $(0, +\infty)$. Conversely, if

$$f(x; \sigma) = \frac{1}{\sqrt{2\pi\rho}} \left\{ e^{-i\frac{\pi}{2}(1-\mu)} \int_0^\sigma e^{-ixy} g^{(+)}(y) y^{\mu-1} dy \right.$$
$$\left. + e^{i\frac{\pi}{2}(1-\mu)} \int_0^\sigma e^{ixy} g^{(-)}(y) y^{\mu-1} dy \right\} \quad (\sigma > 0), \tag{15}$$

then there exists the limit in $L_2(-\infty, +\infty)$ metric

$$\operatorname*{l.\,i.\,m.}_{\sigma \to +\infty} f(x;\sigma) = \tilde{f}(x) = \begin{cases} f(x), & \text{if } x \in (0, +\infty), \\ -\frac{1}{2\pi\rho}\frac{d}{dx}\int_0^{+\infty} \psi_{\rho,\mu}\left(-\frac{y}{x}\right) f(y)dy, & \text{if } x \in (-\infty, 0), \end{cases} \tag{16}$$

where $\psi_{\rho,\mu}$ is the same as in Lemma 1.5-4.

2°. *The functions $g^{(\pm)}(y) \in L_{2,\mu}(0, +\infty)$ and $\tilde{f}(x) \in L_2(-\infty, +\infty)$ are representable almost everywhere in $(0, +\infty)$ and $(-\infty, +\infty)$ in the forms*

$$g^{(\pm)}(y) = \frac{y^{1-\mu}}{\sqrt{2\pi\rho}}\frac{d}{dy}\left\{ y^{\mu}\int_0^{+\infty} E_\rho\left(e^{\pm i\frac{\pi}{2\rho}} y^{\frac{1}{\rho}} x^{\frac{1}{\rho}}; \mu+1\right) x^{\mu-1} f(x)dx \right\} \tag{17}$$

and

$$\tilde{f}(x) = \frac{1}{\sqrt{2\pi\rho}}\left\{ e^{-i\frac{\pi}{2}(1-\mu)}\frac{d}{dx}\int_0^{+\infty}\frac{e^{-ixy}-1}{-iy}g^{(+)}(y)y^{\mu-1}dy \right. $$
$$\left. + e^{i\frac{\pi}{2}(1-\mu)}\frac{d}{dx}\int_0^{+\infty}\frac{e^{ixy}-1}{iy}g^{(-)}(y)y^{\mu-1}dy \right\}. \tag{18}$$

3°. *The following estimates are true:*

$$\int_0^{+\infty} |g^{(\pm)}(y)|^2 y^{2(\mu-1)}dy \le \frac{M_\mu^2}{2\pi\rho}\int_0^{+\infty}|f(x)|^2 dx, \tag{19}$$

and

$$\int_0^{+\infty} |f(x)|^2 dx \le \int_{-\infty}^{+\infty} |\tilde{f}(x)|^2 dx $$
$$= \rho^{-1}\left\{ \int_0^{+\infty} |g^{(+)}(y)|^2 y^{2(\mu-1)}dy + \int_0^{+\infty} |g^{(-)}(y)|^2 y^{2(\mu-1)}dy \right\}, \tag{20}$$

where M_μ is the constant of the estimate 1.5(13).

(d) As was already observed above, any function $f(x) \in L_2(0, +\infty)$ has two transforms generated by means of Fourier kernels $e^{\pm ixy}$ and also two transforms generated by means of Mittag-Leffler type kernels $E_\rho\left(e^{\pm i\pi/2\rho} x^{1/\rho} y^{1/\rho}; \mu\right)$. Let $f_1(x) \in L_{2,\mu}(0, +\infty)$ and $f_2(x) \in L_2(0, +\infty)$ be arbitrary functions whose transforms are

$$g^{(\pm)}(y; f_1; e) = \frac{1}{\sqrt{2\pi\rho}}\frac{d}{dy}\int_0^{+\infty}\frac{e^{\pm iyx}-1}{\pm ix}x^{\mu-1}f_1(x)dx \tag{21}$$

and

$$g^{(\pm)}(y; f_2; E_\rho) = \frac{y^{1-\mu}}{\sqrt{2\pi\rho}}\frac{d}{dy}\left\{ y^{\mu}\int_0^{+\infty} E_\rho\left(e^{\pm i\frac{\pi}{2\rho}} y^{\frac{1}{\rho}} x^{\frac{1}{\rho}}; \mu+1\right) x^{\mu-1} f_2(x)dx \right\}. \tag{22}$$

Then, according to Theorems 1.6-1 and 1.6-3,

$$g^{(\pm)}(y; f_1; e) \in L_2(0, +\infty) \text{ and } g^{(\pm)}(y; f_2; E_\rho) \in L_{2,\mu}(0, +\infty). \tag{23}$$

Moreover, the following analog of Parseval equality is true.

Theorem 1.6-4. *The pairs of functions $f_1(x)$, $f_2(x)$ and $g^{(\pm)}(y; f_1; e)$, satisfy the equality*

$$\int_0^{+\infty} f_1(x) x^{\mu-1} \overline{f_2(x)} dx = e^{-i\frac{\pi}{2}(1-\mu)} \int_0^{+\infty} g^{(-)}(y; f_1; e) \overline{g^{(-)}(y; f_2; E_\rho)} y^{\mu-1} dy$$

$$+ e^{i\frac{\pi}{2}(1-\mu)} \int_0^{+\infty} g^{(+)}(y; f_1; e) \overline{g^{(+)}(y; f_2; E_\rho)} y^{\mu-1} dy.$$

$$(24)$$

(e) Using Theorem 1.6-3 it is possible to generalize essentially the main Theorem 1.6-1, that is, to construct in an explicit form an apparatus of Fourier-Plancherel type integral for any finite system of rays starting from a point of the complex plane.

Let $L\{\vartheta_1, \vartheta_2, \ldots, \vartheta_s\}$ be the system of rays

$$l_k : \arg z = \vartheta_k \qquad (k = 1, 2, \ldots, s; \ s \geq 2),$$
$$-\pi < \vartheta_1 < \vartheta_2 \cdots < \vartheta_s \leq \pi$$

$$(25)$$

starting from the origin. These rays divide the z-plane to s corner domains with a common vertex at $z = 0$. Further, assuming that $\vartheta_{s+1} = 2\pi + \vartheta_1$, we put

$$\rho_s = \rho\{\vartheta_1, \ldots, \vartheta_s\} = \max_{1 \leq k \leq s} \left\{ \frac{\pi}{\vartheta_{k+1} - \vartheta_k} \right\} \qquad (26)$$

and observe that $\rho_s \geq s/2$, and π/ρ_s is the size of the minimal opening of the mentioned corner domains.

Theorem 1.6-5. *Let $\rho \geq \rho_s$, and let $g(z)$ be any function defined on the system of rays $L_s = L\{\vartheta_1, \ldots, \vartheta_s\}$, such that*

$$\mathcal{T}_s(g) \equiv \int_{L_s} |g(z)|^2 |z|^{2(\mu-1)} d|z| = \sum_{k=1}^{s} \int_0^{+\infty} |g\left(re^{i\vartheta_k}\right)|^2 r^{2(\mu-1)} dr < +\infty. \quad (27)$$

Further, let

$$f_k(x; \sigma) = \frac{1}{\sqrt{2\pi\rho}} \int_0^{\sigma} e^{-ixr} g\left(re^{i\vartheta_k}\right) r^{\mu-1} dr \qquad (\sigma > 0, 1 \leq k \leq s). \quad (28)$$

Then:
1°. There exist functions $f_k(x) \in L_2(-\infty, +\infty)$ $(1 \leq k \leq s)$ such that

$$f_k(x) = \underset{\sigma \to +\infty}{\overset{(\mu)}{\text{l. i. m.}}} f_k(x; \sigma) \qquad (1 \leq k \leq s). \quad (29)$$

Conversely, the functions

$$g(\mathrm{re}^{i\varphi};\sigma) = \frac{1}{\sqrt{2\pi\rho}} \sum_{k=1}^{s} \int_{-\sigma}^{\sigma} E_\rho\left((ix)^{1/\rho} r^{1/\rho} e^{i(\varphi-\vartheta_k)};\mu\right) (ix)^{\mu-1} f_k(x) dx \quad (\sigma > 0) \tag{30}$$

converge in mean to $g(\mathrm{re}^{i\varphi})$ *on the system of rays* L_s:

$$\lim_{\sigma\to+\infty} \sum_{j=1}^{s} \int_{0}^{+\infty} \left| g\left(\mathrm{re}^{i\vartheta_j}\right) - g\left(\mathrm{re}^{i\vartheta_j};\sigma\right) \right|^2 r^{2(\mu-1)} dr = 0. \tag{31}$$

2°. *The functions* $g(z)(z \in L_s)$ *and* $f_k(x)$ $(1 \le k \le s)$ *are connected by the formulas*

$$f_k(x) = \frac{1}{\sqrt{2\pi\rho}} \frac{d}{dx} \int_{0}^{+\infty} \frac{e^{-ixr}-1}{-ir} g\left(\mathrm{re}^{i\vartheta_k}\right) r^{\mu-1} dr \quad (1 \le k \le s), \tag{32}$$

$$g(\mathrm{re}^{i\varphi}) = \frac{r^{1-\mu}}{\sqrt{2\pi\rho}} \sum_{k=1}^{s} \frac{d}{dr} \left\{ r^\mu \int_{-\infty}^{+\infty} E_\rho\left((ix)^{1/\rho} r^{1/\rho} e^{i(\varphi-\vartheta_k)};\mu+1\right) (ix)^{\mu-1} f_k(x) dx \right\} \tag{33}$$

which are true for almost every $x \in (-\infty,+\infty)$ *and for almost every* $\mathrm{re}^{i\varphi} \in L_s = L\{\vartheta_1,\ldots,\vartheta_s\}$ *correspondingly.*

3°. *A Parseval type equality is also true:*

$$\mathcal{T}_s(g) = \rho \sum_{k=1}^{s} \int_{-\infty}^{+\infty} |f_k(x)|^2 \, dx. \tag{34}$$

1.7 Some consequences

In this section some particular cases of previous general theorems are observed. These particular cases are directly connected with the classical results of the theory of Fourier transform in L_2.

(a) Theorem 1.7-1. *If* $\mu \in (1/2, 5/2)$, *then:*
1°. *The relations*

$$f(x) = \underset{\sigma\to+\infty}{\overset{(\mu)}{\mathrm{l.\,i.\,m.}}} \sqrt{\frac{2}{\pi}} \int_{0}^{\sigma} \cos\left(xy + \frac{\pi}{2}(1-\mu)\right) g(y) y^{\mu-1} dy, \tag{1}$$

$$g(y) = \underset{\sigma\to+\infty}{\overset{(\mu)}{\mathrm{l.\,i.\,m.}}} \sqrt{\frac{2}{\pi}} \int_{0}^{\sigma} E_{1/2}\left(-y^2 x^2;\mu\right) x^{\mu-1} f(x) dx \tag{2}$$

give invertible and mutually inverse transforms between the spaces of functions $g(y) \in L_{2,\mu}(0, +\infty)$ *and* $f(x) \in L_2(0, +\infty)$, *and the following two-sided estimate is true:*

$$\int_0^{+\infty} |g(y)|^2 y^{2(\mu-1)} dy \asymp \int_0^{+\infty} |f(x)|^2 dx. \tag{3}$$

$2°$. If any functions $f(x) \in L_2(0, +\infty)$ *and* $g(y) \in L_{2,\mu}(0, +\infty)$ *are connected by relations (1) and (2), then*

$$f(x) = \sqrt{\frac{2}{\pi}} \frac{d}{dx} \int_0^{+\infty} \frac{\cos\left(xy - \frac{\pi}{2}\mu\right) - \cos\frac{\pi}{2}\mu}{y} g(y) y^{\mu-1} dy \tag{4}$$

and

$$g(y) = \sqrt{\frac{2}{\pi}} y^{1-\mu} \frac{d}{dy} \left[y^\mu \int_0^{+\infty} E_{1/2}\left(-y^2 x^2; \mu+1\right) x^{\mu-1} f(x) dx \right] \tag{5}$$

almost everywhere in $(0, +\infty)$.

The proof of these assertions will now be briefly outlined.
$1°$. We shall use Theorem 1.6-1, assuming $\rho = 1/2$ and $\mu \in (1/2, 5/2)$. To this end we suppose that $g(y) \in L_{2,\mu}(0, +\infty)$ is an arbitrary function and introduce the following functions depending on $x \in (0, +\infty)$:

$$f^{(\pm)}(x; \sigma) = \frac{1}{\sqrt{\pi}} \int_0^\sigma e^{\pm ixy} g(y) y^{\mu-1} dy \qquad (\sigma > 0),$$

$$f(x; \sigma) = \frac{1}{\sqrt{2}} \left\{ e^{-i\frac{\pi}{2}(1-\mu)} f^{(-)}(x; \sigma) + e^{i\frac{\pi}{2}(1-\mu)} f^{(+)}(x; \sigma) \right\}$$

$$= \sqrt{\frac{2}{\pi}} \int_0^\sigma \cos\left(xy + \frac{\pi}{2}(1-\mu)\right) g(y) y^{\mu-1} dy.$$

Then the existence of the limit function $f(x) \in L_2(0, +\infty)$ in the sense of (1) follows from formulas 1.6(3)–(6). It follows also that

$$f(x) = \frac{1}{\sqrt{2}} \left\{ e^{-i\frac{\pi}{2}(1-\mu)} f^{(-)}(x) + e^{i\frac{\pi}{2}(1-\mu)} f^{(+)}(x) \right\}. \tag{6}$$

Further, by 1.6(5),

$$g(y; \sigma) = \sqrt{\frac{2}{\pi}} \int_0^\sigma E_{1/2}\left(-y^2 x^2; \mu\right) x^{\mu-1} f(x) dx,$$

and the existence of the limit function $g(y) \in L_{2,\mu}(0, +\infty)$, in the sense of (2), follows at once. Finally, formulas 1.6(19)–(20) imply the two-sided estimates (3).

To prove the converse statement, i.e., that formula (1) gives the inversion of transformation (2), we shall assume that $f(x) \in L_2(0, +\infty)$ is any function. Then, according to 1.6(14)–(16), the limit (2) exists, $g(y) = \sqrt{2}g^{(\pm)}(y)$ and it belongs to $L_{2,\mu}(0,+\infty)$ when $\rho = 1/2$ and $\mu \in (1/2, 5/2)$. Besides, it follows that

$$f(x;\sigma) = \frac{2}{\sqrt{\pi}} \int_0^\sigma \cos\left(xy + \frac{\pi}{2}(1-\mu)\right) g^{(\pm)}(y)y^{\mu-1}dy, \qquad x \in (0,+\infty),$$

and hence the limit (1) exists, i.e., the inversion formula of transformation (2) is true.

2°. The pair of formulas (4)–(5) immediately follows, in the same way as above, from the corresponding relations 1.6(17)–(18) where we assume $\rho = 1/2$.

(b) Observe that by formulas 1.2(1)

$$E_{1/2}\left(-x^2y^2; 1\right) = \cos xy \qquad \text{and} \qquad E_{1/2}\left(-x^2y^2; 2\right) = \frac{\sin xy}{xy}. \tag{7}$$

Therefore, the well-known Fourier-Plancherel dual formulas of cos- and sin-transforms in $L_2(0,+\infty)$ follow from Theorem 1.7-1 when $\mu = 1$ and $\mu = 2$.

Theorem 1.7-2. *The dual relations*

$$g_c(y) = \operatorname*{l.i.m.}_{\sigma\to+\infty} \sqrt{\frac{2}{\pi}} \int_0^\sigma \cos(xy)f(x)dx,$$

$$f(x) = \operatorname*{l.i.m.}_{\sigma\to+\infty} \sqrt{\frac{2}{\pi}} \int_0^\sigma \cos(xy)g_c(y)dy \tag{8}$$

and

$$g_s(y) = \operatorname*{l.i.m.}_{\sigma\to+\infty} \sqrt{\frac{2}{\pi}} \int_0^\sigma \sin(xy)f(x)dx,$$

$$f(x) = \operatorname*{l.i.m.}_{\sigma\to+\infty} \sqrt{\frac{2}{\pi}} \int_0^\sigma \sin(xy)g_s(y)dy \tag{9}$$

represent mappings of the whole class of functions $L_2(0,+\infty)$ onto itself. And the Parseval equality

$$\int_0^{+\infty} |f(x)|^2 dx = \int_0^{+\infty} |g_c(y)|^2 dy = \int_0^{+\infty} |g_s(y)|^2 dy \tag{10}$$

is true in both cases.

It should be mentioned that equalities (10) follow from 1.6(7), if we take into account the complementary notations introduced when the proof of Theorem 1.7-1 was briefly outlined. Further, note that the dual formulas (8) and (9) can be written also in the forms (4) and (5) when $\mu = 1$ and $\mu = 2$.

(c) Finally, note that the classical Plancherel theorem on the Fourier transform in $L_2(-\infty, +\infty)$, which is stated below, is contained as a very special case not only in the general Theorem 1.6-5 when $\rho = \mu = 1$, and we have the system of two rays $L\{0, \pi\}$, but also in Theorem 1.6-3.

Theorem 1.7-3.. *The dual formulas*

$$g(y) = \frac{1}{\sqrt{2\pi}} \frac{d}{dy} \int_{-\infty}^{+\infty} \frac{e^{-iyx} - 1}{-ix} f(x)dx,$$

$$f(x) = \frac{1}{\sqrt{2\pi}} \frac{d}{dx} \int_{-\infty}^{+\infty} \frac{e^{ixy} - 1}{iy} g(y)dy \tag{11}$$

and

$$g(y) = \underset{\sigma \to +\infty}{\text{l. i. m.}} \frac{1}{\sqrt{2\pi}} \int_{-\sigma}^{\sigma} e^{-iyx} f(x)dx,$$

$$f(x) = \underset{\sigma \to +\infty}{\text{l. i. m.}} \frac{1}{\sqrt{2\pi}} \int_{-\sigma}^{\sigma} e^{ixy} g(y)dy \tag{12}$$

represent the same one-to-one mapping of the space $L_2(-\infty, +\infty)$ onto itself, and the Parseval relation is true:

$$\int_{-\infty}^{+\infty} |f(x)|^2 dx = \int_{-\infty}^{+\infty} |g(y)|^2 dy. \tag{13}$$

We shall briefly outline a proof based on Theorem 1.6-3, where we put $\rho = \mu = 1$ and consequently have

$$E_\rho(z; \mu) \equiv E_1(z; 1) = e^z, \qquad E_\rho(z; \mu + 1) \equiv E_1(z; 2) = \frac{e^z - 1}{z}.$$

In this case, formulas 1.6(17)–(18) become

$$g^{(\pm)}(y) = \frac{1}{\sqrt{2\pi}} \frac{d}{dy} \int_0^{+\infty} \frac{e^{\pm iyx} - 1}{\pm ix} f(x)dx, \qquad y \in (0, +\infty),$$

$$\tilde{f}(x) = \frac{1}{\sqrt{2\pi}} \left\{ \frac{d}{dx} \int_0^{+\infty} \frac{e^{-ixy} - 1}{-iy} g^{(+)}(y)dy + \frac{d}{dx} \int_0^{+\infty} \frac{e^{ixy} - 1}{iy} g^{(-)}(y)dy \right\}.$$
$$x \in (-\infty, +\infty),$$

But $\Psi_{1,1}(x) \equiv 0$, $x \in (0, +\infty)$ by formula 1.5(21) of Lemma 1.5-4. Therefore,

$$\tilde{f}(x) = \begin{cases} f(x) & \text{when } x \in (0, +\infty), \\ 0 & \text{when } x \in (-\infty, 0), \end{cases}$$

and, as can be easily verified, equality 1.6(20) becomes

$$\int_0^{+\infty} |f(x)|^2 dx = \int_0^{+\infty} |g^{(+)}(y)|^2 dy + \int_0^{+\infty} |g^{(-)}(y)|^2 dy.$$

Further, denote $g(\pm y) \equiv g^{(\mp)}(y)$ $(0 < y < +\infty)$. Then it is easy to obtain the formulas

$$g(y) = \frac{1}{\sqrt{2\pi}} \frac{d}{dy} \int_0^{+\infty} \frac{e^{-ixy} - 1}{-ix} f(x) dx, \quad y \in (-\infty, +\infty),$$

$$\tilde{f}(x) = \frac{1}{\sqrt{2\pi}} \frac{d}{dx} \int_{-\infty}^{+\infty} \frac{e^{ixy} - 1}{iy} g(y) dy = \begin{cases} f(x) & \text{when } x \in (0, +\infty) \\ 0 & \text{when } x \in (-\infty, 0) \end{cases}$$

and equality (13), but only for a subclass of those functions of $L_2(-\infty, +\infty)$ which vanish on the semi-axis $(-\infty, 0)$. Similarly, we consider the case of functions vanishing on the semi-axis $(0, +\infty)$, then we obtain the desired relations (11) for the whole class $L_2(-\infty, +\infty)$. The second pair of dual formulas (12) follows similarly from formulas 1.6(14)–(16) of Theorem 1.6-3.

1.8 Notes

1.1 The function

$$E_{1/\rho}(z) = \sum_{k=0}^{\infty} \frac{z^k}{\Gamma(1 + k/\rho)} (\rho > 0)$$

was first introduced by Mittag-Leffler [1-3] in connection with his new method of the summation of divergent series (see also Buhl [1], Hardy [1, Sec. 8.10] and M.M.Djrbashian [5, Theorems 3.3 and 3.4]). As the earliest investigations where different important properties of $E_{1/\rho}(z)$ were established, the papers of Wiman [1, 2] and Buhl [1] must be mentioned. The function

$$E_\rho(z; \mu) = \sum_{k=0}^{\infty} \frac{z^k}{\Gamma(\mu + k/\rho)} (\rho > 0, \mu > 0)$$

called a Mittag-Leffler type function by the author is the basis of a series of investigations in the 1950s (see M.M.Djrbashian [1-4]) which were improved and summarized in [5]. It must be mentioned that the function $E_\rho(z; \mu)$, which coincides with $E_{1/\rho}(z)$ when $\mu = 1$, has been written in several different forms. For example, Cleota and Hugens [1] wrote it in the form $E_{1/\rho}(z; \mu)$, and the notation $E_{1/\rho,\mu}(z)$, used in the reference book of Bateman [1], was introduced by Agarwal [1]. A series of relations for $E_\rho(z; \mu)$, where $\mu = 1$ is assumed, can be found in Buhl [1], Humbert [1], Agarwal [1] and Humbert-Agarwal [1], but most of these relations are of a formal character. Some of these relations extended for any μ are given by Bateman [1]. These and other simple properties of $E_\rho(z; \mu)$ were improved by M.M.Djrbashian [5, Ch. 3, §1].

1.2 Relations (1)–(9) are given in the above mentioned papers and monographs. Relation (10) was obtained by M.M.Djrbashian-A.B.Nersesian [1]. Identities (14) and (15), obtained by the author, play an important role in the establishment of the main results of this book. Theorem 1.2-1 was proved by Hille-Tamarkin [1] by a direct application of the general formulas of the theory of Volterra integral equations (see also V.I.Smirnov [1, Section 48] and M.M.Djrbashian [5, Theorem 3.1]).

1.3 Theorems 1.3-1 and 1.3-2 were proved by M.M.Djrbashian [3], [5, Chapter 3, §2]. For Theorems 1.3-3 and 1.3-4 see Bateman [1, p. 210] and M.M.Djrbashian [5, Chapter 3, §2]. Theorem 1.3-5 was proved by M.M.Djrbashian-B.A.Saakian [1]. The assertions of Theorems 1.3-6 and 1.3-7, except relations (18), were established in essentially different ways and in different forms by Pollard [1, 2] and Winter [1, 2], only for the special case $\mu = 1$. These theorems, in the way they are formulated here, were proved by a new method in the paper of M.M.Djrbashian-R.A.Bagian [1], where the entire function $\Phi_{\rho,\mu}(s)$ was introduced and relations (18) were established. It should be mentioned that Pollard [1, 2] had established that the function $\Psi_\rho(\tau)$ of the integral representation

$$E_\rho(-x;1) = \int_0^{+\infty} e^{-x\tau} d\Psi_\rho(\tau), \quad o \le x < +\infty, \ \rho \ge 1,$$

proved by him is absolutely continuous, and therefore $\Phi_{\rho,1}(\tau) \equiv \Psi'_\rho(\tau)$, $\tau \in (0, +\infty)$.

1.4 The problem of the distribution of zeros of the function $E_\rho(z;1) \equiv E_{1/\rho}(z)$ was first considered by Wiman [1, 2]. He established Lemmas 1.4-1 and 1.4-2 for $\mu = 1$. The general case when $\mu \ne 1$ was considered by M.M.Djrbashian [5, Chapter 3]. The resuming Theorem 1.4-1 was established by Polya [1], and its consequences Theorems 1.4-2, 1.4-3 and 1.4-4, relating to distribution of zeros of the function

$$\mathcal{E}_\sigma(z;\nu) = E_{1/2}\left(-\sigma^2 z; 1 + \nu\right), \quad \nu \in [0,2),$$

were proved by M.M. Djrbashian [7, §1].

1.5–1.7 For Theorems 1.5-1 and 1.5-2 on Mellin transforms in L_2 see Titchmarsh [1, §3.17] and M.M. Djrbashian [5, Chapter 1, §4.3]. Lemmas 1.5-1–1.5-4 and the assertions of Theorems 1.6-1–1.6-5 were proved by M.M. Djrbashian [5, Chapter 4, §2]. The concluding Theorems 1.7-1–1.7-3 are consequences of Theorems 1.6-1–1.6-5.

2 Further results. Wiener-Paley type theorems

2.1 Introduction

In this chapter we present some other applications of the theory of harmonic analysis in the complex domain. These applications are related to the theory of parametric representations of various classes of entire and analytic functions restricted by additional conditions of weighted integrability on suitable systems of rays. The main results of this chapter will be used later on, but nevertheless, we present them without proofs. The proofs can be found in M. M. Djrbashian's monograph [5].

2.2 Some simple generalizations of the first fundamental Wiener-Paley theorem.

(a) The class of entire functions $f(z)$ of order $\rho = 1/2$ and of type $\leq \sigma$, satisfying the condition

$$\int_0^{+\infty} |f(x)|^2 \, x^\omega dx < +\infty \qquad (-1 < \omega < 1), \tag{1}$$

will be denoted by $W^{2,\omega}_{1/2,\sigma}$. The parametric representation of this class is contained in the following theorem.

Theorem 2.2-1. *1°. The class $W^{2,\omega}_{1/2,\sigma}$ $(-1 < \omega < 1)$ coincides with the set of functions representable in the form*

$$f(z) = \int_0^\sigma E_{1/2}(-z\tau^2; \mu)\tau^{\mu-1}\varphi(\tau)d\tau, \qquad z \in \mathbb{C}, \tag{2}$$

where $\mu = \omega + 3/2$ and $\varphi(\tau) \in L_2(0,\sigma)$ are arbitrary functions.
2°. If $f(z) \in W^{2,\omega}_{1/2,\sigma}$, then the function $\varphi(\tau)$ of the representation (2) is unique, and the inversion formula of the transformation (2)

$$\frac{2}{\pi}\frac{d}{d\tau}\int_0^{+\infty} \frac{\cos(t\tau - \mu\pi/2) - \cos(\mu\pi/2)}{t} f(t^2)t^{\mu-1}dt \equiv \varphi_\sigma(\tau)$$
$$= \begin{cases} \varphi(\tau) & \text{when } \tau \in (0,\sigma), \\ 0 & \text{when } \tau \in (\sigma,+\infty) \end{cases} \tag{3}$$

is true almost everywhere.
3°. The following two-sided estimates are true:

$$\int_0^{+\infty} |f(x)|^2 \, x^\omega dx \asymp \int_0^\sigma |\varphi(\tau)|^2 \, d\tau. \tag{4}$$

Here the suitable constants do not depend on f and φ.

If the particular classes of entire functions

$$W_{1/2,\sigma}^{2,\mp 1/2} : \int_0^{+\infty} \mid f(x) \mid^2 x^{\mp 1/2} dx < +\infty \tag{5}$$

are considered, then the well-known Wiener-Paley theorems on these classes follow from Theorem 2.2-1. Namely, the assertions of the following theorem are obtained.

Theorem 2.2-2. 1°. *The class* $W_{1/2,\sigma}^{2,-1/2}$ $(-1 < \omega < 1)$ *coincides with the set of functions of the form*

$$f(z) = \int_0^\sigma \cos(\sqrt{z}\tau)\varphi(\tau)d\tau, \varphi(\tau) \in L_2(0,\sigma). \tag{6}$$

2°. *The class* $W_{1/2,\sigma}^{2,1/2}$ *coincides with the set of functions of the form*

$$f(z) = \int_0^\sigma \frac{\sin(\sqrt{z}\tau)}{\sqrt{z}}\varphi(\tau)d\tau, \varphi(\tau) \in L_2(0,\sigma). \tag{7}$$

3°. *The inversions of transformations (6) and (7) are given correspondingly by the formulas*

$$\left.\begin{aligned} \frac{2}{\pi}\frac{d}{d\tau}\int_0^{+\infty}\frac{\sin(t\tau)}{t}f\left(t^2\right)dt \\ \frac{2}{\pi}\frac{d}{d\tau}\int_0^{+\infty}\frac{1-\cos(t\tau)}{t}f\left(t^2\right)tdt \end{aligned}\right\} \equiv \varphi_\sigma(\tau) = \left\{\begin{array}{l} \varphi(\tau),\ \tau \in (0,\sigma) \\ 0,\ \tau \in (\sigma,+\infty) \end{array}\right. \tag{8}$$

which are true almost everywhere in $(0,+\infty)$.
4°. *Parseval equalities of the form*

$$\frac{1}{\pi}\int_0^{+\infty} \mid f(x) \mid^2 x^{\mp 1/2}dx = \int_0^\sigma \mid \varphi(\tau) \mid^2 d\tau \tag{9}$$

are true for both cases.

Indeed, if $\omega = -1/2(\mu = 1)$ or $\omega = 1/2(\mu = 2)$, then, by formula 1.2(1),

$$E_{1/2}\left(-z\tau^2; 1\right) = \cos(\sqrt{z}\tau), E_{1/2}\left(-z\tau^2; 2\right) = \frac{\sin(\sqrt{z}\tau)}{\sqrt{z}\tau}.$$

So, assertions 1° to 3° of Theorem 2.2-2 follow from formulas (2) and (3). As to equalities (9), they follow from formulas (8) and 1.7(10).

(b) The following direct generalization of the Wiener-Paley first fundamental theorem is established for the class $W_{1,\sigma}^{2,\omega}$ of entire functions $f(z)$ of order $\rho = 1$ and of type $\leq \sigma$, which satisfy the condition

$$\int_{-\infty}^{+\infty} \mid f(x) \mid^2 \mid x \mid^\omega dx < +\infty \qquad (-1 < \omega < 1). \tag{10}$$

Theorem 2.2-3. *1°. The class $W_{1,\sigma}^{2,\omega}(-1 < \omega < 1)$ coincides with the set of functions of the form*

$$f(z) = \int_{-\sigma}^{\sigma} E_1(iz\tau; \mu) \mid \tau \mid^{\mu-1} \varphi(\tau)d\tau, \qquad z \in \mathbb{C}, \tag{11}$$

where $\mu = 1 + \omega/2$ and $\varphi(\tau) \in L_2(-\sigma, \sigma)$ are arbitrary functions.
2°. If $f(z) \in W_{1,\sigma}^{2,\omega}$, then the function $\varphi(\tau)$ of representation (11) is unique, and the inversion formula

$$\frac{1}{2\pi} \frac{d}{d\tau} \int_{-\infty}^{+\infty} \frac{e^{-i\tau t} - 1}{-it} \exp\left\{i\frac{\pi}{2}(\mu - 1)\,\mathrm{sign}(\tau t)\right\} f(t) \mid t \mid^{\mu-1} dt$$
$$= \begin{cases} \varphi(\tau), & \tau \in (-\sigma, \sigma) \\ 0, & \tau \notin (-\sigma, \sigma) \end{cases} \tag{12}$$

is true almost everywhere.
3°. The following two-sided estimates are true:

$$\int_{-\infty}^{+\infty} \mid f(x) \mid^2 \mid x \mid^{\omega} dx \asymp \int_{-\sigma}^{\sigma} \mid \varphi(\tau) \mid^2 d\tau. \tag{13}$$

If $\omega = 0$ $(\mu = 1)$, then we consider the function

$$E_1(iz\tau; 1) = e^{iz\tau},$$

and the first fundamental Wiener-Paley theorem, which we give below, follows from Theorem 2.2-3.

Theorem 2.2-4. *1°. The class $W_{1,\sigma}^{2,0}$ of entire functions $f(z)$ of order $\rho = 1$ and of type $\leq \sigma$, for which*

$$\int_{-\infty}^{+\infty} \mid f(x) \mid^2 dx < +\infty \tag{14}$$

coincides with the set of functions of the form

$$f(z) = \int_{-\sigma}^{\sigma} e^{iz\tau}\varphi(\tau)d\tau, \qquad \varphi(\tau) \in L_2(-\sigma, \sigma). \tag{15}$$

2°. If $f(z) \in W_{1,\sigma}^{2,0}$, then the function $\varphi(\tau)$ of the representation (15) is unique, and the following inversion formula is true almost everywhere:

$$\frac{1}{2\pi} \frac{d}{d\tau} \int_{-\infty}^{+\infty} \frac{e^{-i\tau t} - 1}{-it} f(t)dt = \begin{cases} \varphi(\tau), & \tau \in (-\sigma, \sigma) \\ 0, & \tau \notin (-\sigma, \sigma) \end{cases}. \tag{16}$$

2.3 A general Wiener-Paley type theorem and some particular results.

(a) First we introduce some notations.

Let $\rho \geq 1/2$ be an arbitrary fixed number and let the integer $s = s(\rho) \geq 0$ be such that

$$s \geq [2\rho] - 1. \tag{1}$$

Further, let the set of numbers $\{\vartheta_1 < \vartheta_2 < \ldots < \vartheta_{s+1}\}$ satisfy the conditions

$$-\pi < \vartheta_1 < \vartheta_2 < \ldots < \vartheta_s \leq \pi < \vartheta_{s+1} = \vartheta_1 + 2\pi$$

$$\max_{1 \leq k \leq s} \{\vartheta_{k+1} - \vartheta_k\} = \frac{\pi}{\rho}. \tag{2}$$

Form the set of pairs $(\vartheta_k, \vartheta_{k+1})_1^s$ from $\{\vartheta_k\}_1^{s+1}$ and then choose the subset of pairs $(\vartheta_{r_k}, \vartheta_{r_k+1})_1^p$ which have the same order of succession and for which

$$\vartheta_{r_k+1} - \vartheta_{r_k} = \frac{\pi}{\rho} \qquad (k = 1, 2, \ldots, p \leq s). \tag{3}$$

In addition, if $p < s$, then the remaining pairs $(\vartheta_{\kappa_k}, \vartheta_{\kappa_k+1})_1^q$ $(q = s - p)$ are supposed to be enumerated in the same order.

Further, denote

$$\Xi_k = \frac{1}{2}[\vartheta_{r_k} + \vartheta_{r_k+1}] \qquad (k = 1, 2, \ldots, p) \tag{4}$$

and suppose that the parameters $\omega \in (-1, 1)$ and $0 \leq \sigma_k \leq \sigma$ $(k = 1, 2, \ldots, p)$ are arbitrary. Then the class

$$W_{\rho,\sigma}^{2,\omega}(\{\vartheta_k\}, \{\sigma_k\})$$

of entire functions $f(z)$ of order $\rho \geq 1/2$ and of type $\leq \sigma$, satisfying the conditions

$$\int_0^{+\infty} |f(e^{-i\vartheta_k}t)|^2 \, t^\omega dt < +\infty \qquad (k = 1, 2, \ldots, s), \tag{5}$$

$$h(-\Xi_k, f) \leq \sigma_k \leq \sigma \qquad (k = 1, 2, \ldots, p), \tag{6}$$

where

$$h(\varphi, f) = \limsup_{r \to \infty} \frac{\log |f(re^{i\varphi})|}{r^\rho} \tag{7}$$

is the indicator of f, will be associated with the set of numbers $\{\vartheta_k\}_1^{s+1}$.

The following most general Wiener-Paley type theorem is true for entire functions of arbitrary finite order.

Theorem 2.3-1. *1°. The class $W_{\rho,\sigma}^{2;\omega}(\{\vartheta_k\},\{\sigma_k\})$ coincides with the set of functions $f(z)$ representable in the form*

$$f(z) = \sum_{k=1}^{p} \int_0^{\sigma_k} E_\rho\left(e^{i\Xi_k}z\tau^{1/\rho};\mu\right)\tau^{\mu-1}\varphi_k(\tau)d\tau, \qquad z \in \mathbb{C}, \tag{8}$$

where $\mu = (1+\omega+\rho)/2\rho$ and $\varphi_k(\tau) \in L_2(0,\sigma_k)(1 \le k \le p)$ are arbitrary functions. 2°. If $f(z) \in W_{\rho,\sigma}^{2;\omega}(\{\vartheta_k\},\{\sigma_k\})$, then the functions $\{\varphi_k(\tau)\}_1^p$ of representation (8) are unique and can be determined almost everywhere by formulas

$$\frac{1}{\sqrt{2\pi\rho}}\left\{e^{i\frac{\pi}{2}(1-\mu)}\Phi_{r_k+1}(-\tau) + e^{-i\frac{\pi}{2}(1-\mu)}\Phi_{r_k}(\tau)\right\} \equiv \tilde{\Phi}_k(\tau)$$
$$= \begin{cases} \varphi_k(\tau), & \tau \in (0,\sigma_k) \\ 0, & \tau \in (\sigma_k,+\infty) \end{cases} \qquad (k = 1,2,\dots,p), \tag{9}$$

where it is supposed that

$$\Phi_k(\tau) = \frac{1}{\sqrt{2\pi}}\frac{d}{d\tau}\int_0^{+\infty}\frac{e^{-i\tau t}-1}{-it}f\left(e^{-i\vartheta_k}t^{1/\rho}\right)t^{\mu-1}dt \qquad (k = 1,2,\dots,s) \tag{10}$$

almost everywhere in $(-\infty,+\infty)$ and $\Phi_{s+1}(\tau) \equiv \Phi_1(\tau)$.

(b) Obviously, the general Theorem 2.3-1 may be reduced to a particular result for any special choice of the system of rays on which conditions (5) and (6) are supposed to be satisfied. For example, it is not difficult to verify that Theorem 2.3-1 contains, as special cases, the simplest generalizations of the Wiener-Paley theorem given in Section 2.2. Now the parametric representations of two other particular classes of entire functions, which follow again from the general Theorem 2.3-1, will be stated.

The class of entire functions $f(z)$ of order $\rho \ge 1/2$ and of type $\le \sigma$, satisfying the condition

$$\sup_{\pi/2\rho\le|\vartheta|\le\pi}\left\{\int_0^{+\infty}|f(e^{-i\vartheta}t)|^2\,t^\omega dt\right\} < +\infty \tag{11}$$

for a given $\omega \in (-1,1)$, will be denoted by $A_{\rho,\sigma}^{2;\omega}\{\pm\pi/2\rho\}$. Then the following theorem is true.

Theorem 2.3-2. *1°. The class $A_{\rho,\sigma}^{2;\omega}\{\pm\pi/2\rho\}$ coincides with the set of functions $f(z)$ representable in the form*

$$f(z) = \int_0^{\sigma} E_\rho\left(z\tau^{1/\rho};\mu\right)\tau^{\mu-1}\varphi(\tau)d\tau, \qquad z \in \mathbb{C}, \tag{12}$$

where $\mu = (1+\omega+\rho)/2\rho$ and $\varphi(\tau) \in L_2(0,\sigma)$ are arbitrary functions.

$2°$. *The function $\varphi(\tau)$ of representation(12) is unique and can be determined almost everywhere by the inversion formula*

$$\frac{1}{2\pi\rho}\left\{e^{i\frac{\pi}{2}(1-\mu)}\frac{d}{d\tau}\int_0^{+\infty}\frac{e^{i\tau t}-1}{it}f\left(e^{-i\frac{\pi}{2\rho}}t^{\frac{1}{\rho}}\right)t^{\mu-1}dt\right.$$

$$\left.+e^{-i\frac{\pi}{2}(1-\mu)}\frac{d}{d\tau}\int_0^{+\infty}\frac{e^{-i\tau t}-1}{-it}f\left(e^{i\frac{\pi}{2\rho}}t^{\frac{1}{\rho}}\right)t^{\mu-1}dt\right\} \qquad (13)$$

$$=\begin{cases}\varphi(\tau), \ \tau\in(0,\sigma), \\ 0, \ \tau\in(\sigma,+\infty).\end{cases}$$

It is easy to observe that the previous theorem is a natural generalization of Theorem 2.2-1, since these theorems merely coincide when we put $\rho=1/2$ and replace z by $-z$.

As the second particular class, we will take $B_{\rho,\sigma}^{2,\omega}$ – the class of entire functions of order $\rho\geq 1$ and of type $\leq\sigma$, satisfying the conditions

$$\sup_{\pi/\rho\leq|\vartheta|\leq\pi}\left\{\int_0^{+\infty}|f(e^{-i\vartheta}t)|^2\,t^\omega dt\right\}<+\infty,\ \int_0^{+\infty}|f(t)|^2\,t^\omega dt<+\infty \qquad (14)$$

for a given $\omega\in(-1,1)$. It is important to notice that $B_{\rho,\sigma}^{2,\omega}$ coincides with $W_{1,\sigma}^{2,\omega}$ when $\rho=1$.

Theorem 2.3-3. $1°$. *The class $B_{\rho,\sigma}^{2,\omega}$ coincides with the set of functions $f(z)$ representable in the form*

$$f(z)=\int_0^\sigma E_\rho\left(e^{i\frac{\pi}{2\rho}}z\tau^{\frac{1}{\rho}};\mu\right)\tau^{\mu-1}\varphi_1(\tau)d\tau+\int_0^\sigma E_\rho\left(e^{-i\frac{\pi}{2\rho}}z\tau^{\frac{1}{\rho}};\mu\right)\tau^{\mu-1}\varphi_2(\tau)d\tau, z\in\mathbb{C},$$

$$(15)$$

where $\mu=(1+\omega+\rho)/2\rho$ and $\varphi_k(\tau)\in L_2(0,\sigma)(k=1,2)$.

$2°$. *The functions φ_1,φ_2 of the representation (15) are unique and can be determined almost everywhere by formulas*

$$\frac{1}{\sqrt{2\pi}\rho}\left\{e^{i\frac{\pi}{2}(1-\mu)}\Psi_{-1}(-\tau)+e^{-i\frac{\pi}{2}(1-\mu)}\Psi_0(\tau)\right\}=\begin{cases}\varphi_1(\tau), \ \tau\in(0,\sigma), \\ 0, \ \tau\in(\sigma,+\infty),\end{cases} \qquad (16)$$

$$\frac{1}{\sqrt{2\pi}\rho}\left\{e^{i\frac{\pi}{2}(1-\mu)}\Psi_0(-\tau)+e^{-i\frac{\pi}{2}(1-\mu)}\Psi_1(\tau)\right\}=\begin{cases}\varphi_2(\tau), \ \tau\in(0,\sigma), \\ 0, \ \tau\in(\sigma,+\infty),\end{cases} \qquad (17)$$

where

$$\Psi_k(\tau)=\frac{1}{\sqrt{2\pi}}\frac{d}{d\tau}\int_0^{+\infty}\frac{e^{-i\tau t}-1}{-it}f\left(e^{-i\frac{\pi}{\rho}k}t^{\frac{1}{\rho}}\right)t^{\mu-1}dt \qquad (k=-1,0,1) \qquad (18)$$

almost everywhere in $(-\infty,+\infty)$.

2.4 Two important cases of the general Wiener-Paley type theorem.

Here we state two important special cases of Theorem 2.3-1, where the orders of entire functions of considered classes are assumed to be

$$\rho_s = s + \frac{1}{2} \ (s = 0, 1, 2, \dots) \ \text{ or } \ \rho_s^* = s \ (s = 1, 2, \dots). \tag{1}$$

These special cases will be used in later chapters.

(a) Let $s \geq 0$ be an integer. Then the set of rays

$$\arg z = \vartheta_j = \frac{\pi}{s + 1/2} j \qquad (0 \leq |z| < +\infty, j = 0, \pm 1, \dots, \pm s) \tag{2}$$

divides the z-plane into $2s + 1$ corner domains of the same opening $\pi/(s + 1/2)$. Thus, if $s = 0$, or, which is the same, $\rho = \rho_0 = 1/2$, then there exists only one ray, $\arg z = \vartheta_0 = 0$, the z-plane is cut along $[0, +\infty)$, and the opening of the corner domain is 2π.

The class of entire functions $\Phi(z)$ of order $s + 1/2(s \geq 0)$ and of type $\leq \sigma$, satisfying the conditions

$$\int_0^{+\infty} |\Phi(e^{-i\frac{\pi}{s+1/2}j}t)|^2 t^\omega dt < +\infty \qquad (-s \leq j \leq s) \tag{3}$$

for a fixed $\omega \in (-1, 1)$, will be denoted by $W^{2,\omega}_{s+1/2,\sigma}$. It can be observed that, in the case considered, $\vartheta_{s+1} = \vartheta_{-s} + 2\pi$ and

$$\Xi_j = \frac{1}{2}[\vartheta_j + \vartheta_{j+1}] = \frac{\pi}{s+1/2}(j + 1/2) \qquad (-s \leq j \leq s), \tag{4}$$

and it is easy to verify, that Theorem 2.3-1 takes the following form.

Theorem 2.4-1. *1°. The class $W^{2,\omega}_{s+1/2,\sigma}$ $(-1 < \omega < 1, s \geq 0)$ coincides with the set of functions representable in the form*

$$\Phi(z) = \sum_{j=-s}^{s} \int_0^\sigma E_{s+1/2}\big(e^{i\pi\frac{j+1/2}{s+1/2}} z\tau^{\frac{1}{s+1/2}}; \mu\big)\varphi_j(\tau)\tau^{\mu-1}d\tau, \quad z \in \mathbb{C}, \tag{5}$$

where $\mu = (3/2 + s + \omega)/(1 + 2s)$ and $\varphi_j(\tau) \in L_2(0, \sigma)$ $(-s \leq j \leq s)$ is an arbitrary function.
2°. If $\Phi(z) \in W^{2,\omega}_{s+1/2,\sigma}$, then the functions $\{\varphi_j(\tau)\}^s_{-s}$ of representation(5) are unique and can be determined by the formulas

$$\frac{1}{\sqrt{2\pi}(s+1/2)}\big\{e^{i\frac{\pi}{2}(1-\mu)}\Phi_{j+1}(-\tau) + e^{-i\frac{\pi}{2}(1-\mu)}\Phi_j(\tau)\big\}$$

$$= \begin{cases} \varphi_j(\tau), \ \tau \in (0, \sigma) \\ 0, \ \tau \in (\sigma, +\infty) \end{cases} \qquad (-s \leq j \leq s), \tag{6}$$

where it is supposed that

$$\Phi_j(\tau) = \frac{1}{\sqrt{2\pi}}\frac{d}{d\tau}\int_0^{+\infty} \frac{e^{-i\tau t} - 1}{-it}\Phi\big(e^{-i\frac{\pi j}{s+1/2}}t^{\frac{1}{s+1/2}}\big)t^{\mu-1}dt \ (-s \leq j \leq s) \tag{7}$$

and $\Phi_{s+1}(\tau) \equiv \Phi_{-s}(\tau)$.

Here it may be noted that Theorem 2.4-1 contains Theorem 2.2-1 on parametric representation of the class of entire functions $W^{2,\omega}_{1/2,\sigma}$, if the simplest case $s = 0$ $(\rho = 1/2)$ is considered.

(b) If $s \geq 1$ is any natural number, then the set of rays

$$\arg z = \vartheta_j = \frac{\pi}{s}j \qquad (0 \leq |z| < +\infty, j = 0, 1, \dots, 2s - 1) \tag{8}$$

divides the z-plane into $2s$ corner domains which have the same opening π/s. So, if $s = 1$, then the set of rays $\arg z = \vartheta_j = \pi j (j = 0, 1)$ divides the plane into two mutually complementary half-planes $G_\pm = \{z : \pm Imz > 0\}$ with the common boundary $(-\infty, +\infty)$.

We denote by $W^{2,\omega}_{s,\sigma}$ the class of entire functions $\Phi(z)$ of order $s(s \geq 1)$ and of type $\leq \sigma$ (where s is assumed, as before, to be a natural number), which satisfy the conditions

$$\int_0^{+\infty} |\Phi(e^{-i\frac{\pi}{s}j}t)|^2 t^\omega dt < +\infty \qquad (j = 0, 1, \dots, 2s - 1) \tag{9}$$

for a given $\omega \in (-1, 1)$.

Observe that, in the case considered, $\vartheta_{2s} = \vartheta_0 + 2\pi = 2\pi$,

$$\Xi_j = \frac{1}{2}[\vartheta_j + \vartheta_{j+1}] = \frac{\pi}{s}(j + 1/2) \qquad (j = 0, 1, \dots, 2s - 1), \tag{10}$$

and Theorem 2.3-1 takes the following form.

Theorem 2.4-2. *1°. The class $W^{2,\omega}_{s,\sigma}$ $(-1 < \omega < 1, s \geq 1)$ coincides with the set of functions representable in the form*

$$\Phi(z) = \sum_{j=0}^{2s-1} \int_0^\sigma E_s\left(e^{i\frac{\pi}{s}(j+1/2)}z\tau^{\frac{1}{s}}; \mu\right)\tau^{\mu-1}\varphi_j(\tau)d\tau, \qquad z \in \mathbb{C}, \tag{11}$$

where $\mu = (1 + \omega + s)/2s$ and $\varphi_j(\tau) \in L_2(0, \sigma)(j = 0, 1, \dots, 2s - 1)$ are arbitrary functions.
2°. If $\Phi(z) \in W^{2,\omega}_{s,\sigma}$, then the functions $\{\varphi_j(\tau)\}_0^{2s-1}$ of representation (11) are unique and can be determined almost everywhere by the formulas

$$\frac{1}{\sqrt{2\pi s}}\left\{e^{i\frac{\pi}{2}(1-\mu)}\Phi_{j+1}(-\tau) + e^{-i\frac{\pi}{2}(1-\mu)}\Phi_j(\tau)\right\}$$
$$= \begin{cases} \varphi_j(\tau), \ \tau \in (0, \sigma) \\ 0, \ \tau \in (\sigma, +\infty) \end{cases} \qquad (j = 0, 1, \dots, 2s - 1), \tag{12}$$

where it is assumed

$$\Phi_j(\tau) = \frac{1}{\sqrt{2\pi}}\frac{d}{d\tau}\int_0^{+\infty}\frac{e^{-i\tau t} - 1}{-it}\Phi\left(e^{-i\frac{\pi}{s}jt^{\frac{1}{s}}}\right)t^{\mu-1}dt \qquad (0 \leq j \leq 2s - 1) \tag{13}$$

and $\Phi_{2s}(\tau) \equiv \Phi_0(\tau)$.

Finally, it should be noted that the last theorem contains the assertions of Theorem 2.2-3 in the particular case $s = 1$.

2.5 Generalizations of the second fundamental Wiener-Paley theorem.

(a) The class of functions $\Phi(z)$, analytic in the half-plane $\mathbb{C}_+ = \{z : \operatorname{Re} z > 0\}$ and satisfying the condition

$$\sup_{0<x<\infty} \int_{-\infty}^{+\infty} |\, \Phi(x+iy)\,|^2 \, dy < +\infty \tag{1}$$

will be denoted by $H^2(\mathbb{C}_+)$. As is well known, Wiener and Paley, along with the famous Theorem 2.2-4, established their second fundamental theorem stating the parametric representation of the class $H^2(\mathbb{C}_+)$. This theorem is given below.

Theorem 2.5-1. $1°$. *The class $H^2(\mathbb{C}_+)$ coincides with the set of functions representable in the form*

$$\Phi(z) = \int_0^{+\infty} e^{-z\tau}\varphi(\tau)d\tau, \qquad z \in \mathbb{C}_+, \tag{2}$$

where $\varphi(\tau) \in L_2(0, +\infty)$ are arbitrary functions.
$2°$. *If $\Phi(z) \in H^2(\mathbb{C}_+)$, then the function $\varphi(\tau)$ of representation (2) is unique and can be determined almost everywhere by means of the Fourier transform*

$$\frac{1}{2\pi}\frac{d}{d\tau}\int_{-\infty}^{+\infty} \frac{e^{i\tau y}-1}{iy}\Phi(iy)dy = \begin{cases} \varphi(\tau), & \tau > 0, \\ 0, & \tau < 0, \end{cases} \tag{3}$$

where $\Phi(iy) \in L_2(-\infty, +\infty)$ are the boundary values of $\Phi(z)$ on $\partial\mathbb{C}_+ = (-i\infty, +i\infty)$.

(b) Theorem 2.5-1 has essential generalizations which will be stated below, but first it is necessary to introduce some notation. We use $H_2\{\alpha; \omega\}$ $(1/2 < \alpha < +\infty, -1 < \omega < 1)$ for the class of functions $\Phi(z)$ analytic in the corner domain

$$\Delta_\alpha = \left\{ z : |\arg z| < \pi/2\alpha \right\}$$

and satisfying the condition

$$\sup_{|\vartheta|<\pi/2\alpha} \left\{ \int_0^{+\infty} |\, \Phi(re^{i\vartheta})\,|^2 \, r^\omega dr \right\} < +\infty. \tag{4}$$

Then the following theorem is true.

Theorem 2.5-2. 1°. *The class $H_2\{\alpha;\omega\}$ coincides with the set of functions representable in the form*

$$\Phi(z) = \int_0^{+\infty} E_\rho\left(e^{i\frac{\pi}{2\gamma}} z\tau^{\frac{1}{\rho}};\mu\right)\tau^{\mu-1}\varphi_-(\tau)d\tau$$

$$+ \int_0^{+\infty} E_\rho\left(e^{-i\frac{\pi}{2\gamma}} z\tau^{\frac{1}{\rho}};\mu\right)\tau^{\mu-1}\varphi_+(\tau)d\tau, \qquad z \in \Delta_\alpha, \qquad (5)$$

where

$$\rho \geq \frac{\alpha}{2\alpha-1}, \qquad \frac{1}{\gamma} = \frac{1}{\rho} + \frac{1}{\alpha}, \qquad \mu = \frac{1+\omega+\rho}{2\rho} \qquad (6)$$

and $\varphi_\pm(\tau) \in L_2(0,+\infty)$ are arbitrary functions.
2°. *If $\Phi(z) \in H_2\{\alpha;\omega\}$ and conditions (6) are satisfied, then the equality*

$$L_\rho(z;\Phi) \equiv \int_0^{+\infty} E_\rho\left(e^{i\frac{\pi}{2\gamma}} z\tau^{\frac{1}{\rho}};\mu\right)\tau^{\mu-1}\varphi_-(\tau;\Phi)d\tau$$

$$+ \int_0^{+\infty} E_\rho\left(e^{-i\frac{\pi}{2\gamma}} z\tau^{\frac{1}{\rho}};\mu\right)\tau^{\mu-1}\varphi_+(\tau;\Phi)d\tau = \Phi(z), \quad z \in \Delta_\alpha \qquad (7)$$

is true. Here the functions $\varphi_\pm(\tau;\Phi) \in L_2(0,+\infty)$ can be determined almost everywhere by formulas

$$\varphi_\pm(\tau;\Phi) = \frac{\exp\left(\pm i\frac{\pi}{2}(1-\mu)\right)}{2\pi\rho}\frac{d}{d\tau}\int_0^{+\infty}\frac{e^{\pm i\tau t}-1}{\pm it}\Phi\left(e^{\pm i\frac{\pi}{2\alpha}}t^{\frac{1}{\rho}}\right)t^{\mu-1}dt, \qquad (8)$$

where $\Phi\left(e^{\pm i\pi/2\alpha}r^{1/\rho}\right) \in L_{2,\mu}(0,+\infty)$ are the boundary values of $\Phi(z)$ on $\partial\Delta_\alpha$.
3°. *The identity*

$$L_\rho^*\left(e^{i\varphi}r^{1/\rho};\Phi\right) \equiv r^{1-\mu}\left\{\frac{d}{dr}\left[r^\mu \int_0^{+\infty} E_\rho\left(e^{i\left(\frac{\pi}{2\gamma}+\varphi\right)}r^{\frac{1}{\rho}}\tau^{\frac{1}{\rho}};\mu+1\right)\tau^{\mu-1}\varphi_-(\tau;\Phi)d\tau\right]\right.$$

$$\left. + \frac{d}{dr}\left[r^\mu \int_0^{+\infty} E_\rho\left(e^{-i\left(\frac{\pi}{2\gamma}-\varphi\right)}r^{\frac{1}{\rho}}\tau^{\frac{1}{\rho}};\mu+1\right)\tau^{\mu-1}\varphi_+(\tau;\Phi)d\tau\right]\right\}$$

$$= \Phi\left(e^{i\varphi}r^{1/\rho}\right) \qquad \left(\mid\varphi\mid\leq \frac{\pi}{2\alpha},\ 0 < r < +\infty\right)$$

$$(9)$$

is true for all $r > 0$, if $\mid\varphi\mid< \pi/2\alpha$, and for almost all $r > 0$, if $\varphi = \pm\pi/2\alpha$ (in the last case, the notation $\Phi\left(e^{\pm i\pi/2\alpha}r^{1/\rho}\right)$ means the boundary values of Φ).

The following theorem is also true.

Theorem 2.5-3. 1°. *If*

$$\rho > \frac{2\alpha}{2\alpha-1}, \qquad \kappa = \frac{\alpha\rho}{(2\alpha-1)\rho-2\alpha}, \qquad (10)$$

and $\Delta(\kappa;\pi) = \{z :\mid Argz - \pi\mid< \pi/2\kappa\}$, then

$$L_\rho(z;\Phi) \equiv 0, z \in \Delta(\kappa,\pi). \qquad (11)$$

2°. *The identity*

$$L_\rho^*\left(e^{i\varphi}r^{1/\rho};\Phi\right) \equiv 0 \qquad (12)$$

is true for all $r > 0$, if $\mid\pi-\varphi\mid< \pi/2\kappa$, and for almost all $r > 0$, if $\mid\pi-\varphi\mid= \pi/2\kappa$.

(c) The main Theorem 2.5-2 takes the following simple form in the case when $\rho \geq \alpha/(2\alpha - 1)$ takes its minimal value.

Theorem 2.5-4. 1°. *The class* $H_2\{\alpha; \omega\}(1/2 < \alpha < +\infty, -1 < \omega < 1)$ *coincides with the set of functions representable in the form*

$$\Phi(z) = \int_0^{+\infty} E_\rho\left(-z\tau^{1/\rho}; \mu\right) \tau^{\mu-1}\varphi(\tau)d\tau, z \in \Delta_\alpha, \tag{13}$$

where

$$\rho = \frac{\alpha}{2\alpha - 1}, \mu = \frac{1}{2} + (1 + \omega)\left(1 - \frac{1}{2\alpha}\right), \tag{14}$$

and $\varphi(\tau) \in L_2(0, +\infty)$ *are arbitrary functions.*
2°. *If* $\Phi(z) \in H_2\{\alpha; \omega\}$ *and conditions* (14) *are satisfied, then*

$$\Phi(z) = \int_0^{+\infty} E_\rho\left(-z\tau^{1/\rho}; \mu\right) \tau^{\mu-1}\varphi(\tau; \Phi)d\tau, \qquad z \in \Delta_\alpha, \tag{15}$$

where

$$\varphi(\tau; \Phi) = \frac{1}{2\pi\rho}\frac{d}{d\tau}\int_{-\infty}^{+\infty}\frac{e^{-i\tau t} - 1}{-it}\Phi\left(e^{-i\frac{\pi}{2\alpha}\,\mathrm{sign}\,t}\,|\,t\,|^{\frac{1}{\rho}}\right)\left(e^{i\frac{\pi}{2}\,\mathrm{sign}\,t}\,|\,t\,|\right)^{\mu-1}dt. \tag{16}$$

3°. *The representation*

$$\Phi\left(e^{i\varphi}r^{\frac{1}{\rho}}\right) = r^{1-\mu}\frac{d}{dr}\left[r^\mu\int_0^{+\infty} E_\rho\left(-e^{i\varphi}r^{\frac{1}{\rho}}\tau^{\frac{1}{\rho}}; \mu + 1\right)\tau^{\mu-1}\varphi(\tau; \Phi)d\tau\right] \tag{17}$$

is true for all $r > 0$, *if* $|\varphi| < \pi/2\alpha$, *and for almost all* $r > 0$, *if* $\varphi = \pm\pi/2\alpha$.

(d) The following theorem is obtained from Theorem $2.5 - 4(1^\circ)$, where we assume $\alpha = 1$ and $\omega = 0$.

Theorem 2.5-5. *The class* $H_2\{1; 0\} \equiv \overset{*}{H}{}^2(\mathbb{C}_+)$ *of functions* $\Phi(z)$, *analytic in the half-plane* $\mathbb{C}_+$ *and satisfying the condition*

$$\sup_{|\vartheta| < \pi/2}\left\{\int_0^{+\infty}|\Phi(re^{i\vartheta})|^2\,dr\right\} < +\infty, \tag{18}$$

coincides with the set of functions representable by the Laplace integral

$$\Phi(z) = \int_0^{+\infty} e^{-z\tau}\varphi(\tau)d\tau, \qquad z \in \mathbb{C}_+, \tag{19}$$

where $\varphi(\tau) \in L_2(0, +\infty)$ *are arbitrary functions.*

The following important statement is the result of Wiener-Paley Theorem 2.5-1 and Theorem 2.5-5.

Theorem 2.5-6. *The classes $H^2(\mathbb{C}_+)$ and $\overset{*}{H}{}^2(\mathbb{C}_+)$ coincide:*

$$H^2(\mathbb{C}_+) \equiv \overset{*}{H}{}^2(\mathbb{C}_+). \tag{20}$$

Note that an essential generalization of the last identity is contained in Theorem 3.2-1 of the next chapter.

(e) Now we state an additional result connected with the last two theorems. To this end, assume

$$\rho > \frac{1}{2}, \ \mu = \frac{1 + \omega + \rho}{2\rho} \qquad (-1 < \omega < 1) \tag{21}$$

and introduce the mutually complementary corner domains

$$\Delta_\rho = \{z : |\arg z| < \pi/2\rho\}, \qquad \Delta_\rho^* = \mathbb{C} \setminus \overline{\Delta_\rho}. \tag{22}$$

Then the following theorem is true.

Theorem 2.5-7. *If $\varphi(\tau) \in L_2(0, +\infty)$ is any function, then the integral*

$$\Phi(z) = \int_0^{+\infty} E_\rho \left(z\tau^{1/\rho}; \mu \right) \tau^{\mu-1} \varphi(\tau) d\tau, \qquad z \in \Delta_\rho^*, \tag{23}$$

represents a function analytic in Δ_ρ^* and satisfying the condition

$$\sup_{\pi/2\rho < |\vartheta| \le \pi} \left\{ \int_0^{+\infty} |\Phi(re^{i\vartheta})|^2 \, r^\omega dr \right\} \le M_\mu \int_0^{+\infty} |\varphi(\tau)|^2 \, d\tau, \tag{24}$$

where the constant $M_\mu > 0$ depends only on μ.

2.6 Notes

In should be mentioned that the fundamental Theorems 2.2-4 and 2.5-1 were established by Wiener and Paley [1] in the 1930s.

2.2-2.4 The results of these sections relating to parametric representations of different classes of entire functions of finite order are improvements of the first fundamental Wiener-Paley Theorem 2.2-4 (see M.M. Djrbashian [5, Chapter 6]).

2.5 The results of this section are generalizations of the second fundamental Wiener-Paley Theorem 2.5-1. The first step of investigations establishing such generalizations was the paper of M.M. Djrbashian and A.E. Avetisian [1]. Further, the frames of these investigations were extended in M.M. Djrbashian [5, Chapter 7]. More detailed notes relating to the results of these investigations are given in M.M. Djrbashian [5, pp. 664–665].

3 Some estimates in Banach spaces of analytic functions

3.1 Introduction

The series of lemmas and theorems proved in this chapter establishes some estimates of norms in different weighted spaces of functions analytic in a half-plane and also in different weighted spaces of entire functions of exponential type. Later chapters of the book are based on these results and the results of Chapters 1 and 2.

3.2 Some estimates in Hardy classes over a half-plane

First we introduce some classes of analytic functions and mention a series of their most important properties.

(a) Let $H_\pm^p (0 < p < +\infty)$ be the Hardy classes over the half-planes

$$G_\pm = \{z : \pm \operatorname{Im} z > 0\}. \tag{1}$$

In other words, let $H_\pm^p$ be the classes of functions $\omega_\pm(z)$ analytic in $G_\pm$ and satisfying, correspondingly, the conditions

$$\sup_{y>0} \left\{ \int_{-\infty}^{+\infty} |\, \omega_\pm(x \pm iy)\, |^p \, dx \right\}^{1/p} < +\infty. \tag{2}$$

The following assertions are well known.

$1°$. If $\omega_\pm(z) \in H_\pm^p (0 < p < +\infty)$, then its non-tangential boundary values $\omega_\pm(x)$ exist for almost all $x \in (-\infty, +\infty)$. Moreover,

$$\lim_{y \to +0} \omega_\pm(x \pm iy) = \omega_\pm(x) \in L_p(-\infty, +\infty) \tag{3}$$

and

$$\|\omega_\pm\|_p = \sup_{y>0} \left\{ \int_{-\infty}^{+\infty} |\, \omega_\pm(x \pm iy)\, |^p \, dx \right\}^{1/p},$$

if it is assumed that

$$\|\omega_\pm\|_p = \left\{ \int_{-\infty}^{+\infty} |\, \omega_\pm(x)\, |^p \, dx \right\}^{1/p}.$$

$2°$. The following relations are true:

$$\|\omega_\pm\|_p = \lim_{y \to +0} \|\omega_\pm(x \pm iy)\|_p, \quad \lim_{y \to +0} \|\omega_\pm(x) - \omega_\pm(x \pm iy)\|_p = 0, \tag{4}$$

$$\|\omega_\pm(x \pm iy)\|_p \leq \|\omega_\pm\|_p (0 < y < +\infty). \tag{5}$$

It should be noted that if $1 \leq p < +\infty$, then the classes $H_\pm^p$ are Banach spaces with norm (2).

(b) It is necessary now to introduce the classes $\overset{*}{H}{}^p{}_\pm$ $(0 < p < +\infty)$ which are more general than the classes $\overset{*}{H}{}^2$ $(\mathbb{C}_+)$ introduced by 2.5(18). We shall say that a function $\omega_\pm(z)$, analytic in a half-plane $G_\pm$ is of class $\overset{*}{H}{}^p{}_\pm$ $(0 < p < +\infty)$, if

$$\|\omega_\pm\|_p^* = \sup_{0 < \vartheta < \pi} \left\{ \int_0^{+\infty} |\,\omega_\pm(re^{\pm i\vartheta})\,|^p \, dr \right\}^{1/p} < +\infty. \tag{6}$$

If z is replaced in Theorem 2.5-6 by $\mp iz$, then it easily follows that the classes $H_\pm^p$ and $\overset{*}{H}{}^p{}_\pm$ coincide when $p = 2$. Later a theorem will be proved where the equality $H_\pm^p = \overset{*}{H}{}^p{}_\pm$ is stated for any $p \in (0, +\infty)$.

(c) First we prove some auxiliary assertions.

Lemma 3.2-1. *If* $\omega(z) \in \overset{*}{H}{}_+^2$ *and*

$$\int_0^{+\infty} |\,\omega(x)\,|^2 \, dx \leq M_1 < +\infty, \qquad \int_{-\infty}^0 |\,\omega(x)\,|^2 \, dx \leq M_2 < +\infty, \tag{7}$$

then

$$\int_0^{+\infty} |\,\omega(re^{i\vartheta})\,|^2 \, dr \leq M_1^{1-\vartheta/\pi} M_2^{\vartheta/\pi}, \qquad 0 < \vartheta < \pi. \tag{8}$$

Proof. As it is known, the limits in the metric of $L_2(0, +\infty)$

$$\mathop{\text{l.i.m.}}_{\vartheta \to 0} \omega(re^{i\vartheta}) = \omega(r), \qquad \mathop{\text{l.i.m.}}_{\vartheta \to \pi - 0} \omega(re^{i\vartheta}) = \omega(-r) \tag{9}$$

exist for any function $\omega(z) \in \overset{*}{H}{}_+^2$ $(\equiv H_+^2)$ and, what is more, $\omega(\pm r) \in L_2(0, +\infty)$. Thus, according to Theorem 1.5-1, the Mellin transforms of the functions $\omega(\pm r)$ and $\omega(re^{i\vartheta})$ $(0 \leq \vartheta \leq \pi)$

$$\begin{aligned}
\Omega_\pm(s) &= \mathop{\text{l.i.m.}}_{\alpha \to +\infty} \int_{1/a}^a \omega(\pm r) r^{s-1} dr, && \operatorname{Re} s = 1/2, \\
\Omega(s; \vartheta) &= \mathop{\text{l.i.m.}}_{\alpha \to +\infty} \int_{1/a}^a \omega(re^{i\vartheta}) r^{s-1} dr, && \operatorname{Re} s = 1/2 \quad (0 \leq \vartheta \leq \pi)
\end{aligned} \tag{9'}$$

converging in the norm of $L_2(1/2 - i\infty, 1/2 + i\infty)$, also exist. In addition, it is obvious that

$$\Omega(s; 0) = \Omega_+(s), \qquad \Omega(s; \pi) = \Omega_-(s).$$

It is also known that the functions of (9′) are connected by the equalities

$$\Omega(s;\vartheta) = e^{is(\pi-\vartheta)}\Omega_-(s) = e^{-is\vartheta}\Omega_+(s) \quad (0 \le \vartheta \le \pi) \tag{9''}$$

which are true almost everywhere on the line $s = 1/2 + it$ $(-\infty < t < +\infty)$. It follows that for almost all $t \in (-\infty, +\infty)$

$$\left|\Omega\left(\frac{1}{2}+it;\vartheta\right)\right|^2 = e^{2(\vartheta-\pi/2)t}\left|\Omega_-\left(\frac{1}{2}+it\right)\Omega_+\left(\frac{1}{2}+it\right)\right|, 0 \le \vartheta \le \pi.$$

So, using the Parseval equality 1.5(6), we obtain

$$M(\vartheta) \equiv \int_0^{+\infty}\left|\omega\left(re^{i\vartheta}\right)\right|^2 dr = \frac{1}{2\pi}\int_{-\infty}^{+\infty}\left|\Omega\left(\frac{1}{2}+it;\vartheta\right)\right|^2 dt$$

$$= \frac{1}{2\pi}\int_{-\infty}^{+\infty}e^{2(\vartheta-\pi/2)t}\left|\Omega_-\left(\frac{1}{2}+it\right)\Omega_+\left(\frac{1}{2}+it\right)\right|dt < +\infty, \quad 0 \le \vartheta \le \pi.$$

$$\tag{10}$$

Hence, it follows, in particular, that $M(\vartheta)$ is continuous on $[0,\pi]$. In addition, by (7),

$$M(0) = \int_0^{+\infty}|\omega(r)|^2 dr \le M_1, \qquad M(\pi) = \int_0^{+\infty}|\omega(-r)|^2 dr \le M_2. \tag{11}$$

Now let ϑ_1 and ϑ_2 be any numbers, such that $0 \le \vartheta_1 < \vartheta_2 \le \pi$. Then, applying the Schwarz inequality, it can be obtained from (10) that

$$M\left(\frac{\vartheta_1+\vartheta_2}{2}\right) = \frac{1}{2\pi}\int_{-\infty}^{+\infty}e^{(\vartheta_1+\vartheta_2-\pi)t}\left|\Omega_-\left(\frac{1}{2}+it\right)\Omega_+\left(\frac{1}{2}+it\right)\right|dt$$

$$\le \sqrt{M(\vartheta_1)M(\vartheta_2)},$$

and it follows that the function $\log M(\vartheta)$ is convex on $[0,\pi]$. So, according to (11),

$$\log M(\vartheta) \le \frac{\pi-\vartheta}{\pi}\log M_1 + \frac{\vartheta}{\pi}\log M_2$$

and

$$M(\vartheta) \le M_1^{1-\frac{\vartheta}{\pi}}M_2^{\frac{\vartheta}{\pi}}.$$

Thus the estimate (8) and the lemma are proved.

(d) Now we prove a series of lemmas concerning a class $\overset{*}{H}{}_+^{p,\epsilon}$ $(0 < p < +\infty,$ $0 < \epsilon < \pi/2)$ rather different from $\overset{*}{H}{}_+^p$ considered above. A function $\omega(z)$ analytic in the corner domain $-\epsilon < \mathrm{Arg}\, z < \pi + \epsilon$ is said to be of $\overset{*}{H}{}_+^{p,\epsilon}$, if

$$\|\omega\|_{p,\epsilon}^* \equiv \sup_{-\epsilon<\vartheta<\pi+\epsilon}\left\{\int_0^{+\infty}\left|\omega(re^{i\vartheta})\right|^p dr\right\}^{1/p} < +\infty. \tag{12}$$

Lemma 3.2-2. *If* $\omega(z) \in \overset{*}{H}{}^{p,\epsilon}_+$ $(0 < p < +\infty, 0 < \epsilon < \pi/2)$, *then*

$$| \omega(z) |^p \le C/ | z |, \qquad z \in G_+, \tag{13}$$

where the constant $C > 0$ *does not depend on* z.

Proof. We shall denote the disk with its center at the point $z \in G_+$ and with the radius $| z |\sin \epsilon$ by $K(z)$. Further, we assume that Arg $\zeta = \alpha$ and Arg $\zeta = \beta$ $(0 < \beta - \alpha < 2\pi)$ are the rays tangent to the disk $K(z)$, and $R(z)$ is the intersection of the corner domain $\alpha < \text{Arg } \zeta < \beta$ and the domain

$$r_1 = | z | (1 - \sin \epsilon) < | \zeta | < | z | (1 + \sin \epsilon) = r_2.$$

Then the geometric consideration obviously shows that the domains $K(z)$ and $R(z)$ are contained in the initial domain of analyticity of the function $\omega(z)$. Since the function $| \omega(\zeta) |^p$ $(\zeta = \xi + i\eta)$ is subharmonic in the disk $K(z)$, the following inequalities can be obtained easily:

$$\pi \sin^2 \epsilon \, | z |^2 | \omega(z) |^p \le \iint_{K(z)} | \omega(\zeta) |^p \, d\xi d\eta$$

$$\le \int_\alpha^\beta d\vartheta \int_{r_1}^{r_2} | \omega(re^{i\vartheta}) |^p \, r dr \le r_2 \int_\alpha^\beta d\vartheta \int_{r_1}^{r_2} | \omega(re^{i\vartheta}) |^p \, dr$$

$$\le r_2(\beta - \alpha) \left[\|\omega\|^*_{p,\epsilon} \right]^p \le (\beta - \alpha)(1 + \sin \epsilon) \, | z | \left[\|\omega\|^*_{p,\epsilon} \right]^p .$$

Hence the estimate (13) follows at once.

Lemma 3.2-3. *Let* $\omega(z) \in \overset{*}{H}{}^{p,\epsilon}_+$ $(0 < p < +\infty, 0 < \epsilon < \pi/2)$, *and let* $m > 1/p$ *be any integer. Also, let*

$$\omega_\delta(z) = \left(\frac{z}{z + i\delta} \right)^m \frac{e^{i\delta z} - 1}{i\delta z} \omega(z), \qquad z \in G_+, \tag{14}$$

where $\delta > 0$ *is any number. Then* $\omega_\delta(z) \in H^p_+$.

Proof. First observe that, if $\Pi^+ = \{z : -1 < \text{Re} \, z < 1, \, 0 < \text{Im} \, z < 1\}$, is a rectangle, then

$$\sup_{\Pi^+} \left| \frac{e^{i\delta z} - 1}{i\delta z} \right| \le C_1, \qquad \sup_{\Pi^+} \left| \frac{1}{z + i\delta} \right|^m \le C_2,$$

where the constants $C_1 > 0$, $C_2 > 0$ do not depend on z (such constants will be denoted further by C_k $(k \ge 1)$). Hence it follows, according to (13) and (14), that

$$| \omega_\delta(z) | \le C_3 | z |^{m - 1/p} \le C_4, \qquad z \in \Pi^+,$$

and, therefore,

$$\sup_{0<y<1}\left\{\int_{-1}^{1}|\,\omega_\delta(x+iy)\,|^p\,dx\right\}<+\infty. \tag{15}$$

On the other hand, it is obvious that

$$\left|\frac{z}{z+i\delta}\right|^{mp}\left|\frac{e^{i\delta z}-1}{i\delta z}\right|^{p}\le C_5\,|z|^{-p}\quad(0<\operatorname{Im}z<1/2,|\operatorname{Re}z|\ge 1).$$

Thus the estimate

$$|\omega_\delta(z)|^p\le C_5\,|z|^{-p-1}\qquad(0<\operatorname{Im}z<1/2,|\operatorname{Re}z|\ge 1)$$

also follows from (13) and (14). Consequently,

$$\int_{1}^{+\infty}|\omega_\delta(x+iy)|^p\,dx\le C_5\int_{1}^{+\infty}\frac{dx}{(x-y)^{p+1}}=\frac{C_6}{(1-y)^p},\qquad 0<y<\frac{1}{2},$$

and

$$\sup_{0<y<1/2}\left\{\int_{1}^{+\infty}|\omega_\delta(x+iy)|^p\,dx\right\}<+\infty. \tag{16}$$

The estimate

$$\sup_{0<y<1/2}\left\{\int_{-\infty}^{-1}|\omega_\delta(x+iy)|^p\,dx\right\}<+\infty \tag{17}$$

can also be obtained by similar arguments. Further, it is obvious that

$$\left|\frac{z}{z+i\delta}\right|^{mp}\left|\frac{e^{i\delta z}-1}{i\delta z}\right|^{p}\le C_7\,|z|^{-p}\qquad(\operatorname{Im}z\ge 1/2).$$

Consequently,

$$|\omega_\delta(z)|^p\le C_8\,|z|^{-p-1}\qquad(\operatorname{Im}z\ge 1/2)$$

and

$$\sup_{y\ge 1/2}\left\{\int_{-\infty}^{+\infty}|\omega_\delta(x+iy)|^p\,dx\right\}<+\infty. \tag{18}$$

The comparison of estimates (15)-(18) gives

$$\sup_{y\ge 0}\left\{\int_{-\infty}^{+\infty}|\omega_\delta(x+iy)|^p\,dx\right\}<+\infty,$$

i.e., $\omega_\delta(z)\in H_+^p$.

Lemma 3.2-4. *If $\omega(z)\in \overset{*}{H}{}_+^{p,\epsilon}$ $(0<p<+\infty,0<\epsilon<\pi/2)$, then $\omega(z)\in H_+^p$ and*

$$\|\omega\|_p\le 2^{1/p}\|\omega\|_p^{*}. \tag{19}$$

Proof. The auxiliary function

$$\tilde{\omega}_\delta(x) = \left(\frac{x}{x+i\delta}\right)^m \frac{e^{i\delta x}-1}{i\delta x}, \qquad x \in (-\infty,+\infty)$$

will be considered now, along with $\omega_\delta(z)$ introduced above, assuming again that $m > 1/p$ is an integer and $\delta > 0$ is any number. It is obvious that

$$|\omega(x) - \omega_\delta(x)|^p = |\omega(x)|^p \, |1 - \tilde{\omega}_\delta(x)|^p \tag{20}$$

and

$$|\tilde{\omega}_\delta(x)| \le C_9, \qquad \lim_{\delta \to +0} \tilde{\omega}_\delta(x) = 1, \qquad x \in (-\infty,0) \cup (0,+\infty). \tag{21}$$

But $\omega(x) \in L_p(-\infty,+\infty)$, according to (12), so the relation

$$\lim_{\delta \to +0} \|\omega - \omega_\delta\|_p = 0 \tag{22}$$

follows from (20) and (21) by the Lebesgue theorem on bounded convergence. Further, as is well known, the space H_+^p is complete. Therefore $\omega(z) \in H_+^p$ by Lemma 3.2-3 and (22). On the other hand, $\omega(z)$ is analytic in the corner domain $-\epsilon < \mathrm{Arg}\, z < \pi + \epsilon$. Thus

$$\int_{1/\sigma}^{\sigma} |\omega(x)|^p \, dx = \lim_{\vartheta \to +0} \int_{1/\sigma}^{\sigma} \left|\omega(re^{i\vartheta})\right|^p \, dr \le \left(\|\omega\|_p^*\right)^p,$$

$$\int_{1/\sigma}^{\sigma} |\omega(-x)|^p \, dx = \lim_{\vartheta \to \pi-0} \int_{1/\sigma}^{\sigma} \left|\omega(re^{i\vartheta})\right|^p \, dr \le \left(\|\omega\|_p^*\right)^p$$

for any $\sigma > 1$. Adding these inequalities and letting $\sigma \to +\infty$, we obtain the estimate (19).

(e) Now we are ready to prove the main result of this section.

Theorem 3.2-1. *Let $p \in (0,+\infty)$ be an arbitrary number. Then the following assertions are true:*

1°. *If $\omega(z) \in \overset{*}{H}_+^p$ and*

$$\int_0^{+\infty} |\omega(x)|^p \, dx \le M_1 < +\infty, \quad \int_{-\infty}^0 |\omega(x)|^p \, dx \le M_2 < +\infty, \tag{23}$$

then

$$\int_0^{+\infty} \left|\omega(re^{i\vartheta})\right|^p \, dr \le M_1^{1-\vartheta/p} M_2^{\vartheta/p}, \qquad 0 < \vartheta < \pi. \tag{24}$$

2°. $\overset{*}{H}_+^p = H_+^p.$ \hfill (25)

3°. *If $\omega(z) \in \overset{*}{H}_+^p = H_+^p$ is any function, then*

$$2^{-1/p}\|\omega\|_p \le \|\omega\|_p^* \le \|\omega\|_p. \tag{26}$$

Proof. 1°. As is well known, any function $\omega(z) \in H_+^p$ is representable in the form $\omega(z) = B(z)\omega_*(z)$, $z \in G_+$, where $B(z)$ is the Blaschke product containing its zeros, and $\omega_*(z) \in H_+^p$ is a non-vanishing function. Obviously, $\tilde{\omega}(z) = [\omega_*(z)]^{p/2} \in H_+^2$ and also $\tilde{\omega}(z) \in \overset{*}{H}{}_+^2$, because the classes H_+^2 and $\overset{*}{H}{}_+^2$ coincide, as was mentioned in Section 3.2(b). On the other hand, $|B(z)| \leq 1$, $z \in G_+$ and, consequently, $|\omega(z)|^p \leq |\omega_*(z)|^p$, $z \in G_+$. Hence

$$\int_0^{+\infty} |\omega(re^{i\vartheta})|^p \, dr \leq \int_0^{+\infty} |\omega_*(re^{i\vartheta})|^p \, dr = \int_0^{+\infty} |\tilde{\omega}(re^{i\vartheta})|^2 \, dr, \quad 0 \leq \vartheta \leq \pi. \tag{27}$$

And, since $\tilde{\omega}(z) \in H_+^2 = \overset{*}{H}{}_+^2$,

$$\sup_{0 < \vartheta < \pi} \left\{ \int_0^{+\infty} |\omega(re^{i\vartheta})|^p \, dr \right\} < +\infty.$$

Thus the inclusion $H_+^p \subseteq \overset{*}{H}{}_+^p$ is proved. In addition, it is clear that

$$\int_0^{+\infty} |\tilde{\omega}(x)|^2 \, dx = \int_0^{+\infty} |\omega(x)|^p \, dx \leq M_1 < +\infty,$$
$$\int_{-\infty}^0 |\tilde{\omega}(x)|^2 \, dx = \int_{-\infty}^0 |\omega(x)|^p \, dx \leq M_2 < +\infty.$$

And the inclusion $\tilde{\omega}(z) \in \overset{*}{H}{}_+^2$, together with Lemma 3.2-1, leads to the estimate

$$\int_0^{+\infty} |\tilde{\omega}(re^{i\vartheta})|^2 \, dr \leq M_1^{1-\vartheta/\pi} M_2^{\vartheta/\pi}, \qquad 0 < \vartheta < \pi,$$

which proves, by virtue of inequality (27), assertion 1°, but only for functions $\omega(z) \in H_+^p \subseteq \overset{*}{H}{}_+^p$.

2°. The inclusion $H_+^p \subseteq \overset{*}{H}{}_+^p$ is already proved, so it remains to prove the converse inclusion, i.e., if $\omega(z) \in \overset{*}{H}{}_+^p$, then $\omega(z) \in H_+^p$. To this end we note that the suitably selected branch of the function

$$w = i \left(e^{-i\pi/2} z \right)^{1+2\epsilon/\pi}$$

maps conformally the half-plane G_+ onto the corner domain $-\epsilon < \text{Arg } z < \pi + \epsilon$. The image of the ray Arg $z = \vartheta$ will be denoted by Arg $w = t$. Then, if $\gamma = 2\epsilon/\pi$, the equality

$$\int_0^{e^{i\vartheta}\infty} |\omega(z)|^p \, |dz| = \frac{1}{1+\gamma} \int_0^{e^{it}\infty} \left| \omega \left(i \left(e^{-i\pi/2} w \right)^{\frac{1}{1+\gamma}} \right) \right|^p |w|^{-\frac{\gamma}{1+\gamma}} \, |dw|$$

implies that the function

$$\tilde{\omega}_\epsilon(w) = (1+\gamma)^{-\frac{1}{p}} w^{-\frac{\gamma}{p(1+\gamma)}} \omega\left(i\left(e^{-i\pi/2}w\right)^{\frac{1}{1+\gamma}}\right) \tag{28}$$

is of class $\overset{*}{H}{}_{+}^{p,\epsilon}$ and, what is more, $\|\tilde{\omega}_\epsilon\|_{p,\epsilon}^* = \|\omega\|_p^*$. Thus, by Lemma 3.2-4,

$$\tilde{\omega}_\epsilon(w) \in H_+^p, \qquad \|\tilde{\omega}_\epsilon\|_p \le 2^{1/p}\|\omega\|_p^* \tag{29}$$

or, what is the same,

$$\int_{-\infty}^{+\infty} |\tilde{\omega}_\epsilon(x+iy)|^p \, dx \le 2\left(\|\omega\|_p^*\right)^p, \qquad y \in (0, +\infty). \tag{29'}$$

Now note that

$$\lim_{\epsilon \to +0} \tilde{\omega}_\epsilon(w) = \omega(w), \qquad w \in G_+,$$

as follows from (28). On the other hand, Fatou's lemma gives

$$\int_{-\infty}^{+\infty} \liminf_{\epsilon \to +0} |\tilde{\omega}_\epsilon(x+iy)|^p \, dx \le \liminf_{\epsilon \to +0} \int_{-\infty}^{+\infty} |\tilde{\omega}_\epsilon(x+iy)|^p \, dx.$$

Therefore, by (29')

$$\int_{-\infty}^{+\infty} |\omega(x+iy)|^p \, dx \le 2\left(\|\omega\|_p^*\right)^p, \qquad y \in (0, +\infty).$$

Hence the inclusion $\omega(z) \in H_+^p$ and the estimate

$$\|\omega\|_p \le 2^{1/p}\|\omega\|_p^*$$

follow. Thus, $\overset{*}{H}{}_{+}^p \subseteq H_+^p$, and the assertion $2°$ is proved, and assertion $1°$ is completely proved.

$3°$. As the last estimate is proved, it remains to prove only the right-hand side inequality of (26). To this end, introduce the function

$$\tilde{\Omega}(\vartheta) = \left\{\int_0^{+\infty} |\omega(x)|^p \, dx\right\}^{1-\vartheta/\pi} \left\{\int_{-\infty}^0 |\omega(x)|^p \, dx\right\}^{\vartheta/\pi}, \qquad 0 \le \vartheta \le \pi$$

and observe that

$$\int_0^{+\infty} |\omega(re^{i\vartheta})|^p \, dr \le \tilde{\Omega}(\vartheta), \qquad 0 \le \vartheta \le \pi, \tag{30}$$

as follows from assertion $1°$. The function $\tilde{\Omega}(\vartheta)$ is not only continuous, but it is also monotonic or constant on $[0, \pi]$. Thus $\tilde{\Omega}(\vartheta) \le \max\{\tilde{\Omega}(0); \tilde{\Omega}(\pi)\} \le \|\omega\|_p^p$, $0 \le \vartheta \le \pi$, and the inequality $\|\omega\|_p^* \le \|\omega\|_p$ follows from (30). So the proof of Theorem 3.2-1 is complete.

3.3 Some estimates in weighted Hardy classes over a half-plane.

(a) Let $K(t)$ be an arbitrary measurable function on $(-\infty, +\infty)$, satisfying the condition

$$\frac{K(t)}{1 + |t|} \in L_1(-\infty, +\infty). \tag{1}$$

Then the integral

$$\mathcal{K}(z) = \frac{1}{2\pi i} \int_{-\infty}^{+\infty} \frac{K(t)}{t - z} dt = \begin{cases} \mathcal{K}^+(z), z \in G_+, \\ \mathcal{K}^-(z), z \in G_-, \end{cases} \tag{2}$$

called a *Cauchy type integral*, uniformly converges in $G_+ \cup G_-$, and so the functions $\mathcal{K}^+(z)$ and $\mathcal{K}^-(z)$ are analytic in the half-planes G_+ and G_- correspondingly. If we suppose that $K(t)$ is an arbitrary function of class $L_p(-\infty, +\infty)$ $(1 < p < +\infty)$, then, using Hölder's inequality, it can be easily verified that $K(t)$ satisfies condition (1). The well-known Riesz-Titchmarsh theorem stated below relates to the projection of $L_p(-\infty, +\infty)$ onto $H_{\pm}^p$, which is realized by integral (2).

Theorem 3.3-1. *Let $K(t) \in L_p(-\infty, +\infty)$ $(1 < p < +\infty)$. Then the Cauchy type integral (2) satisfies the conditions*

$$\mathcal{K}^{\pm}(z) \in H_{\pm}^p, \|\mathcal{K}^{\pm}\|_p \le A_p \|K\|_p, \tag{3}$$

where the constant $A_p > 0$ depends only on p.

The comparison of this theorem and Theorem 3.2-1 shows that the estimates

$$\sup_{0 < \vartheta < \pi} \left\{ \int_0^{+\infty} \left| \mathcal{K}(re^{\pm i\vartheta}) \right|^p dr \right\}^{1/p} \le A_p \left| K \right|_p \tag{4}$$

are true provided that $K(t) \in L_p(-\infty, +\infty)$ $(1 < p < +\infty)$.

(b) Now we shall prove a weighted analog of Theorem 3.3-1.

Theorem 3.3-2. *Let $K(t)$ be a measurable function on $(-\infty, +\infty)$, such that*

$$\|K\|_{p,\kappa} = \left\{ \int_{-\infty}^{+\infty} |K(t)|^p |t|^\kappa dt \right\}^{1/p} < +\infty \quad (1 < p < +\infty, -1 < \kappa < p - 1). \tag{5}$$

Then the Cauchy type integral (2) is analytic in half-planes G_+ and G_-, and, in addition,

$$\sup_{0 < \vartheta < \pi} \left\{ \int_0^{+\infty} |\mathcal{K}(re^{\pm i\vartheta})|^p r^\kappa dr \right\}^{1/p} \le A_{p,\kappa} \|K\|_{p,\kappa}, \tag{6}$$

where the constant $A_{p,\kappa} > 0$ depends only on p and κ.

Proof. Using Hölder's inequality we arrive at the estimate

$$\int_{-\infty}^{+\infty} \frac{|K(t)|}{1 + |t|} dt \leq \|K\|_{p,\kappa} \left\{ \int_{-\infty}^{+\infty} \frac{|t|^{-\kappa q/p}}{(1 + |t|)^q} \right\}^{1/q} < +\infty,$$

where $q = p/(p - 1)$ and the boundedness of the right-hand side quantity follows from the boundedness of the integral (5), since $1 < p < +\infty$ and $-1 < \kappa < p - 1$. Thus, $K(t)$ satisfies condition (1), and the Cauchy type integral (2) is analytic in both half-planes G_+ and G_-. Further, we denote

$$\mathcal{K}_1(z) = \frac{1}{2\pi i} \int_0^{+\infty} \frac{K(t)}{t - z} dt, \, z \in G_+ \cup G_- \tag{7}$$

and prove the inequalities

$$\sup_{0 < \vartheta < \pi} \left\{ \int_0^{+\infty} |\mathcal{K}_1(re^{\pm i\vartheta})|^p r^\kappa dr \right\}^{1/p} \leq B_{p,\kappa} \|K\|_{p,\kappa}, \tag{8}$$

where the constant $B_{p,\kappa} > 0$ is assumed to depend only on p and κ, as the first step of the proof of the estimate (6). To this end we denote

$$H(\pm\vartheta; r; r - t) = \frac{|r - t|}{t - re^{\pm i\vartheta}}, \qquad \vartheta \in (0, \pi),$$

and introduce the operators $T^{(\pm\vartheta)}$ by the formula

$$T^{(\pm\vartheta)}(K_+)(r) = \frac{1}{2\pi i} \int_0^{+\infty} \left\{ H(\pm\vartheta; r; r - t) \, |r - t|^{-1} \right\} K_+(t) dt, \tag{9}$$

where $K_+(t) \in L_p(0, +\infty)$. One can prove that $T^{(\pm\vartheta)}$ are bounded operators on $L_p(0, +\infty)$ and, moreover, if $K_+(t) \in L_p(0, +\infty)$ is any function, and if

$$\|K_+\|_p^+ = \left\{ \int_0^{+\infty} |K_+(t)|^p \, dt \right\}^{1/p}, \tag{10}$$

then

$$\|T^{(\pm\vartheta)}(K_+)\|_p^+ \leq A_p \|K_+\|_p^+, \tag{11}$$

where the constant $A_p > 0$ depends only on p. Indeed, it is enough to put

$$K(t) = \begin{cases} K_+(t), & t \in (0, +\infty) \\ 0, & t \in (-\infty, 0) \end{cases} \tag{12}$$

in Theorem 3.3-1 and to observe that, in this case, $\mathcal{K}(z) = \mathcal{K}_1(z)$ and

$$T^{(\pm\vartheta)}(K_+)(r) = \mathcal{K}_1(re^{\pm i\vartheta}) \qquad (0 < \vartheta < \pi, 0 < r < +\infty). \tag{13}$$

Therefore, inequality (4) precisely takes the form (11). On the other hand, it is obvious that $|H(\pm\vartheta; r; r - t)| \leq 1 (0 < \vartheta < \pi, 0 < r, t < +\infty)$. Hence, a theorem proved by Stein [1] leads to the conclusion that the estimate

$$\|T^{(\pm\vartheta)}(K_+)(r)r^{\kappa/p}\|_p^+ \leq B_{p,\kappa}\|K_+(t)t^{\kappa/p}\|_p^+, \tag{14}$$

where the constant $B_{p,\kappa} > 0$ is of the same type as in (8), is true for any function $K_+(t)$ measurable on $(0, +\infty)$. Comparing (7), (12) and (13) and using the estimate (14) we arrive at equality (8). Now observe that, if similar to (7),

$$\mathcal{K}_2(z) = \frac{1}{2\pi i}\int_{-\infty}^0 \frac{K(z)}{t - z}dt, \qquad z \in G_+ \cup G_-, \tag{15}$$

then the same arguments evidently lead to the estimate

$$\sup_{0 < \vartheta < \pi}\left\{\int_0^{+\infty}\left|\mathcal{K}_2(re^{\pm i\vartheta})\right|^p r^\kappa dr\right\}^{1/p} \leq B_{p,\kappa}\|K\|_{p,\kappa}. \tag{16}$$

But, by (2),(7) and (15), $\mathcal{K}(z) = \mathcal{K}_1(z) + \mathcal{K}_2(z), \qquad z \in G_+ \cup G_-$. Thus, (6) follows from (8) and (16).

(c) It is necessary now to introduce the norm

$$\|K\|_{p,\kappa,1} = \left\{\int_{-\infty}^{+\infty}|K(t)|^p\left(1 + t^2\right)^{\kappa/2}dt\right\}^{1/p}, \tag{17}$$

which differs from $\|.\|_{p,\kappa}$ considered above, and to prove the following theorem, which is not a simple consequence of Theorems 3.2-1 and 3.3-1.

Theorem 3.3-3. *Let $K(t)$ be a measurable function on $(-\infty, +\infty)$, having finite norm $\|K\|_{p,\kappa,1}$ $(1 < p < +\infty, -1 < \kappa < p - 1)$. Then the following assertions are true for the Cauchy type integral (2):*
1°. If we single out any univalent branches of functions $(z \pm i)^{\kappa/p}(z \in G_\pm)$ in the z-plane cut along $(-i\infty, -i) \subset G_-$ or $(+i, +i\infty) \subset G_+$ correspondingly, then

$$(z \pm i)^{\kappa/p}\mathcal{K}^\pm(z) \in H_\pm^p. \tag{18}$$

$$2°. \qquad \|(x \pm i)^{\kappa/p}\mathcal{K}^\pm(x)\|_p \leq A_{p,\kappa}\|K\|_{p,\kappa,1}, \tag{19}$$

where the constant $A_{p,\kappa} > 0$ depends only on p and κ.

Proof. First we introduce the integrals

$$\mathcal{K}_1^\pm(z) = \frac{1}{2\pi i}\int_0^1 \frac{K(t)}{t - z}dt, \quad \mathcal{K}_2^\pm(z) = \frac{1}{2\pi i}\int_1^{+\infty}\frac{K(t)}{t - z}dt, \quad z \in G_\pm, \tag{20}$$

and prove the estimates

$$\sup_{0<\vartheta<\pi}\left\{\int_0^{+\infty}\left|\mathcal{K}_j^+(re^{i\vartheta})\right|^p\left|re^{i\vartheta}+i\right|^\kappa dr\right\}^{1/p}\le C_1\|K\|_{p,\kappa,1}\,(j=1,2),\quad(21)$$

$$\sup_{0<\vartheta<\pi}\left\{\int_0^{+\infty}\left|\mathcal{K}_j^-(re^{-i\vartheta})\right|^p\left|re^{-i\vartheta}-i\right|^\kappa dr\right\}^{1/p}\le C_2\|K\|_{p,\kappa,1}\,(j=1,2),\quad(22)$$

which will be used later. Here and elsewhere $C_n>0$ $(n=1,2,\dots)$ will mean constants depending only on p and κ. To this end, using Hölder's inequality and the simple estimate $1\le|t+i|\le\sqrt{2}$ $(0\le t\le1)$ we obtain

$$\int_0^1|K(t)|\,dt\le\left\{\int_0^1|K(t)|^p\,dt\right\}^{1/p}$$

$$\le C_3\left\{\int_0^1|K(t)|^p\,|t+i|^\kappa\,dt\right\}^{1/p}\le C_3\|K\|_{p,\kappa,1}.\qquad(23)$$

Hence the inequalities

$$\sup_{0<\vartheta<\pi}\left\{\int_0^2\left|\mathcal{K}_1^+(re^{i\vartheta})\right|^p\left|re^{i\vartheta}+i\right|^\kappa dr\right\}^{1/p}$$

$$\le C_4\sup_{0<\vartheta<\pi}\left\{\int_0^2\left|\mathcal{K}_1^+(re^{i\vartheta})\right|^p dr\right\}^{1/p}\le C_5\left\{\int_0^1|K(t)|^p\,dt\right\}^{1/p}$$

$$\le C_6\left\{\int_0^1|K(t)|^p\,(1+t^2)^{\kappa/2}\,dt\right\}^{1/p}\le C_6\|K\|_{p,\kappa,1}\qquad(24)$$

follow from Theorem 3.3-2 (where we take $\kappa=0$) and from the simple estimates $1\le\left|re^{i\vartheta}+i\right|\le3$ $(0\le r\le2,0\le\vartheta\le\pi)$. Observe now that

$$U_1(\vartheta)=\left\{\int_2^{+\infty}\left|\mathcal{K}_1^+(re^{i\vartheta})\right|^p\left|re^{i\vartheta}+i\right|^\kappa dr\right\}^{1/p}$$

$$\le\left\{\int_2^{+\infty}\left(\int_0^1\frac{|K(t)|\,\left|re^{i\vartheta}+i\right|^{\kappa/p}}{|t-re^{i\vartheta}|}dt\right)^p dr\right\}^{1/p},\qquad 0<\vartheta<\pi.\quad(25)$$

So, if the well-known inequality

$$\left\{\int_{a_1}^{b_1}\left(\int_{a_2}^{b_2}|F(x,y)|\,dx\right)^p dy\right\}^{1/p}\le\int_{a_2}^{b_2}\left(\int_{a_1}^{b_1}|F(x,y)|^p\,dy\right)^{1/p}dx,\qquad(26)$$

where $a_1 < b_1, a_2 < b_2$, and the simple estimate $\left|t - re^{i\vartheta}\right| \geq r/2$ ($r \geq 2, 0 \leq t \leq 1, 0 \leq \vartheta \leq \pi$) are used, then we obtain

$$U_1(\vartheta) \leq \int_0^1 \left(\int_2^{+\infty} \frac{|K(t)|^p \left|re^{i\vartheta} + i\right|^\kappa}{|t - re^{i\vartheta}|^p} dr \right)^{1/p} dt$$

$$\leq 2 \int_0^1 |K(t)| \left(\int_2^{+\infty} \frac{\left|re^{i\vartheta} + i\right|^\kappa}{r^p} dr \right)^{1/p} dt, \qquad 0 < \vartheta < \pi. \tag{25'}$$

But, obviously,

$$r \leq \left|re^{i\vartheta} + i\right| \leq 3r \qquad (r \geq 1/2, 0 \leq \vartheta \leq \pi) \tag{27}$$

and $\int_2^{+\infty} r^{\kappa - p} dr < +\infty$ since $\kappa < p - 1$. Thus, the estimate

$$\sup_{0 < \vartheta < \pi} \left\{ \int_2^{+\infty} \left|\mathcal{K}_1^+(re^{i\vartheta})\right|^p \left|re^{i\vartheta} + i\right|^\kappa dr \right\}^{1/p} \leq C_7 \|K\|_{p,\kappa,1} \tag{28}$$

follows from $(25'), (27)$ and (23). And the estimate (21), for $j = 1$, follows from (24) and (28). The proof of the case $j = 1$ of the estimate (22) is similar. So, it remains to prove the cases $j = 2$ of estimates (21) and (22). To this end, observe first that the estimate

$$\left|\mathcal{K}_2^+(z)\right| \leq \frac{1}{2\pi} \left\{ \int_1^{+\infty} |K(t)|^p |t + i|^\kappa dt \right\}^{1/p} \left\{ \int_1^{+\infty} \frac{dt}{|t - z|^q |t + i|^{\kappa q/p}} \right\}^{1/q},$$

where $z \in G_+, |z| \leq 1/2$ and $q = p/(p - 1)$, follows directly from definition (20) of the function $\mathcal{K}_2^+(z)$ and Hölder's inequality. But $|t - z| \geq t/2$ and $t \leq |t + i| \leq 2t$ when $|z| \leq 1/2$ and $1 \leq t < +\infty$. On the other hand, $\kappa > -1$. Hence it follows that

$$\left\{ \int_1^{+\infty} \frac{dt}{|t - z|^q |t + i|^{q\kappa/p}} \right\}^{1/q} \leq \left\{ C_8 \int_1^{+\infty} \frac{dt}{t^{q(1 + \kappa/p)}} \right\}^{1/q} < +\infty,$$

so

$$\sup_{|z| \leq 1/2, z \in G_+} \left\{ \left|\mathcal{K}_2^+(z)\right| \right\} \leq C_9 \|K\|_{p,\kappa,1}.$$

Consequently, the simple inequalities $1/2 \leq \left|re^{i\vartheta} + i\right| \leq 3/2$ ($0 \leq r \leq 1/2, 0 \leq \vartheta \leq \pi$) lead to the estimate

$$\left\{ \int_0^{1/2} \left|\mathcal{K}_2^+(re^{i\vartheta})\right|^p \left|re^{i\vartheta} + i\right|^\kappa dr \right\}^{1/p} \leq C_{10} \|K\|_{p,\kappa,1}, \quad 0 < \vartheta < \pi. \tag{29}$$

Observe also, that the estimate

$$\left\{ \int_1^{+\infty} |K(t)|^p \, t^\kappa dt \right\}^{1/p} \leq C_{11} \|K\|_{p,\kappa,1}$$

is true along with (27). Therefore, by Theorem 3.3-2,

$$\sup_{0<\vartheta<\pi} \left\{ \int_{1/2}^{+\infty} \left| \mathcal{K}_2^+(re^{i\vartheta}) \right|^p \left| re^{i\vartheta} + i \right|^\kappa dr \right\}^{1/p} \leq C_{12} \|K\|_{p,\kappa,1}, \tag{30}$$

and the desired estimate (21) ($j = 2$) follows from (29) and (30). Estimate (22) ($j = 2$) can be proved in the same way.

So, estimates (21) and (22) are already proven, and, if the function

$$\mathcal{K}_1(z) = \mathcal{K}_1^\pm(z) + \mathcal{K}_2^\pm(z) = \frac{1}{2\pi i} \int_0^{+\infty} \frac{K(t)}{t-z} dt, z \in G_\pm$$

is introduced, then the estimates

$$\sup_{0<\vartheta<\pi} \left\{ \int_0^{+\infty} \left| \mathcal{K}_1(re^{\pm i\vartheta}) \right|^p \left| re^{\pm i\vartheta} \pm i \right|^\kappa dr \right\}^{1/p} \leq C_{13} \|K\|_{p,\kappa,1} \tag{31}$$

easily follow. In the same way, if

$$\mathcal{K}_2(z) = \frac{1}{2\pi i} \int_{-\infty}^{\circ} \frac{K(t)}{t-z} dt, z \in G_\pm,$$

then the estimates

$$\sup_{0<\vartheta<\pi} \left\{ \int_0^{+\infty} \left| \mathcal{K}_2(re^{\pm i\vartheta}) \right|^p \left| re^{\pm i\vartheta} \pm i \right|^\kappa dr \right\}^{1/p} \leq C_{14} \|K\|_{p,\kappa,1} \tag{32}$$

hold, since the simple change of variables t and z transforms the integral $\mathcal{K}_2(z)$ into an integral similar to $\mathcal{K}_1(z)$. However, the function $\mathcal{K}(z)$ of the theorem can be represented in the form $\mathcal{K}(z) = \mathcal{K}_1(z) + \mathcal{K}_2(z)$, $z \in G_\pm$, So the following estimates remain true for $\mathcal{K}(z)$:

$$\sup_{0<\vartheta<\pi} \left\{ \int_0^{+\infty} \left| \mathcal{K}(re^{\pm i\vartheta}) \right|^p \left| re^{\pm i\vartheta} \pm i \right|^\kappa dr \right\}^{1/p} \leq C_{15} \|K\|_{p,\kappa,1} \tag{33}$$

Consequently, $(z+i)^{\kappa/p}\mathcal{K}^+(z) \in \overset{*}{H}{}_+^p$, and it follows from Theorem 3.2-1 that $(z+i)^{\kappa/p}\mathcal{K}^+(z) \in H_+^p$. It also follows that estimate (19) is true for the function $\mathcal{K}^+(z)$. In the same way, we can conclude that assertions $1°$ and $2°$ are true for the function $\mathcal{K}^-(z)$. This completes the proof of Theorem 3.3-3.

3.4 Some estimates in Banach spaces of entire functions of exponential type.

(a) First we formulate some well-known statements used throughout this section.

$1°$. If $\varphi(x) \in L_p(-\infty, +\infty)$ $(1 \le p < +\infty)$ and , as always,

$$\|\varphi\|_p = \left\{ \int_{-\infty}^{+\infty} |\varphi(x)|^p \, dx \right\}^{1/p},$$

then Steklov's function,

$$\varphi_h(x) = \frac{1}{2h} \int_{x-h}^{x+h} \varphi(t) dt (h > 0)$$

is continuous and uniformly bounded on the whole axis $-\infty < x < +\infty$. In addition, it satisfies the conditions

$$\|\varphi_h\|_p \le \|\varphi\|_p, \lim_{h \to +0} \|\varphi - \varphi_h\|_p = 0, \lim_{h \to +0} \|\varphi_h\|_p = \|\varphi\|_p.$$

$2°$. Let $f(z)$ be analytic in the half-plane $G_+ = \{z : \operatorname{Im} z > 0\}$ and continuous up to its boundary. Also let $f(z)$ be of exponential type $\le \sigma$, i.e.,

$$\limsup_{r \to +\infty} \left\{ r^{-1} \log M(r; f) \right\} \le \sigma, M(r; f) = \max_{0 \le \vartheta \le \pi} \left| f(re^{i\vartheta}) \right| \quad (r > 0).$$

Then the inequality

$$|f(x)| \le M < +\infty, \qquad x \in (-\infty, +\infty)$$

is sufficient to conclude that

$$|f(z)| \le M e^{\sigma \operatorname{Im} z}, \qquad z \in G_+.$$

Indeed, it is necessary to observe only that the last inequality holds, if the refined Phragmen-Lindelöf principle is applied to the function

$$f_\epsilon(z) = f(z) \exp \left\{ i(\sigma + \epsilon)z \right\}, \qquad z \in G_+, \ (\epsilon > 0)$$

and the passage to the limit $\epsilon \to +0$ is done.

$3°$. Let $U(z)$ be a subharmonic function in the half-plane G_+, continuous up to its boundary, and let

$$\sup_{z \in G_+} \{U(z)\} < +\infty.$$

Then

$$U(z) \le M < +\infty, z \in G_+,$$

if
$$U(x) \le M < +\infty, \quad -\infty < x < +\infty.$$

This is the most simple form of the maximum principle for subharmonic functions.

(b) Now we introduce some notations. Let
$$W_{1,\sigma}^{p,\kappa} \equiv W_{\sigma}^{p,\kappa} \qquad (1 < p < +\infty, -1 < \kappa < +\infty) \tag{1}$$

be the class of entire functions $f(z)$ of exponential type $\le \sigma$ satisfying the condition
$$\|f\|_{p,\kappa} = \left\{ \int_{-\infty}^{+\infty} |f(x)|^p \, |x|^\kappa \, dx \right\}^{1/p} < +\infty. \tag{2}$$

Along with this, let
$$W_{1,\sigma}^{p,\kappa}[\gamma] \equiv W_{\sigma}^{p,\kappa}[\gamma] \qquad (1 < p < +\infty, -1 < \kappa < +\infty, 0 \le \gamma < +\infty) \tag{3}$$

be the class of entire functions $f(z)$ of exponential type $\le \sigma$, satisfying the condition
$$\|f\|_{p,\kappa,\gamma} = \left\{ \int_{-\infty}^{+\infty} |f(x)|^p \, (x^2 + \gamma^2)^{\kappa/2} dx \right\}^{1/p} < +\infty. \tag{4}$$

It is easy to observe that the classes $W_{\sigma}^{p,\kappa}$ and $W_{\sigma}^{p,\kappa}[\gamma]$ coincide in spite of being differently normed when $\gamma > 0$, and it is obvious that
$$W_{\sigma}^{p,\kappa}[0] \equiv W_{\sigma}^{p,\kappa} \quad \text{and} \quad \|f\|_{p,\kappa,0} \equiv \|f\|_{p,\kappa}. \tag{5}$$

Later on, the branches of the multivalent functions $z^{\kappa/p}$ and $(z \pm i\gamma)^{\kappa/p}$ are supposed to be chosen as follows: if the function $z^{\kappa/p}$ is considered in the half-plane G_+(or G_-), then we choose a univalent branch analytic outside the cut $(-i\infty, -i0)$(or $(i0, +i\infty)$). Further, if $\gamma > 0$ is any number, then we suppose that $(z + i\gamma)^{\kappa/p}$ and $(z - i\gamma)^{\kappa/p}$ are some univalent branches analytic outside the cuts $(-i\infty, -i\gamma) \subset G_-$ and $(i\gamma, +i\infty) \subset G_+$ correspondingly.

Now let $f(z) \in W_{\sigma}^{p,\kappa}[\gamma]$ $(0 \le \gamma < +\infty)$ be any function. We associate with it the following group of functions:
$$\begin{aligned} F^{\pm}(z) &= z^{\kappa/p} f(z), & z \in G_{\pm}, \\ F_{\gamma}^{\pm}(z) &= (z \pm i\gamma)^{\kappa/p} f(z), & z \in G_{\pm} \end{aligned} \tag{6}$$

and note that $F_0^{\pm}(z) \equiv F^{\pm}(z), z \in G_{\pm}$. All these functions are obviously analytic in corresponding half-planes and are continuous up to their common boundary $(-\infty, +\infty)$ (excluding only the point $x = 0$ when $\gamma = 0$). In addition, all these functions are of exponential type $\le \sigma$.

(c) Two functions associated with an arbitrary $f(z) \in W_{\sigma}^{p,\kappa}[\gamma]$ $(1 < p < +\infty, -1 < \kappa < +\infty, 0 \le \gamma < +\infty)$ will be now introduced by the formulas
$$\omega_{\gamma}^{\pm}(z; f) = e^{\pm i\sigma z}(z \pm i\gamma)^{\kappa/p} f(z), \qquad z \in G_{\pm}, \tag{7}$$

and the following lemma will be proved.

Lemma 3.4-1. *If $f(z) \in W^{p,\kappa}_{\sigma}[\gamma]$, then*

$$1°. \qquad \omega^{\pm}_{\gamma}(z; f) \in H^{p}_{\pm},$$
$$2°. \qquad \|f(x \pm i\gamma)\|_{p,\kappa,\gamma} \le e^{\sigma\gamma}\|f\|_{p,\kappa}. \tag{8}$$

Proof. 1°. It suffices to prove this assertion only for $\omega^{+}_{\gamma}(z; f)$. The proof for $\omega^{-}_{\gamma}(z; f)$ will obviously be the same. To this end, we introduce for any $h > 0$ the function

$$\mathcal{F}^{+}_{h}(z) = \frac{1}{2h}\int_{z-h}^{z+h} F^{+}_{\gamma}(t)dt = \frac{1}{2h}\int_{z-h}^{z+h} f(t)(t + i\gamma)^{\kappa/p}dt, \quad z \in G_{+}. \tag{9}$$

Then the properties of $F^{+}_{\gamma}(z)$, which were mentioned above, and the properties of the integrated functions of Section 3.4(a)1° lead to the following conclusions.

(i) The function $\mathcal{F}^{+}_{h}(z)$ is analytic in G_{+} and continuous up to its boundary $(-\infty, +\infty)$ where it is uniformly bounded. Besides, by (6),

$$\|\mathcal{F}^{+}_{h}\|_{p} \le \|F^{+}_{\gamma}\|_{p} = \|f\|_{p,\kappa,\gamma}$$

for any $h > 0$. It should also be noticed that the inequality

$$\left|\mathcal{F}^{+}_{h}(x)\right| \le (2h)^{-1/p}\|f\|_{p,\kappa,\gamma}, \qquad x \in (-\infty, +\infty)$$

follows from (9) and from Hölder's inequality.

(ii) Also, it follows from (9) that

$$M(r; \mathcal{F}^{+}_{h}) = \max_{\substack{|z|=r \\ z\in G_+}} \left|\mathcal{F}^{+}_{h}(z)\right| \le \max_{\substack{|z|=r+h \\ z\in G_+}} \left|F^{+}_{\gamma}(z)\right| = M\left(r + h; F^{+}_{\gamma}\right).$$

Thus, the functions $F^{+}_{\gamma}(z)$ and $\mathcal{F}^{+}_{h}(z)$ are both of exponential type $\le \sigma$ in G_{+}. The inequality

$$\left|e^{i\sigma z}\mathcal{F}^{+}_{h}(z)\right| \le (2h)^{-1/p}\|f\|_{p,\kappa,\gamma}, \qquad z \in G_{+} \tag{10}$$

can be easily obtained from properties (i) and (ii) of the function $\mathcal{F}^{+}_{h}(z)$ mentioned above and statement 3.4(a)2°. Further, we introduce for any $R > 0$ the function

$$U_{h}(z; R) = \int_{-R}^{R} \left|e^{i\sigma(z+t)}\mathcal{F}^{+}_{h}(z + t)\right|^{p} dt, \quad z \in G_{+}. \tag{11}$$

This function is subharmonic and bounded in G_{+}. Besides, it is continuous up to the boundary $(-\infty, +\infty)$ of G_{+}, since $\mathcal{F}^{+}_{h}(z)$ is of the same kind. Further,

$$U_{h}(x; R) = \int_{-R}^{R} |\mathcal{F}^{+}_{h}(x + \tau)|^{p}d\tau$$

$$= (2h)^{-p}\int_{-R}^{R} \left|\int_{x+\tau-h}^{x+\tau+h} f(t)(t + i\gamma)^{\kappa/p}dt\right|^{p} d\tau, \quad x \in (-\infty, +\infty).$$

Hence it follows that for any $x \in (-\infty, +\infty)$ and $R > 0$

$$U_h(x; R) \leq \frac{1}{2h} \int_{-R}^{R} d\tau \int_{-h}^{h} |f(x + \tau + t)|^p \left[(x + \tau + t)^2 + \gamma^2\right]^{\kappa/2} dt$$

$$= \frac{1}{2h} \int_{-h}^{h} dt \int_{x+t-R}^{x+t+R} |f(\tau)|^p (\tau^2 + \gamma^2)^{\kappa/2} d\tau \leq \|f\|_{p,\kappa,\gamma}^p,$$

if Hölder's inequality is used. Thus the function $U_h(z, R)$ is subharmonic and bounded in G_+, continuous up to its boundary $(-\infty, +\infty)$ and

$$U_h(x; R) \leq \|f\|_{p,\kappa,\gamma}^p, \qquad R > 0$$

for any $x \in (-\infty, +\infty)$. Consequently,

$$U_h(z; R) \leq \|f\|_{p,\kappa,\gamma}^p, \quad z \in G_+, \qquad R > 0,$$

as it follows from 3.4(a)3°. Hence we obtain the estimate

$$\int_{-\infty}^{+\infty} |\mathcal{F}_h^+(x + t + iy)|^p \, dt = \int_{-\infty}^{+\infty} |\mathcal{F}_h^+(x + iy)|^p \, dx \leq e^{\sigma p y} \|f\|_{p,\kappa,\gamma}^p, \ y > 0, \quad (12)$$

if we use the definition (11) of $U_h(z; R)$ and do the passage $R \to +\infty$. For any $y > 0$

$$\|F_\gamma^+(x + iy)\|_p^p = \int_{-\infty}^{+\infty} \liminf_{h \to +0} |\mathcal{F}_h^+(x + iy)|^p \, dx \leq \liminf_{h \to +0} \int_{-\infty}^{+\infty} |\mathcal{F}_h^+(x + iy)|^p \, dx$$

by Fatou's lemma. Therefore, the passage $h \to +0$ in (12) gives

$$\int_{-\infty}^{+\infty} \left| f(x + iy)(x + iy + i\gamma)^{\kappa/p} \right|^p \, dx \leq e^{\sigma p y} \|f\|_{p,\kappa,\gamma}^p, y > 0.$$

Hence, it follows that

$$\|\omega_\gamma^+(x + iy; f)\|_p \leq \|f\|_{p,\kappa,\gamma}, \ y > 0,$$

since $|\exp\{i\sigma(x + iy)\}| = e^{-\sigma y}$. Thus, assertion 1° is proved.

2°. Observe that

$$\|f(x \pm i\gamma)\|_{p,\kappa,\gamma} = \|f(x \pm i\gamma)(x \pm i\gamma)^{\kappa/p}\|_p \tag{13}$$

according to the definitions of these norms. In addition, the inclusions

$$e^{\pm i\sigma z} z^{\kappa/p} f(z) \in H_\pm^p$$

follow particularly from the case $\gamma = 0$ of assertion 1°, which is already proved. Therefore, the properties 3.2(5) imply

$$e^{-\sigma\gamma} \|f(x \pm i\gamma)(x \pm i\gamma)^{\kappa/p}\|_p = \|e^{\pm i\sigma(x \pm i\gamma)} f(x \pm i\gamma)(x \pm i\gamma)^{\kappa/p}\|_p$$

$$\leq \|e^{\pm i\sigma x} f(x) x^{\kappa/p}\|_p \leq \|f\|_{p,\kappa},$$

and assertion 2° follows from the equality (13).

(d) Lemma 3.4-2. $1°$. *If $f(z) \in W_\sigma^{p,\kappa}[\gamma]$ $(0 \leq \gamma < +\infty)$ is any function, then*

$$|f(z)| \leq C(1+|z|)^{-\kappa/p} e^{\sigma|\operatorname{Im} z|} \|f\|_{p,\kappa,\gamma}, \qquad z \in \mathbb{C}, \tag{14}$$

where the constant $C > 0$ does not depend on f and z.

$2°$. *The class $W_\sigma^{p,\kappa}[\gamma]$ is a Banach space with the norm $\|\cdot\|_{p,\kappa,\gamma}$.*

Proof. $1°$. If $G_\gamma^\pm = \{z : \pm Imz \geq 2(1+\gamma)\}, \gamma \in [0,+\infty)$, then, clearly,

$$G_{\gamma+1/2}^\pm \subset G_\gamma^\pm \subset \mathbb{C}_\gamma = \{z : |z| \geq 2(1+\gamma)\}.$$

Further, if $K_z = \{\zeta : |\zeta - z| < 1\}$, then it is easy to verify that

$$|z|/2 \leq |\zeta \pm i\gamma| \leq 3|z|/2 \text{ when } z \in \mathbb{C}_\gamma \text{ and } \zeta \in K_z, \tag{15}$$

and, of course, when $z \in G_\gamma^\pm$ and $\zeta \in K_z$. Now note that, since $|f(z)|^p$ is a subharmonic function,

$$|f(z)|^p \leq \frac{1}{\pi} \iint_{|\zeta-z|<1} |f(\zeta)|^p |d\zeta|^2, \qquad z \in \mathbb{C}, \tag{16}$$

where $|d\zeta|^2$ is the Lebesgue plane measure. From (15) it follows that $|\zeta \pm i\gamma|^\kappa \geq (\pi C_\kappa)^{-1} |z|^\kappa$ $(z \in \mathbb{C}_\gamma, |\zeta - z| < 1)$, where $C_\kappa > 0$ is a constant, and therefore (16) gives

$$|f(z)|^p \leq C_\kappa |z|^{-\kappa} \iint_{|\zeta-z|<1} |f(\zeta)|^p |\zeta \pm i\gamma|^\kappa |d\zeta|^2, \qquad z \in \mathbb{C}_\gamma. \tag{17}$$

Observe now that

$$\mathcal{Y}^\pm(z) = \iint_{|\zeta-z|<1} |f(\zeta)|^p |\zeta \pm i\gamma|^\kappa |d\zeta|^2$$

$$\leq \int_{y-1}^{y+1} d\eta \int_{x-1}^{x+1} |f(\xi + i\eta)|^p |\xi + i\eta \pm i\gamma|^\kappa \, d\xi \leq \int_{y-1}^{y+1} J^\pm(\eta) d\eta, \tag{18}$$

$$J^\pm(\eta) = \int_{-\infty}^{+\infty} |f(\xi + i\eta)|^p |\xi + i\eta \pm i\gamma|^\kappa \, d\xi \tag{19}$$

for arbitrary $z = x + iy \in \mathbb{C}$. The right-hand side integrals of (19) are finite and bounded in any segment $[-R, R]$, as it follows from the inequalities

$$J^\pm(\eta) \leq \exp\{p\sigma |\eta \pm \gamma|\} \|f(x \mp i\gamma)\|_{p,\kappa}^p \tag{20}$$

which are consequences of assertion $2°$ of Lemma 3.4-1. Further, since

$$\omega_\gamma^\pm(z; f) = e^{\pm i\sigma z}(z \pm i\gamma)^{\kappa/p} f(z) \in H_\pm^p$$

by assertion 1° of Lemma 3.4-1, the use of inequality 3.2(5) gives

$$e^{-\sigma p\eta}J^{+}(\eta) = \|\omega_\gamma^{+}(\xi + i\eta; f)\|_p^p \leq \|\omega_\gamma^{+}(\xi; f)\|_p^p = \|f\|_{p,\kappa,\gamma}^p, \quad \text{if } \eta \geq 0, \qquad (20')$$

$$e^{\sigma p\eta}J^{-}(\eta) = \|\omega_\gamma^{-}(\xi + i\eta; f)\|_p^p \leq \|\omega_\gamma^{-}(\xi; f)\|_p^p = \|f\|_{p,\kappa,\gamma}^p, \quad \text{if } \eta \leq 0. \qquad (20'')$$

Using these inequalities and the inequalities of (18) we obtain

$$\mathcal{Y}^{\pm}(z) \leq \frac{e^{\sigma p} - e^{-\sigma p}}{\sigma p} e^{\sigma p|\operatorname{Im} z|}\|f\|_{p,\kappa,\gamma}^p, \qquad z \in G_{\gamma+1/2}^{\pm}, \qquad (21)$$

which together with (17) imply

$$|f(z)| \leq C_{\kappa,p}(1 + |z|)^{-\kappa/p}e^{\sigma|\operatorname{Im} z|}\|f\|_{p,\kappa,\gamma}, \qquad z \in G_{\gamma+1/2}^{\pm}, \qquad (22)$$

where the constant $C_{\kappa,p} > 0$ is independent of f and z.

Now let

$$\Pi_\gamma = \{z : |\operatorname{Im} z| < 3 + 2\gamma\} = \mathbb{C} \setminus \left\{G_{\gamma+1/2}^{+} \cup G_{\gamma+1/2}^{-}\right\}$$

be a horizontal strip, with the width $2(3 + 2\gamma)$, symmetric with respect to the real axis; also, let

$$\Pi_\gamma^{\pm} = \{z : |\operatorname{Im} z| < 3 + 2\gamma, \pm \operatorname{Re} z \geq 2(1 + \gamma)\} \subset \mathbb{C}_\gamma$$

be half-strips, and let

$$\square_\gamma = \{z : |\operatorname{Re} z| \leq 2(1 + \gamma), |\operatorname{Im} z| < 3 + 2\gamma\}$$

be a rectangle. Then it is clear that

$$\Pi_\gamma = \Pi_\gamma^{+} \cup \Pi_\gamma^{-} \cup \square_\gamma. \qquad (23)$$

To prove the inequality

$$\sup_{z \in \Pi_\gamma} \left\{(1 + |z|)^{\kappa/p}|f(z)|\right\} < +\infty, \qquad (24)$$

first observe that obviously

$$\sup_{z \in \square_\gamma} \left\{(1 + |z|)^{\kappa/p}|f(z)|\right\} < +\infty.$$

Next, it follows from the inclusions $\Pi_\gamma^{\pm} \subset \mathbb{C}_\gamma$, estimates (17), the boundedness of functions $J^{\pm}(\eta)$ on any segment $-R \leq \eta \leq R$ (see (20'), (20'')) and estimates (18) that

$$\sup_{z \in \Pi_\gamma^{\pm}} \left\{(1 + |z|)^{\kappa/p}|f(z)|\right\} < +\infty.$$

Thus (24) holds. Further, estimates (22) imply particularly

$$\sup_{z \in \partial \Pi_\gamma} \left\{ (1 + |z|)^{\kappa/p} |f(z)| \right\} \leq C_1 \|f\|_{p,\kappa,\gamma}, \tag{25}$$

where $\partial \Pi_\gamma$ is the boundary of the strip Π_γ, which is the sum of boundaries of the half-planes $G^{\pm}_{\gamma+1/2}$. Now denote by $\{z + i2(2 + 2\gamma)\}^{\kappa/p}$ any univalent branch of the same function, chosen to be analytic outside the cut $(-i\infty, -i2(2+2\gamma)) \subset G^-_{\gamma+1/2}$, and introduce the function

$$f_\gamma(z) = \{z + i2(2 + 2\gamma)\}^{\kappa/p} f(z).$$

Evidently

$$|z + i2(2 + 2\gamma)| \asymp 1 + |z|, z \in \Pi_\gamma. \tag{26}$$

Thus, by inequalities (24) and (25), the function $f_\gamma(z)$ is bounded in the strip Π_γ and satisfies the condition

$$\sup_{z \in \partial \Pi_\gamma} \left\{ |f_\gamma(z)| \right\} \leq C_2 \|f\|_{p,\kappa,\gamma}$$

on its boundary. Hence, by Phragmen-Lindelöf principle,

$$|f_\gamma(z)| \leq C_2 \|f\|_{p,\kappa,\gamma}, \qquad z \in \Pi_\gamma,$$

and, if we return to the function $f(z)$, then (26) implies

$$|f(z)| \leq C_3 (1 + |z|)^{-\kappa/p} \|f\|_{p,\kappa,\gamma}, \qquad z \in \Pi_\gamma.$$

Estimate (14) follows from (22).

2°. The completeness of the space $W^{p,\kappa}_\sigma[\gamma]$ follows from inequality (22), which is already proved.

(e) The theorem proved below shows that the classes $W^{p,\kappa}_\sigma$ and $W^{p,\kappa}_\sigma[\gamma](0 < \gamma < +\infty)$ not only coincide, but they also have equivalent norms.

Theorem 3.4-1. *If* $f(z) \in W^{p,\kappa}_\sigma(1 < p < +\infty, -1 < \kappa < +\infty)$ *is an arbitrary function and* $\gamma \in (0, +\infty)$ *is any number, then*

$$1^\circ. \qquad \|f\|_{p,\kappa,\gamma} \asymp \|f\|_{p,\kappa}. \tag{27}$$

$$2^\circ. \qquad \|f(x \pm i\gamma)\|_{p,\kappa} \asymp \|f\|_{p,\kappa}. \tag{28}$$

Proof. $1°$. Observe that the classes $W_\sigma^{p,\kappa}[0]$ and $W_\sigma^{p,\kappa}$ merely coincide and consider for any $\gamma \in (0, +\infty)$ the operator T_γ defined in the following way: $T_\gamma f = f \in W_\sigma^{p,\kappa}[\gamma]$, $\quad \forall f \in W_\sigma^{p,\kappa}[0]$. Evidently, T_γ^{-1} also exists, and it is a mapping of $W_\sigma^{p,\kappa}[\gamma]$ onto $W_\sigma^{p,\kappa}[0]$. Two cases are possible:

1. If $\kappa \geq 0$, then $|x|^\kappa \leq |x + i\gamma|^\kappa$. Hence

$$\|f\|_{p,\kappa} \leq \|f\|_{p,\kappa,\gamma} = \|T_\gamma f\|_{p,\kappa,\gamma}, \qquad f \in W_\sigma^{p,\kappa}[0], \tag{29}$$

and consequently

$$\|T_\gamma^{-1} f\|_{p,\kappa} \leq \|f\|_{p,\kappa,\gamma}, \qquad f \in W_\sigma^{p,\kappa}[\gamma].$$

So T_γ^{-1} is a bounded linear operator, and the operator T_γ is also bounded by Banach's theorem, i.e.,

$$\|T_\gamma f\|_{p,\kappa,\gamma} = \|f\|_{p,\kappa,\gamma} \leq A\|f\|_{p,\kappa}, \qquad f \in W_\sigma^{p,\kappa}[0], \tag{30}$$

where $A > 0$ is a constant independent of f, and (29) and (30) give the two-sided estimates (27).

2. If $-1 < \kappa < 0$, then $|x + i\gamma|^\kappa \leq |x|^\kappa$. Hence

$$\|T_\gamma f\|_{p,\kappa,\gamma} = \|f\|_{p,\kappa,\gamma} \leq \|f\|_{p,\kappa}, \qquad f \in W_\sigma^{p,\kappa}[0], \tag{31}$$

i.e., the operator T_γ is bounded. Evidently, T_γ^{-1} is also bounded, i.e.,

$$\|T_\gamma^{-1} f\|_{p,\kappa} = \|f\|_{p,\kappa} \leq A^*\|f\|_{p,\kappa,\gamma}, \qquad f \in W_\sigma^{p,\kappa}[\gamma], \tag{32}$$

where $A^* > 0$ is a constant independent of f, and (31) and (32) give the two-sided estimates (27).

$2°$. If the function $f(z)$ in (27) is replaced by $f(z \pm i\gamma)$, then, using the second estimate of (8) we conclude that

$$\|f(x \pm i\gamma)\|_{p,\kappa} \leq C\|f(x \pm i\gamma)\|_{p,\kappa,\gamma} \leq Ce^{\sigma\gamma}\|f\|_{p,\kappa} \tag{33}$$

or, which is the same,

$$\|f\|_{p,\kappa} \leq Ce^{\sigma\gamma}\|f(x \pm i\gamma)\|_{p,\kappa}. \tag{34}$$

The two-sided estimate (28) follows from (33) and (34).

(f) The sequence $\{x_k\}_{-\infty}^{+\infty}$ of real numbers will be said to be of class Δ_δ if

$$\begin{aligned}
&a) \quad x_0 = 0,\ x_{-k} = -x_k\ (1 \leq k < \infty),\ x_k < x_{k+1}\ (-\infty < k < +\infty),\\
&b) \quad |x_k| \asymp 1 + |k|\ (k = \pm 1, \pm 2, \ldots),\\
&c) \quad \inf_{-\infty < k < +\infty} |x_{k+1} - x_k| > \delta > 0.
\end{aligned} \tag{35}$$

Theorem 3.4-2. *If $f(z) \in W_\sigma^{p,\kappa}$ $(1 < p < +\infty, -1 < \kappa < +\infty)$, $\{x_k\}_{-\infty}^{+\infty} \in \Delta_\delta$ and h is any real number, then*

$$\sum_{k=-\infty}^{+\infty} |f(x_k + ih)|^p (1 + |k|)^\kappa \leq C\|f\|_{p,\kappa}^p, \tag{36}$$

where $C > 0$ is a constant independent of f.

Proof. The function $f(z)z^{\kappa/p}$ is analytic in the z-plane, outside the cut $(-\infty, 0)$. Therefore, $|f(z)|^p |z|^\kappa$ is subharmonic in the half-plane $Re\, z > 0$, and

$$|f(x_k)|^p |x_k|^\kappa \leq \frac{1}{2\pi\rho} \int_{|\zeta - x_k| = \rho} |f(\zeta)|^p |\zeta|^\kappa |d\zeta| \quad (k = 1, 2, \dots)$$

for any $\rho \in (0, \delta)$, since $x_k > 0$ when $k \geq 1$. Thus, if we multiply both sides of these inequalities by $\rho d\rho$ and integrate them along $(0, \delta/2)$, then we shall obtain

$$|f(x_k)|^p |x_k|^\kappa \leq \frac{4\delta^{-2}}{\pi} \iint_{|\zeta - x_k| < \delta/2} |f(\zeta)|^p |\zeta|^\kappa |d\zeta|^2$$

$$\leq \frac{4\delta^{-2}}{\pi} \int_{-\delta/2}^{\delta/2} d\eta \int_{x_k - \delta/2}^{x_k + \delta/2} |f(\xi + i\eta)|^p |\xi + i\eta|^\kappa d\xi \quad (k = 1, 2, \dots).$$

A similar consideration for the function $f(-z)z^{\kappa/p}$ leads to the same inequalities for $k = -1, -2, \dots$, since $-x_k = x_{-k}$. If we sum up all these inequalities, we obtain

$$\sum_{1 \leq |k| < +\infty} |f(x_k)|^p |x_k|^\kappa \leq \frac{4\delta^{-2}}{\pi} \int_{-\delta/2}^{\delta/2} d\eta \int_{-\infty}^{+\infty} |f(\xi + i\eta)|^p |\xi + i\eta|^\kappa d\xi, \tag{37}$$

but assertion $2°$ of Lemma 3.4-1 implies

$$\int_{-\infty}^{+\infty} |f(\xi + i\eta)|^p |\xi + i\eta|^\kappa d\xi = \|f(\xi + i\eta)\|_{p,\kappa,|\eta|}^p \leq \exp\{p\sigma |\eta|\}\|f\|_{p,\kappa}^p.$$

Therefore, by (37),

$$\sum_{1 \leq |k| < +\infty} |f(x_k)|^p |x_k|^\kappa \leq C_{\sigma,\delta}\|f\|_{p,\kappa}^p, \tag{38}$$

where the constant $C_{\sigma,\delta} > 0$ is independent of f. Applying this estimate and Theorem $3.4 - 1(2°)$ to the function $f(z + ih) \in W_\sigma^{p,\kappa}$ we obtain

$$\sum_{1 \leq |k| < +\infty} |f(x_k + ih)|^p |x_k|^\kappa \leq C_{\sigma,\delta}(h)\|f\|_{p,\kappa}^p, \tag{38'}$$

where the constant $C_{\sigma,\delta}(h) > 0$ is also independent of f. And now, to complete the proof, it remains to observe that, by Lemma $3.4 - 2(1°$, when $\gamma = 0)$,

$$|f(x_0 + ih)| = |f(ih)| \leq C_h\|f\|_{p,\kappa}$$

and also, that $|x_k|^\kappa \asymp (1 + |k|)^\kappa$ $(k \neq 0)$ by (35). Then estimate (36) follows from (38') and from the last inequality.

Theorem 3.4-3. *Let $\omega_{\pm}(z) \in H_{\pm}^p$ and $\{x_k\}_{-\infty}^{+\infty} \in \Delta_\delta$. Then*

$$\sum_{k=-\infty}^{+\infty} |\omega_{\pm}(x_k \pm i)|^p \leq C_\delta \|\omega_{\pm}\|_p^p, \tag{39}$$

where $C_\delta > 0$ is a constant independent of $\omega_{\pm}$.

Proof. The functions $|\omega_{\pm}(z)|^p$ are subharmonic in the half-planes $G_{\pm}$. Hence

$$|\omega_{\pm}(x_k \pm i)|^p \leq \frac{4\delta_0^{-2}}{\pi} \int_0^{2\pi} \int_0^{\delta_0/2} |\omega_{\pm}(x_k \pm i + \rho e^{i\vartheta})|^p \, \rho d\rho d\vartheta \quad (-\infty < k < +\infty) \tag{40}$$

for any $\rho \in (0, \delta_0)(\delta_0 = \min\{1, \delta\})$. But, since $\{x_k\}_{-\infty}^{+\infty} \in \Delta_\delta$, the disks $|z - (x_k \pm i)| < \delta_0/2$ are situated inside the parallel strips $0 < \pm \operatorname{Im} z < 1 + \delta_0/2$, and they do not intersect. Therefore,

$$\sum_{k=-\infty}^{+\infty} |\omega_{\pm}(x_k \pm i)|^p \leq \frac{4\delta_0^{-2}}{\pi} \int_0^{1+\delta_0/2} dy \int_{-\infty}^{+\infty} |\omega_{\pm}(x \pm iy)|^p \, dx \leq C_\delta \|\omega_{\pm}\|_p^p$$

by inequalities (40) and properties 3.2(5) of functions of classes $H_{\pm}^p$. Thus, (39) is proved.

(g) The final lemma of this chapter relating to the case of entire functions of order $1/2$ is similar to Lemma 3.4-2. Here the notation

$$W_{1/2,\sigma}^{p,\omega} \qquad (1 < p < +\infty, -1 < \omega < +\infty, 0 < \sigma < +\infty) \tag{41}$$

will be used for the classes of entire functions $\Phi(z)$ of order $1/2$ and of type $\leq \sigma$, which satisfy the condition

$$\|\Phi\|_{p,\omega}^+ = \left\{ \int_{-\infty}^{+\infty} |\Phi(x)|^p \, x^\omega dx \right\}^{1/p} < +\infty. \tag{42}$$

Lemma 3.4-3. *1°. If $\Phi(z) \in W_{1/2,\sigma}^{p,\omega}$ is an arbitrary function, then*

$$|\Phi(z)| \leq A(1 + |z|)^{-\frac{2\omega+1}{2p}} e^{\sigma|z|^{1/2}} \|\Phi\|_{p,\omega}^+, \qquad z \in \mathbb{C}, \tag{43}$$

where the constant $A > 0$ is independent of Φ and z.
2°. The class $W_{1/2,\sigma}^{p,\omega}$ is a Banach space with norm (42).

Proof. 1°. Suppose

$$f(z) = \Phi(z^2), \qquad z \in \mathbb{C}. \tag{44}$$

Then obviously $f(z)$ is an entire function of exponential type $\leq \sigma$, and if $\kappa = 1+2\omega$, then it is easy to verify that

$$\|f\|_{p,\kappa}^p = \int_{-\infty}^{+\infty} |f(x)|^p |x|^\kappa \, dx = \int_0^{+\infty} |\Phi(x)|^p x^\omega dx = \left\{ \|\Phi\|_{p,\omega}^+ \right\}^p . \tag{45}$$

Hence $f(z) \in W_\sigma^{p,\kappa}$. Therefore, by Lemma 3.4-2 ($1°$) (the case $\gamma = 0$) and by formulas (44) and (45)

$$\left| \Phi(z^2) \right| \leq C(1 + |z|)^{-\kappa/p} e^{\sigma|z|} \|\Phi\|_{p,\omega}^+, z \in \mathbb{C}, \tag{46}$$

where $C > 0$ is a constant independent of Φ and z. And, if we use the inequalities

$$(1 + |z|)/2 \leq (1 + |z|^{1/2})^2 \leq 2(1 + |z|), \qquad z \in \mathbb{C},$$

then estimate (43) follows from (46).

$2°$. The completeness of the space $W_{1/2,\sigma}^{p,\omega}$ follows from the standard arguments based on estimate (43) and on the completeness of the space $L_p(0, +\infty)$ with the weight x^ω. Thus $W_{1/2,\sigma}^{p,\omega}$ is a Banach space, and the proof is complete.

3.5 Notes

3.2 For properties (3), (4) and (5) of the functions of classes $H_\pm^p$ over the half-planes $G_\pm$ see, for example, Hoffman [1, Chapter 8] and Garnett [1, Chapter 2, Theorem 3.1]. Lemma 3.2-1 was proved by S.A. Akopian [1]. For assertions (9) and relations ($9''$) see M.M. Djrbashian [5, pp. 414–415, 508]. The assertions of Lemmas 3.2-2–3.2-4 were established by A.M. Sedletski [1]. With respect to Theorem 3.2-1, it should be mentioned that the identity of the classes H_+^p and $\overset{*}{H}{}_+^p$ when $p = 2$ was established in M.M. Djrbashian and A.E. Avetisian [1]. For any $p \in (0, +\infty)$ the identity of these classes was established by S.A.Akopian [1] who proved the inclusion $H_+^p \subseteq \overset{*}{H}{}_+^p$ and by A.M. Sedletski who proved the inverse inclusion.

3.3 For Theorem 3.3-1 see Titchmarsh [1, Chapter 5]. An assertion more general than Theorem 3.3-2 was established by V.M. Martirosian [1]. Theorem 3.3-3 was proved in the papers of M.M. Djrbashian-S.G. Raphaelian [1] and M.M. Djrbashian [7, Theorem 0]. For inequality (26) see Garnett [1, Chapter 1].

3.4 The property of integrated functions is given by I.P. Natanson [1, Chapter 18, Lemmas 1–4]. Lemmas 3.4-1, 3.4-2, Theorem 3.4-1 and a particular case of Theorem 3.4-2 were established by M.M. Djrbashian-S.G. Raphaelian [1]; see also M.M. Djrbashian [7]. Theorem 3.4-3 was proved in a different way by Sh.A. Grigorian [1].

Finally, note that several results of this chapter were first announced by S.G. Raphaelian [1, 2] in a different form. The results of this chapter were proved in §2 and, partially, in §3 of M.M. Djrbashian [7].

4 Interpolation series expansions in spaces $W_{1/2,\sigma}^{p,\omega}$ of entire functions

4.1 Introduction

In this chapter we deduce the expansions in certain interpolation series for the classes $W_{1/2,\sigma}^{p,\omega}(1 < p < +\infty, -1 < \omega < p-1)$ of entire functions of order $\rho < 1/2$ or of order $\rho = 1/2$ and of type $\leq \sigma(0 < \sigma < +\infty)$, for which

$$\|\Phi\|_{p,\omega}^{+} = \left\{ \int_0^{+\infty} |\Phi(x)|^p \, x^\omega dx \right\}^{1/p} < +\infty. \tag{1}$$

As points of interpolation of the above mentioned interpolation series we take the sequence of zeros $\{\lambda_k = \lambda_k(\sigma,\nu)\}_1^\infty$ of the Mittag-Leffler type entire function

$$\mathcal{E}_\sigma(z;\nu) = E_{1/2}\left(-\sigma^2 z; 1+\nu\right) \qquad (0 \leq \nu < 2). \tag{2}$$

The distribution of its zeros was illustrated earlier in Theorems 1.4-3 and 1.4-4 of Chapter 1.

4.2 Lemmas on special Mittag-Leffler type functions

We consider here the properties of two pairs of functions. One of these pairs is

$$\begin{aligned}
e_\nu(z) &= E_{1/2}(-z; 1+\nu), \\
s_\nu(z) &= z E_{1/2}\left(-z^2; 1+\nu\right) = z e_\nu\left(z^2\right).
\end{aligned} \tag{1}$$

The other pair is

$$\begin{aligned}
\mathcal{E}_\sigma(z;\nu) &= E_{1/2}\left(-\sigma^2 z; 1+\nu\right) = e_\nu\left(\sigma^2 z\right), \\
S_\sigma(z;\nu) &= z E_{1/2}\left(-\sigma^2 z^2; 1+\nu\right) = \sigma^{-1} s_\nu(\sigma z).
\end{aligned} \tag{2}$$

It should be noted that the function $s_\nu(z)$ can also be expressed as the sum

$$s_\nu(z) = \frac{1}{2i} \left\{E_1(iz;\nu) - E_1(-iz;\nu)\right\}. \tag{3}$$

This follows strictly from definition 1.1(1) of $E_\rho(z;\mu)$. In the same way, one can obtain that

$$S_\sigma(z;\nu) = \frac{\sigma^{-1}}{2i} \left\{E_1(i\sigma z;\nu) - E_1(-i\sigma z;\nu)\right\}. \tag{3'}$$

(a) Let $\alpha \in (\pi/2, \pi)$ be any number. Consider the mutually complementary corner domains

$$\Delta_\alpha = \{\zeta : |\operatorname{Arg}\zeta| < \alpha\}, \qquad \Delta_\alpha^* = \{\zeta : |\pi - \operatorname{Arg}\zeta| < \pi - \alpha\}. \tag{4}$$

Their common boundary is the sum of two rays, $\Gamma(\alpha) = \{\zeta : \arg\zeta = \pm\alpha\}$ directed negatively with respect to Δ_α. Further, if we use Theorem 1.3-3 relating to integral representations of the function $E_\rho(z;\nu)$, then the case $\rho = 1$ gives, for any $\nu \in [0,2)$,

$$E_1(z;\nu) = \begin{cases} z^{1-\nu}e^z + R_\nu(z;\alpha) & \text{when } z \in \Delta_\alpha, \\ R_\nu(z;\alpha) & \text{when } z \in \Delta_\alpha^*, \end{cases} \tag{5}$$

where

$$R_\nu(z;\alpha) = \frac{1}{2\pi i}\int_{\Gamma(\alpha)} \frac{e^\zeta \zeta^{1-\nu}}{\zeta - z}\,d\zeta, \qquad z \in \Delta_\alpha^* \cup \Delta_\alpha. \tag{5'}$$

Now we are ready to prove the following lemma.

Lemma 4.2-1. 1°. *The function $s_\nu(x)(0 \leq \nu < 2)$ is representable in the form*

$$\begin{aligned} s_\nu(x) &= x^{1-\nu}\cos\left(x - \frac{\pi}{2}\nu\right) + r_\nu(x), & x \in (0,+\infty), \\ s_\nu(x) &= -|x|^{1-\nu}\cos\left(x + \frac{\pi}{2}\nu\right) + r_\nu(x), & x \in (-\infty,0), \end{aligned} \tag{6}$$

where

$$r_\nu(x) = \frac{x}{2\pi i}\int_{\Gamma_0} \frac{e^\zeta \zeta^{1-\nu}}{\zeta^2 + x^2}\,d\zeta, \qquad 0 < |x| < +\infty, \tag{6'}$$

and $\Gamma_0 = \Gamma(3\pi/4)$ is the union of rays $\arg\zeta = \pm 3\pi/4$ directed in the suitable order.
2°. *The asymptotic formulas*

$$s_\nu(x) = \begin{cases} x^{1-\nu}\cos(x - \pi\nu/2) + O(x^{-1}), & x \to +\infty \\ -|x|^{1-\nu}\cos(x + \pi\nu/2) + O(x^{-1}), & x \to -\infty \end{cases} \tag{7}$$

are true for any $\nu \in [0,2)$.

Proof. 1°. Observe that both semi-axes $(i0, +i\infty)$ and $(i0, -i\infty)$ lie in the corner domain $\Delta_{3\pi/4}$. Thus, if $z = x(0 < |x| < +\infty)$, representations (5) and (5') of $E_1(\pm ix;\nu)$, where $\alpha = 3\pi/4$, may be inserted in (3). Then we arrive at the formula

$$s_\nu(x) = \frac{1}{2i}\left\{(ix)^{1-\nu}e^{ix} - (-ix)^{1-\nu}e^{-ix}\right\} + r_\nu(x), \qquad 0 < |x| < +\infty, \tag{8}$$

where

$$r_\nu(x) = \frac{1}{2i}\left\{R_\nu(ix; 3\pi/4) - R_\nu(-ix; 3\pi/4)\right\}.$$

Both the formulas (6) and (6') follow, if we observe that the first term of the right-hand side of (8) can be written as $x^{1-\nu}\cos(x - \pi\nu/2)$ when $x > 0$ and as $-|x|^{1-\nu}\cos(x + \pi\nu/2)$ when $x < 0$.

$2°$. To obtain an estimate for $r_\nu(x)$, note that the inequality

$$\left| \frac{e^\zeta \zeta^{1-\nu}}{\zeta^2 + x^2} \right| \le \frac{e^{-r/\sqrt{2}} r^{1-\nu}}{|\mp ir^2 + x^2|}, \qquad 0 < |\zeta| = r < +\infty$$

is true when $x(0 < |x| < +\infty)$ is any number and ζ is on rays $\arg \zeta = \pm 3\pi/4$ forming the contour of integration Γ_0 in $(6')$. Therefore,

$$|r_\nu(x)| \le \frac{|x|^{-1}}{\pi} \int_0^{+\infty} e^{-r\sqrt{2}} r^{1-\nu} dr < +\infty, \qquad 0 < |x| < +\infty,$$

and the asymptotic formulas (7) follow from (6) and $(6')$.

(b) Denote by

$$\Delta_\pm = \left\{ z : \left| \operatorname{Arg} z \mp \frac{\pi}{2} \right| < \frac{\pi}{4} \right\} \subset G_\pm$$

two corner domains of openings $\pi/2$ lying correspondingly in the half-planes $G_\pm = \{z : \pm \operatorname{Im} z > 0\}$. Let $\Delta = \Delta_+ \cup \Delta_-$ be their sum.

Lemma 4.2-2. *If $\nu \in [0, 2)$ is any number, then the following two-sided estimates are true:*

$$|s_\nu(z)| \asymp (1 + |z|)^{1-\nu} e^{|\operatorname{Im} z|},$$

$$|S_\sigma(z; \nu)| \asymp (1 + |z|)^{1-\nu} e^{\sigma|\operatorname{Im} z|} \qquad (9)$$

$$(z \in \overline{\Delta}, 1 \le |z| < +\infty),$$

where the suitable constants are independent of z.

Proof. The mapping $w = iz$ transforms the domains $\Delta_\pm$ into domains $\Delta_\pm^* = i\Delta_\mp$ which lie correspondingly in the right $(\operatorname{Re} w > 0)$ and in the left $(\operatorname{Re} w < 0)$ half-planes of w-plane. And, obviously,

$$\Delta_-^* = \Delta_{3\pi/4}^*, \qquad \Delta_+^* = \Delta_{3\pi/4}. \qquad (10)$$

Now note that the parameter $\alpha \in (\pi/2, \pi)$ is arbitrary in the representation (5)-$(5')$ of the function $E_1(z; \nu)$, so it may vary, if we desire to find an estimate for the integral $R_\nu(z; \alpha)$. To this end first observe that, if $\alpha \in (\pi/2, 3\pi/4)$, then clearly, $|\zeta - w| \ge |w| \sin(3\pi/4 - \alpha)$ when $w \in \overline{\Delta_-^*} \cup \overline{\Delta_+^*}$ and $\zeta \in \Gamma(\alpha)$. Also the estimate

$$|R_\nu(w; \alpha)| \le \frac{a(\alpha)}{|w|}, \qquad w \in \overline{\Delta^*} \setminus \{0\}, \qquad (11)$$

where $\Delta^* = \Delta_+^* \cup \Delta_-^*$ and

$$a(\alpha) = \frac{1}{\pi} \left[\sin\left(\frac{3\pi}{4} - \alpha \right) \right]^{-1} \int_0^{+\infty} e^{-|\cos \alpha| r} r^{1-\nu} dr < +\infty$$

is a constant, follows easily from representation (5$'$). On the other hand, by (10) and (5),

$$E_1(w;\nu) = \begin{cases} w^{1-\nu}e^w + R_\nu(w;\alpha) & \text{when } w \in \overline{\Delta^*_+} \setminus \{0\}, \\ R_\nu(w;\alpha) & \text{when } w \in \overline{\Delta^*_-} \setminus \{0\} \end{cases}$$

and

$$E_1(-w;\nu) = \begin{cases} (-w)^{1-\nu}e^{-w} + R_\nu(-w;\alpha) & \text{when } w \in \overline{\Delta^*_-} \setminus \{0\}, \\ R_\nu(-w;\alpha) & \text{when } w \in \overline{\Delta^*_+} \setminus \{0\}. \end{cases}$$

Thus we have

$$\tilde{s}_\nu(w) \equiv \frac{1}{2i} \left[E_1(w;\nu) - E_1(-w;\nu) \right] = s_\nu(-iw)$$

$$= \begin{cases} \frac{1}{2i}\left[w^{1-\nu}e^w + R_\nu(w;\alpha) - R_\nu(-w;\alpha) \right] & \text{when } w \in \overline{\Delta^*_+} \setminus \{0\}, \\ \frac{1}{2i}\left[-(-w)^{1-\nu}e^{-w} + R_\nu(w;\alpha) - R_\nu(-w;\alpha) \right] & \text{when } w \in \overline{\Delta^*_-} \setminus \{0\}, \end{cases}$$

$$\tag{12}$$

and the estimate

$$|\tilde{s}_\nu(w)| \leq \frac{1}{2}|w|^{1-\nu}e^{|\operatorname{Re} w|} + \frac{a(\alpha)}{|w|}, \qquad w \in \overline{\Delta^*} = \overline{\Delta^*_+} \bigcup \overline{\Delta^*_-}, \qquad w \neq 0$$

follows from representation (12) and inequality (11). But $0 \leq \nu < 2$, and hence there exists a constant $C_\nu > 0$, such that

$$|\tilde{s}_\nu(w)| \leq C_\nu(1+|w|)^{1-\nu}e^{|\operatorname{Re} w|}, \qquad w \in \overline{\Delta^*}, \ |w| \geq 1. \tag{13}$$

The inequalities

$$\frac{1}{2}|w|^{1-\nu}e^{\operatorname{Re} w} \leq |\tilde{s}_\nu(w)| + \frac{a(\alpha)}{|w|}, \qquad w \in \overline{\Delta^*_+} \setminus \{0\},$$

$$\frac{1}{2}|w|^{1-\nu}e^{-\operatorname{Re} w} \leq |\tilde{s}_\nu(w)| + \frac{a(\alpha)}{|w|}, \qquad w \in \overline{\Delta^*_-} \setminus \{0\}$$

can also be obtained from representation (12) by use of (11). But $\operatorname{Re} w \geq 0$ when $w \in \overline{\Delta^*_+}$, and $\operatorname{Re} w \leq 0$ when $w \in \overline{\Delta^*_-}$, thus the last two inequalities may be united as follows:

$$|\tilde{s}_\nu(w)| \geq \frac{1}{2}|w|^{1-\nu}e^{|\operatorname{Re} w|} - \frac{a(\alpha)}{|w|}, \qquad w \in \overline{\Delta^*} \setminus \{0\}.$$

Now the condition $0 \leq \nu < 2$ leads to the estimate

$$|\tilde{s}_\nu(w)| \geq C^*_\nu(1+|w|)^{1-\nu}e^{|\operatorname{Re} w|}, \qquad w \in \overline{\Delta^*}, |w| \geq 1,$$

where $C^*_\nu > 0$ is a constant independent of w. So, by (13),

$$|\tilde{s}_\nu(w)| \asymp (1+|w|)^{1-\nu}e^{|\operatorname{Re} w|}, \qquad w \in \overline{\Delta^*}, |w| \geq 1. \tag{14}$$

Hence the first of the two-sided estimates (9) follows, if we observe that the inverse mapping $z = -iw$ transforms the domains $\Delta_{\mp}^*$ into the initial corner domains $\Delta_{\pm} = -i\Delta_{\mp}^*$ of the z-plane; it also transforms the function $\tilde{s}_\nu(w)$ into $s_\nu(z) = \tilde{s}_\nu(iz)$ and the two-sided estimate (14) to the first of the two-sided estimates (9). The second two-sided estimate of (9) follows from the first one and from connection (2) between the functions $S_\sigma(z;\nu)$ and $s_\nu(\sigma z)$. This completes the proof.

The next lemma is proved in a similar way.

Lemma 4.2-3. *If $0 \leq \nu < 2$ and $\gamma \in (0, +\infty)$ are any numbers, then*

$$\left.\begin{array}{r} s_\nu(x \pm i\gamma) \\ S_\sigma(x \pm i\gamma; \nu) \end{array}\right\} \asymp |x \pm i\gamma|^{1-\nu}, \qquad x \in (-\infty, +\infty), \tag{15}$$

where the suitable constants depend only on ν, γ and σ.

Proof. First observe that the mapping $w = iz$ transforms the lines $z = x \pm i\gamma(-\infty < x < +\infty)$ into the lines $w = \mp\gamma + ix$ lying correspondingly in the left $(\operatorname{Re} w < 0)$ and in the right $(\operatorname{Re} w > 0)$ half-planes. Then introduce the notation

$$L_\pm(\gamma) = \{\pm\gamma + iv : 2\gamma \leq |v| < +\infty\} \tag{16}$$

for the pairs of rays lying on the lines $\pm\gamma + iv$ $(-\infty < v < +\infty)$ and also introduce the notation

$$l_\pm(\gamma) = \{\pm\gamma + iv : 0 \leq |v| \leq 2\gamma\} \tag{17}$$

for their complementary segments. Then, obviously, $L_\pm(\gamma) \subset \Delta_{3\pi/4}$, and it is clear that

$$|w| = \sqrt{\gamma^2 + (\operatorname{Im} w)^2} \geq \sqrt{5}\gamma, \quad |\operatorname{Im} w| = |w|\sqrt{1 - \frac{\gamma^2}{|w|^2}} \geq \frac{2}{\sqrt{5}}|w|,$$

if $w \in L_\pm(\gamma)$. Hence the geometric consideration gives

$$|w - \zeta| \geq \frac{|\operatorname{Im} w| - \gamma}{\sqrt{2}} \geq \frac{1}{\sqrt{2}}\left(\frac{2}{\sqrt{5}}|w| - \gamma\right) \geq \frac{\gamma}{\sqrt{2}},$$

when $w \in L_\pm(\gamma)$ and $\zeta \in \Gamma_0 = \Gamma(3\pi/4)$, so the estimate

$$\left|R_\nu\left(w; \frac{3\pi}{4}\right)\right| \leq \frac{b}{2|w|/\sqrt{5} - \gamma}, \qquad w \in L(\gamma) = L_+(\gamma) \bigcup L_-(\gamma), \tag{18}$$

where

$$b = \frac{\sqrt{2}}{\pi}\int_0^{+\infty} e^{-r/\sqrt{2}}r^{1-\nu}dr < +\infty \qquad (0 \leq \nu < 2),$$

follows from representation (5′) of the function $R_\nu(z;\alpha)$, if we set $\alpha = 3\pi/4$. On the other hand, (5) implies $E_1(\pm w;\nu) = (\pm w)^{1-\nu} e^{\pm w} + R_\nu(\pm w; 3\pi/4)$, $w \in L(\gamma)$. Hence (12) passes to

$$\tilde{s}_\nu(w) = \frac{1}{2i}\left[w^{1-\nu}e^w - (-w)^{1-\nu}e^{-w} + R_\nu(w;3\pi/4) - R_\nu(-w;3\pi/4)\right], w \in L(\gamma),$$

$$(19)$$

and so, by (18),

$$|\tilde{s}_\nu(w)| \leq |w|^{1-\nu}\, e^\gamma + \frac{b}{2\,|w|\,/\sqrt{5} - \gamma}, \qquad w \in L(\gamma) = L_+(\gamma)\bigcup L_-(\gamma).$$

But $2|w|/\sqrt{5} - \gamma \geq \gamma$ and $0 \leq \nu < 2$, and therefore, it follows from the last estimate that there exists a constant $C^{(1)}_{\nu,\gamma} > 0$, independent of w, such that

$$|\tilde{s}_\nu(w)| \leq C^{(1)}_{\nu,\gamma}\,|w|^{1-\nu}, \qquad w \in L(\gamma).$$

Further, if we use the representation (19) and estimate (18), we obtain

$$|\tilde{s}_\nu(w)| \geq \frac{1}{2}\left(e^\gamma - 1\right)|w|^{1-\nu} - \frac{b}{2\,|w|\,/\sqrt{5} - \gamma}, \; w \in L(\gamma).$$

Similarly, since $0 \leq \nu < 2$, we obtain

$$|\tilde{s}_\nu(w)| \geq C^{(2)}_{\nu,\gamma}\,|w|^{1-\nu}, \qquad w \in L(\gamma),$$

where $C^{(2)}_{\nu,\gamma} > 0$ is a constant independent of w. Thus the two-sided estimates $|\tilde{s}_\nu(w)| \asymp |w|^{1-\nu}$ ($w \in L(\gamma)$) are true. On the other hand, according to the second of identities (2), to the identity $\tilde{s}_\nu(w) = s_\nu(-iw)$ and to Theorem 1.4-2, the function $|\tilde{s}_\nu(w)w^{-1+\nu}|$ is non-vanishing on the parallel lines $\mathrm{Re}\,w = \pm\gamma$. Besides, this function is continuous on the segments $l_\pm(\gamma)$. Therefore, $|\tilde{s}_\nu(w)| \asymp |w|^{1-\nu}$ ($w \in l_\pm(\gamma)$), and we come to the conclusion that the two-sided estimate

$$|\tilde{s}_\nu(w)| \asymp |w|^{1-\nu} \qquad (20)$$

is true on both parallel lines $w = \pm\gamma + iv$ $(0 < \gamma < +\infty, -\infty < v < +\infty)$. It remains to observe that the inverse mapping $z = -iw$ transforms the lines $w = \mp\gamma + ix$ $(-\infty < x < +\infty)$ into the initial lines $z = x \pm i\gamma$ $(-\infty < x < +\infty)$, and the function $\tilde{s}_\nu(w)$ passes to $s_\nu(z) = \tilde{s}_\nu(iz)$, so the two-sided estimate (20) turns into the first estimate of (15). The second of the desired estimates (15) follows from the connection (2) between the functions $s_\nu(z)$ and $S_\sigma(z;\nu)$.

(c) Now we return to the function

$$\mathcal{E}_\sigma(z;\nu) = E_{1/2}\left(-\sigma^2 z; 1 + \nu\right) \qquad (0 \leq \nu < 2)$$

and note that, according to Theorem 1.4-3, its zeros are simple and positive. Let $\{\lambda_k = \lambda_k(\sigma,\nu)\}_1^\infty$ $(0 < \lambda_k < \lambda_{k+1}, 1 \leq k < +\infty)$ be the zeros of this function. Then the following lemma is true.

Lemma 4.2-4. *Let $1 < p < +\infty$ and $-1 < \omega < p - 1$. Then the following assertions are true:*
1°. If $\nu \in [0, 2)$, then

$$\frac{\mathcal{E}_\sigma(z; \nu)}{z - \lambda_k} \in W^{p,\omega}_{1/2,\sigma} \qquad (1 \le k < +\infty), \tag{21}$$

and if $0 \le \nu < 2(1+\omega)/p$, then

$$\mathcal{E}_\sigma(z; \nu) \notin W^{p,\omega}_{1/2,\sigma}. \tag{22}$$

2°. If $2(1+\omega)/p < \nu < 2$ and we put $\lambda_0 = 0$, then

$$\frac{z\mathcal{E}_\sigma(z; \nu)}{z - \lambda_k} \in W^{p,\omega}_{1/2,\sigma} \qquad (0 \le k < +\infty), \tag{23}$$

and if $0 \le \nu < 2$, then

$$z\mathcal{E}_\sigma(z; \nu) \notin W^{p,\omega}_{1/2,\sigma}. \tag{24}$$

Proof. First note that all the functions we are concerned with are obviously entire, of order $1/2$ and of type σ.
1°. Let $\nu \in [0, 2)$ be any number. Then, by the asymptotic formula 1.3(5),

$$|\mathcal{E}_\sigma(x; \nu)| = \left|E_{1/2}\left(-\sigma^2 x; 1 + \nu\right)\right| \le C_1(1 + x)^{-\nu/2}, \qquad x \in [0, +\infty), \tag{25}$$

where the constant $C_1 > 0$ is independent of x. Hence

$$\int_0^1 \left|\frac{\mathcal{E}_\sigma(x; \nu)}{x - \lambda_k}\right|^p x^\omega dx < +\infty \qquad (1 \le k < +\infty). \tag{26}$$

From the estimate (25) it also follows that

$$\left|\frac{\mathcal{E}_\sigma(x; \nu)}{x - \lambda_k}\right|^p x^\omega \le C_2(k)x^{\omega-(1+\nu/2)p}, x \in [1, +\infty), \qquad 1 \le k < +\infty,$$

where the constant $C_2(k) > 0$ is independent of x. But $\omega - (1+\nu/2)p < -1$, since $\nu \ge 0$ and $-1 < \omega < p - 1$. So we have

$$\int_1^{+\infty} \left|\frac{\mathcal{E}_\sigma(x; \nu)}{x - \lambda_k}\right|^p x^\omega dx < +\infty \qquad (1 \le k < +\infty). \tag{27}$$

Inclusion (21) now follows as the result of (26) and (27).
To prove (22), first note that the asymptotic formula 1.3(5) implies

$$(\sigma^2 x)^{-\nu/2} \left|\cos(\sigma\sqrt{x} - \pi\nu/2)\right| \le |\mathcal{E}_\sigma(x; \nu)| + \varphi_\nu(x), \ 1 \le x < +\infty, \ \nu \in [0, 2), \tag{28}$$

where

$$\varphi_\nu(x) = \begin{cases} c_\nu x^{-2}, & \text{if } \nu = 0,1 \\ d_\nu x^{-1}, & \text{if } \nu \neq 0,1 \end{cases} \qquad (c_\nu, d_\nu > 0). \qquad (28')$$

Hence, using the well-known inequality $(a+b)^p \le 2^p(a^p + b^p)$ $(0 \le a,b < +\infty)$, we obtain

$$\begin{aligned}
(2\sigma^\nu)^{-p} &\int_1^R \left| \cos\left(\sigma\sqrt{x} - \frac{\pi\nu}{2}\right) \right|^p x^{\omega - p\nu/2} dx \\
&= 2(2\sigma^\nu)^{-p} \int_1^{\sqrt{R}} \frac{|\cos(\sigma\tau - \pi\nu/2)|^p}{\tau^\alpha} d\tau \\
&\le \int_1^R |\mathcal{E}_\sigma(x;\nu)|^p x^\omega dx + \int_1^R \varphi_\nu^p(x) x^\omega dx, \qquad R \ge 1,
\end{aligned} \qquad (29)$$

where $\alpha = p\nu - 2\omega - 1 < 1$, since $\nu < 2(1+\omega)/p$. But $\omega < p - 1$ and $\alpha < 1$, so it follows from $(28')$ that the last integral of (29) is bounded and the left integral of (29) is unbounded as $R \to +\infty$. Thus (22) is proved.

$2°$. From (25) it follows that

$$\int_0^1 \left| \frac{x\mathcal{E}_\sigma(x;\nu)}{x - \lambda_k} \right|^p x^\omega dx < +\infty \qquad (0 \le k < +\infty),$$

and it is clear that

$$\left| \frac{x\mathcal{E}_\sigma(x;\nu)}{x - \lambda_k} \right|^p x^\omega \le C_3(k) x^{\omega - p\nu/2}, 1 \le x < +\infty, \qquad 0 \le k < +\infty,$$

where the constant $C_3(k) > 0$ is independent of x. But $p\nu/2 - \omega > 1$ since $\nu > 2(1+\omega)/p$. Therefore,

$$\int_0^{+\infty} \left| \frac{x\mathcal{E}_\sigma(x;\nu)}{x - \lambda_k} \right|^p x^\omega dx < +\infty \qquad (0 \le k < +\infty),$$

which implies inclusion (23).

Now, to complete the proof, it remains to prove only (24). To this end we return to the inequality $(28)-(28')$ which easily passes to the integral estimate

$$\begin{aligned}
(2\sigma^\nu)^{-p} &\int_1^R \left| \cos\left(\sigma\sqrt{x} - \frac{\pi\nu}{2}\right) \right|^p x^{\omega + p(1 - \nu/2)} dx \\
&= 2(2\sigma^\nu)^{-p} \int_1^{\sqrt{R}} \frac{|\cos(\sigma\tau - \pi\nu/2)|^p}{\tau^\beta} d\tau \\
&\le \int_1^R |x\mathcal{E}_\sigma(x;\nu)|^p x^\omega dx + \int_1^R \varphi_\nu^p(x) x^{\omega + p} dx, \qquad R \ge 1,
\end{aligned} \qquad (30)$$

where $\beta = p\nu - 2p - 2\omega - 1 < -2\omega - 1 < 1$ since $\nu < 2$ and $-1 < \omega < p - 1$. Representation $(28')$ of the function $\varphi_\nu(x)$ shows that

$$U_\nu(R) = \int_1^R \varphi_\nu^p(x) x^{\omega+p} dx = \begin{cases} O(1), & \text{if } \nu = 0,1 \\ O(R^{1+\omega}), & \text{if } \nu \neq 0,1 \end{cases} \quad \text{as } R \to +\infty. \tag{31}$$

On the other hand, if

$$\Delta_k = \left\{ \tau : \pi k - \frac{\pi}{3} \leq \sigma\tau - \frac{\pi}{2}\nu \leq \pi k + \frac{\pi}{3} \right\} \quad (1 \leq k < +\infty),$$

then mes $\Delta_k = 2\pi/3\sigma \ (k \geq 1)$, and it is clear that

$$\min_{\tau \in \Delta_k} \left| \cos\left(\sigma\tau - \frac{\pi}{2}\nu \right) \right| \geq \frac{1}{2} \quad (1 \leq k < +\infty).$$

Next denote $R_k^\pm = [(\pi k \pm \pi/3 + \pi\nu/2)/\sigma]^2$; then, evidently, Δ_k coincides with the segment $\left[\sqrt{R_k^-}, \sqrt{R_k^+} \right] \ (1 \leq k < +\infty)$, and $\Delta_k \subset [1, +\infty)$ when $k \geq k_0$. Hence the inequality (30) can be written down for $R = R_n^+ (n \geq k_0)$:

$$\int_1^{R_n^+} |x\mathcal{E}_\sigma(x;\nu)|^p x^\omega dx \geq - \int_1^{R_n^+} \varphi_\nu^p(x) x^{\omega+p} dx$$
$$+ 2(2\sigma^\nu)^{-p} \int_1^{\sqrt{R_n^+}} \frac{|\cos(\sigma\tau - \pi\nu/2)|^p}{\tau^\beta} d\tau \equiv -U_\nu(R_n^+) + V_\nu(R_n^+). \tag{32}$$

But

$$V_\nu(R_n^+) > 2(2\sigma^\nu)^{-p} \sum_{k=k_o}^n \int_{\Delta_k} \frac{|\cos(\sigma\tau - \pi\nu/2)|^p}{\tau^\beta} d\tau$$
$$\geq \begin{cases} \sigma^{-\nu p} 2^{-2p+1} \sum_{k=k_o}^n \frac{\text{mes } \Delta_k}{(R_k^+)^{\beta/2}}, & \text{if } 0 \leq \beta < 1, \\ \sigma^{-\nu p} 2^{-2p+1} \sum_{k=k_o}^n \frac{\text{mes } \Delta_k}{(R_k^-)^{\beta/2}}, & \text{if } \beta < 0. \end{cases}$$

Therefore,

$$V_\nu(R_n^+) > C \sum_{k=k_o}^n \frac{1}{k^\beta} \asymp n^{1-\beta}, \qquad k_0 \leq n < +\infty, \tag{33}$$

and (24) follows from (31)-(33) as $n \to +\infty$, since $1 - \beta = 2(1 + \omega) + p(2 - \nu)$.

(d) Concluding this section, we prove one more lemma.

Lemma 4.2-5. *Let* $\nu \in [0,2)$ *and let* $\{\lambda_k\}_1^\infty \ (0 < \lambda_k < \lambda_{k+1}, 1 \leq k < +\infty)$ *be the sequence of zeros of the function* $\mathcal{E}_\sigma(z;\nu)$. *Then the following two-sided estimates are true:*

$$\mathcal{E}_\sigma'(\lambda_k;\nu) \asymp (1+k)^{-1-\nu}(1 \leq k < +\infty). \tag{34}$$

Proof. From the power expansion

$$\mathcal{E}_\sigma(z;\nu) = \sum_{k=0}^{\infty} (-1)^k \sigma^{2k} \frac{z^k}{\Gamma(1+\nu+2k)}$$

it follows that $\mathcal{E}_\sigma(z;\nu-1) = \nu\mathcal{E}_\sigma(z;\nu) + 2z\mathcal{E}'_\sigma(z;\nu)$. So, if we take $z = \lambda_k(1 \le k < +\infty)$, we obtain

$$2\lambda_k\mathcal{E}'_\sigma(\lambda_k;\nu) = \mathcal{E}_\sigma(\lambda_k;\nu-1) = E_{1/2}\left(-\sigma^2\lambda_k;\nu\right) \qquad (1 \le k < +\infty). \qquad (35)$$

Now, using the asymptotic formulas 1.3(5) and 1.4(12)-(14), which were established for the function $E_{1/2}\left(-\sigma^2\lambda_k;\nu\right)$ and for its zeros $\{\lambda_k\}_1^\infty$, we obtain

$$2\lambda_k\mathcal{E}'_\sigma(\lambda_k;\nu) = (\sigma^2\lambda_k)^{(1-\nu)/2} \cos\left(\sigma\sqrt{\lambda_k} + \frac{\pi}{2}(1-\nu)\right) + O(1/\lambda_k)$$
$$= (\sigma^2\lambda_k)^{(1-\nu)/2} \cos\left(\pi k + O(k^{\nu-2})\right) + O(1/\lambda_k)$$

as $k \to +\infty$. This completes the proof since $\lambda_k \asymp (1+k)^2$ and $\nu < 2$.

4.3 Two special interpolation series

Bearing in mind the definition

$$\mathcal{E}_\sigma(z;\nu) = E_{1/2}\left(-\sigma^2 z; 1+\nu\right) \qquad (1)$$

of the function $\mathcal{E}_\sigma(z;\nu)$, note once again that all its zeros $\{\lambda_k = \lambda_k(\sigma,\nu)\}_1^\infty$ ($0 < \lambda_k < \lambda_{k+1}$, $1 \le k < +\infty$) are simple and positive and that their distribution is illustrated in Theorems 1.4-3 and 1.4-4, in the case when $\nu \in [0,2)$. Note also that the two-sided estimates

$$\lambda_k \asymp (1+k)^2 \qquad (1 \le k < +\infty) \qquad (2)$$

are true uniformly with respect to k. In this section we consider the following two series:

$$\Phi(z) = \sum_{k=1}^{\infty} a_k \frac{\mathcal{E}_\sigma(z;\nu)}{\mathcal{E}'_\sigma(\lambda_k;\nu)(z-\lambda_k)}, \qquad (3)$$

$$\Psi(z) = b_0\Gamma(1+\nu)\mathcal{E}_\sigma(z;\nu) + \sum_{k=1}^{\infty} b_k \frac{z\mathcal{E}_\sigma(z;\nu)}{\lambda_k\mathcal{E}'_\sigma(\lambda_k;\nu)(z-\lambda_k)}. \qquad (4)$$

Here $\{a_k\}_1^\infty$ and $\{b_k\}_0^\infty$ are sequences of complex numbers, and $\mathcal{E}'_\sigma(\lambda_k;\nu) \ne 0$ ($1 \le k < +\infty$) since zeros of this function are simple. Evidently, the sums of these series have the interpolation data

$$\Phi(\lambda_k) = a_k \quad (1 \le k < +\infty), \qquad \Psi(\lambda_k) = b_k \quad (\lambda_0 = 0, 0 \le k < +\infty), \qquad (5)$$

if they uniformly converge in $\mathbb{C}$. Our aim is to find conditions in which the series (3) and (4) converge to entire functions $\Phi(z)$ and $\Psi(z)$ of class $W_{1/2,\sigma}^{p,\omega}$ $(1 < p < +\infty, -1 < \omega < p - 1)$.

(a) To this end, introduce the Banach space $l^{p,\kappa}$ $(1 < p < +\infty, -1 < \kappa < +\infty)$ of sequences $\{c_k\} \subset \mathbb{C}$ for which

$$\|\{c_k\}\|_{p,\kappa} = \left\{ \sum_k |c_k|^p (1 + |k|)^\kappa \right\}^{1/p} < +\infty. \tag{6}$$

Then the following lemma is true.

Lemma 4.3-1. *Let the sequences $\{a_k\}_1^\infty$ and $\{b_k\}_0^\infty$ belong to the class $l^{p,\kappa}$, where*

$$1 < p < +\infty, \quad -1 < \omega < p - 1, \quad \kappa = 1 + 2\omega. \tag{7}$$

Then series (3) and (4) converge absolutely and uniformly in any compact $K \subset \mathbb{C}$ to entire functions $\Phi(z)$ and $\Psi(z)$, if, correspondingly,

$$0 \le \nu < 2(1 + \omega)/p \quad \text{or} \quad 0 \le \nu < 2. \tag{8}$$

Besides, these functions have the interpolation data (5).

Proof. Consider the partial sum

$$\Phi_{n,m}(z) = \sum_{k=n}^{m} a_k \frac{\mathcal{E}_\sigma(z;\nu)}{\mathcal{E}_\sigma'(\lambda_k;\nu)(z - \lambda_k)}, \quad 1 \le n < m < +\infty. \tag{9}$$

If $K \subset \mathbb{C}$ is compact, then, obviously,

$$|\Phi_{n,m}(z)| \le A_1 \sum_{k=n}^{m} \frac{|a_k|}{\lambda_k |\mathcal{E}_\sigma'(\lambda_k;\nu)|}, \quad z \in K, \tag{10}$$

where

$$A_1 = A_1(K) = \max_{1 \le k < +\infty} \left\{ \sup_{z \in K} \left| \frac{\lambda_k \mathcal{E}_\sigma(z;\nu)}{z - \lambda_k} \right| \right\} < +\infty.$$

But $\lambda_k |\mathcal{E}_\sigma'(\lambda_k;\nu)| \asymp (1 + k)^{1-\nu}$ $(1 \le k < +\infty)$, according to 4.2(34) and (2). Therefore, by estimate (10) and Hölder's inequality,

$$|\Phi_{n,m}(z)| \le A_2 \sum_{k=n}^{m} |a_k|(1 + k)^{\nu-1}$$

$$\le A_2 \left(\sum_{k=n}^{m} |a_k|^p (1 + k)^\kappa \right)^{1/p} \left(\sum_{k=n}^{m} (1 + k)^{q(\nu-1-\kappa/p)} \right)^{1/q}, \quad z \in K, \tag{11}$$

where $\kappa = 1 + 2\omega$, $q = p/(p-1)$, and the constant $A_2 > 0$ is independent of z, n and m. If the first of conditions (8) is satisfied, i.e. if $0 \leq \nu < (1 + \kappa)/p$, then $q(\nu - 1 - \kappa/p) < -1$. Thus, a partial sum of a convergent series is written in the last brackets of (11) and, consequently,

$$|\Phi_{n,m}(z)| \leq A_3 \|\{a_k\}^m_n\|_{p,\kappa}, \qquad z \in K,$$

where the constant $A_3 > 0$ is also independent of z, n and m. We have also $\{a_k\}^\infty_1 \in l^{p,\kappa}$. Hence the convergence of the series (3) and the desired properties of its sum $\Phi(z)$ hold. The assertion on convergence of the sum (4), in the case of the second hypothesis of (8) (i.e., when $0 \leq \nu < 2$), can be established in a similar way. For this purpose we consider the partial sum

$$\Psi_{n,m}(z) = \sum_{k=n}^{m} b_k \frac{z\mathcal{E}_\sigma(z; \nu)}{\lambda_k \mathcal{E}'_\sigma(\lambda_k; \nu)(z - \lambda_k)} \qquad (1 \leq n < m < +\infty). \tag{12}$$

Observe that

$$A_4 = A_4(K) = \max_{1 \leq k < +\infty} \left\{ \sup_{z \in K} \left| \frac{\lambda_k z\mathcal{E}_\sigma(z; \nu)}{z - \lambda_k} \right| \right\} < +\infty$$

and, consequently,

$$|\Psi_{n,m}(z)| \leq A_4 \sum_{k=n}^{m} \frac{|b_k|}{\lambda_k^2 |\mathcal{E}'_\sigma(\lambda_k; \nu)|}, \qquad z \in K.$$

But $\lambda_k^2 |\mathcal{E}'_\sigma(\lambda_k; \nu)| \asymp (1 + k)^{3-\nu}$ by 4.2(34) and (2). Thus,

$$|\Psi_{n,m}(z)| \leq A_5 \sum_{k=n}^{m} |b_k|(1 + k)^{\nu - 3}$$

$$\leq A_5 \left(\sum_{k=n}^{m} |b_k|^p (1 + k)^\kappa \right)^{1/p} \left(\sum_{k=n}^{m} (1 + k)^{q(\nu - 3 - \kappa/p)} \right)^{1/q}, \qquad z \in K. \tag{13}$$

We also have $\nu < 2$ and $\kappa = 1 + 2\omega > -1$; therefore, $q(\nu - 3 - \kappa/p) < -q(1 + \kappa/p) < -1$, so the last brackets of (13) contain a partial sum of a convergent series. The desired conclusion follows, and the proof is complete.

    Below we shall see that considerably more can be said about the entire functions $\Phi(z)$ and $\Psi(z)$ which are the sums of series (3) and (4), if the parameter $\nu \in [0, 2)$ is restricted by conditions stronger than (8).

(b) Now we prove the first main theorem of this type.

Theorem 4.3-1. *Let $\{a_k\}_1^\infty \in l^{p,\kappa}$ $(1 < p < +\infty,\ \kappa = 1 + 2\omega,\ -1 < \omega < p - 1)$, and let the parameter $\nu \in [0, 2)$ satisfy the complementary condition*

$$\nu \in \Delta(\kappa, p) = \left(\frac{1 + \kappa}{p} - 1, \frac{1 + \kappa}{p}\right) \cap [0, 2). \tag{14}$$

Then:
1°. The series

$$\Phi(z) = \sum_{k=1}^\infty a_k \frac{\mathcal{E}_\sigma(z; \nu)}{\mathcal{E}_\sigma'(\lambda_k; \nu)(z - \lambda_k)}$$

converges in the norm $\|\cdot\|_{p,\omega}^+$ (4.1(1)) to an entire function $\Phi(z) \in W_{1/2,\sigma}^{p,\omega}$.
2°. This function is such that

$$\Phi(\lambda_k) = a_k(1 \le k < +\infty),\quad \|\Phi\|_{p,\omega}^+ \asymp \|\{a_k\}_1^\infty\|_{p,\kappa}, \tag{15}$$

where the suitable constants of two-sided estimates are independent of the sequence $\{a_k\}_1^\infty$.

Proof. Let $\Phi_{n,m}(z)$ be the sum (9). Then $\Phi_{n,m}(z) \in W_{1/2,\sigma}^{p,\omega}$ $(1 \le n < m < +\infty)$ by Lemma $4.2 - 4(1°)$. But

$$\|\Phi_{n,m}\|_{p,\omega}^+ = \left(\int_0^{+\infty} |\Phi_{n,m}(x)|^p\, x^\omega dx\right)^{1/p}$$

$$= \left(\int_{-\infty}^{+\infty} \left|\Phi_{n,m}(x^2)\right|^p |x|^\kappa\, dx\right)^{1/p} < +\infty \tag{16}$$

by 4.1(1). Hence the entire functions $\Phi_{n,m}(z^2)$ of exponential type $\le \sigma$ are of class $W_\sigma^{p,\kappa}$ for any $1 \le n < m < +\infty$. We now denote

$$\varphi_{n,m}(z) = \sum_{k=n}^m \frac{a_k}{\mathcal{E}_\sigma'(\lambda_k; \nu)(z^2 - \lambda_k)}. \tag{17}$$

Then, using Theorem 3.4-1, we can obtain from (16) that

$$\{\|\Phi_{n,m}\|_{p,\omega}^+\}^p \le C_1 \int_{-\infty}^{+\infty} \left|\Phi_{n,m}\left((x + i)^2\right)\right|^p |x + i|^\kappa\, dx$$

$$= C_1 \int_{-\infty}^{+\infty} \left|\mathcal{E}_\sigma\left((x + i)^2; \nu\right)\varphi_{n,m}(x + i)\right|^p |x + i|^\kappa\, dx,\qquad 1 \le n < m < +\infty, \tag{18}$$

where $C_1 > 0$ is a constant independent of n and m (such constants will be denoted further in the proof by $C_j(j \ge 2)$). Now note that $\mathcal{E}_\sigma(z^2; \nu) = (\sigma z)^{-1} s_\nu(\sigma z)$ according to 4.2(2). Hence, by the estimates 4.2(15),

$$\left|\mathcal{E}_\sigma\left((x + i)^2; \nu\right)\right| \asymp |x + i|^{-\nu},\ -\infty < x < +\infty, \tag{19}$$

and it follows from (18) that

$$\left\{\|\Phi_{n,m}\|_{p,\omega}^{+}\right\}^{p} \leq C_{2} \int_{-\infty}^{+\infty} |\varphi_{n,m}(x+i)|^{p} |x+i|^{\kappa_{1}} dx$$

$$\equiv C_{2}\|\varphi_{n,m}(x+i)\|_{p,\kappa_{1},1}^{p}, \quad (19')$$

where $\kappa_{1} = \kappa - \nu p$ and $-1 < \kappa_{1} < p-1$ by condition (14). The well-known Han-Banach theorem implies

$$\|\varphi_{n,m}(x+i)\|_{p,\kappa_{1},1} = \sup_{\|K\|_{q}\leq 1} \left| \int_{-\infty}^{+\infty} \varphi_{n,m}(x+i)K(x) |x+i|^{\kappa_{1}/p} dx \right|.$$

Consequently, there exists a function $K_{0}(x)$ such that $K_{0}(x)|x+i|^{-\kappa_{1}/p} \in L_{q}(-\infty,+\infty)$ $(q = p/(p-1))$ and

$$\|K_{0}\|_{q,\kappa_{2},1} = \left\{ \int_{-\infty}^{+\infty} |K_{0}(x)|^{q} |x+i|^{\kappa_{2}} dx \right\}^{1/q} \leq 1,$$

where $\kappa_{2} = -\kappa_{1}q/p, -1 < \kappa_{2} < q-1$ and

$$\|\varphi_{n,m}(x+i)\|_{p,\kappa_{1},1} \leq C_{3} \left| \int_{-\infty}^{+\infty} \varphi_{n,m}(x+i)K_{0}(x)dx \right|. \quad (20)$$

Now consider the Cauchy type integral

$$\mathcal{K}_{0}^{-}(z) = \frac{1}{2\pi i} \int_{-\infty}^{+\infty} \frac{K_{0}(t)}{t-z}dt, z \in G_{-} = \{z : \operatorname{Im} z < 0\}, \quad (21)$$

and note that, by assertions 1° and 2° of Theorem 3.3-3,

$$(z-i)^{\kappa_{2}/q}\mathcal{K}_{0}^{-}(z) \in H_{-}^{q}, \|(x-i)^{\kappa_{2}/q}\mathcal{K}_{0}^{-}(x)\|_{q} \leq C_{4}\|K_{0}\|_{q,\kappa_{2},1} \leq C_{4}. \quad (22)$$

Further, taking $\lambda_{k} = x_{k}^{2}$ $(1 \leq k < +\infty)$, insert representation (17) into the right-hand side integral of (20). Then, by (21),

$$\|\varphi_{n,m}(x+i)\|_{p,\kappa_{1},1} \leq C_{3} \left| \sum_{k=n}^{m} \frac{a_{k}}{\mathcal{E}_{\sigma}'(\lambda_{k};\nu)} \int_{-\infty}^{+\infty} \frac{K_{0}(t)dt}{(t+i)^{2} - x_{k}^{2}} \right|$$

$$\leq \pi C_{3} \left| \sum_{k=n}^{m} \frac{a_{k}}{\sqrt{\lambda_{k}}\mathcal{E}_{\sigma}'(\lambda_{k};\nu)} \left[\mathcal{K}_{0}^{-}(x_{k}-i) - \mathcal{K}_{0}^{-}(-x_{k}-i)\right] \right|.$$

But $\sqrt{\lambda_{k}}|\mathcal{E}_{\sigma}'(\lambda_{k};\nu)| \asymp (1+k)^{-\nu}(1 \leq k < +\infty)$. Thus, the use of Hölder's inequality gives

$$\|\varphi_{n,m}(x+i)\|_{p,\kappa_{1},1} \leq C_{5}\left\{ \left(\sum_{k=1}^{\infty} |\mathcal{K}_{0}^{-}(x_{k}-i)|^{q} (1+k)^{\kappa_{2}} \right)^{1/q} \right.$$

$$\left. + \left(\sum_{k=1}^{\infty} |\mathcal{K}_{0}^{-}(-x_{k}-i)|^{q} (1+k)^{\kappa_{2}} \right)^{1/q} \right\}\|\{a_{k}\}_{n}^{m}\|_{p,\kappa}, \quad 1 \leq n < m < +\infty.$$

$$(23)$$

Now observe, that, according to Theorem 1.4-4,

$$\sqrt{\lambda_k} = x_k = \frac{\pi}{\sigma}k + \frac{\pi}{2\sigma}(\nu - 1) + O(k^{\nu-2}) \text{ as } k \to +\infty$$

for any $\nu \in [0,2)$. Therefore, supposing that $\lambda_0 = x_0 = 0$, we obtain

$$\inf_{0 \le k < +\infty}[x_{k+1} - x_k] > \delta > 0$$

and $|x_k - 2i| \asymp 1 + k (0 \le k < +\infty)$. Thus, by Theorem 3.4-3 and by the first of relations (22),

$$\sum_{k=1}^{\infty} \left|\mathcal{K}_0^-(\pm x_k - i)\right|^q (1 + k)^{\kappa_2} \le C_6 < +\infty.$$

This, together with (23) and (19′), gives

$$\|\varphi_{n,m}(x + i)\|_{p,\kappa_1,1} \le C_7\|\{a_k\}_n^m\|_{p,\kappa}, \quad \|\Phi_{n,m}\|_{p,\omega}^+ \le C_8\|\{a_k\}_n^m\|_{p,\kappa},$$

and since $\{a_k\}_1^\infty \in l^{p,\kappa}$ $(\kappa = 1 + 2\omega)$, assertion 1° of the theorem follows from the last inequality, letting $n, m \to \infty$.

Let $n = 1$ in the last inequalities. Then the passage $m \to \infty$ gives

$$\|\Phi\|_{p,\omega}^+ \le C_8\|\{a_k\}_1^\infty\|_{p,\kappa}. \tag{24}$$

But, as was proved earlier, in Lemma 4.3-1, $\Phi(\lambda_k) = a_k$ $(k \ge 1)$.

On the other hand, if $\Phi(z) \in W_{1/2,\sigma}^{p,\omega}$, then, obviously, $f(z) = \Phi(z^2) \in W_\sigma^{p,\kappa}$ $(\kappa = 1 + 2\omega)$, and the inequality

$$\|\{f(x_k)\}_1^\infty\|_{p,\kappa} = \|\{\Phi(\lambda_k)\}_1^\infty\|_{p,\kappa} \le C_9\|f\|_{p,\kappa} = C_9\|\Phi\|_{p,\omega}^+ \tag{25}$$

follows from Theorem 3.4-2 (when $h = 0$). The two-sided estimate (15) follows from (24) and (25). This completes the proof.

(c) The next theorem is similar to the previous one, but its corresponding assertions concern the interpolation series (4).

Theorem 4.3-2. *Let* $\{b_k\}_0^\infty \in l^{p,\kappa}$ $(1 < p < +\infty, \kappa = 1 + 2\omega, -1 < \omega < p - 1)$ *and let the parameter* $\nu \in [0,2)$ *satisfy the additional condition*

$$\nu \in \Delta^*(\kappa, p) = \left(\frac{1+\kappa}{p}, 1 + \frac{1+\kappa}{p}\right) \cap [0,2). \tag{26}$$

Then:
1°. The series

$$\Psi(z) = b_0\Gamma(1+\nu)\mathcal{E}_\sigma(z;\nu) + \sum_{k=1}^{\infty} b_k \frac{z\mathcal{E}_\sigma(z;\nu)}{\lambda_k\mathcal{E}_\sigma'(\lambda_k;\nu)(z - \lambda_k)}$$

converges in the norm $\|\cdot\|_{p,\omega}^+$ *(4.1(1)) to an entire function* $\Psi(z) \in W_{1/2,\sigma}^{p,\omega}$.
2°. This function satisfies the relations

$$\Psi(\lambda_k) = b_k \quad (\lambda_0 = 0, 0 \le k < +\infty), \quad \|\Psi\|_{p,\omega}^+ \asymp \|\{b_k\}_0^\infty\|_{p,\kappa}, \tag{27}$$

where the suitable constants of the two-sided estimates do not depend on the sequence $\{b_k\}_0^\infty$.

Proof. It is similar to one of the previous theorem. Therefore, we give only the most important points of it. First we introduce the partial sums

$$\Psi_0(z) = b_0\Gamma(1+\nu)\mathcal{E}_\sigma(z;\nu), \qquad \Psi_{0,m}(z) = b_0\Gamma(1+\nu)\mathcal{E}_\sigma(z;\nu)$$
$$+ \sum_{k=1}^{m} b_k \frac{z\mathcal{E}_\sigma(z;\nu)}{\lambda_k\mathcal{E}'_\sigma(\lambda_k;\nu)(z-\lambda_k)} \quad (1 \le m < +\infty), \tag{28}$$

$$\Psi_{n,m}(z) = \sum_{k=n}^{m} b_k \frac{z\mathcal{E}_\sigma(z;\nu)}{\lambda_k\mathcal{E}'_\sigma(\lambda_k;\nu)(z-\lambda_k)} \quad (1 \le n < m < +\infty) \tag{29}$$

and note that all these functions are of class $W^{p,\omega}_{1/2,\sigma}$ by Lemma 4.2-4 ($2°$). Further, we introduce the function

$$\psi_{n,m}(z) = \sum_{k=n}^{m} b_k \frac{z}{\lambda_k\mathcal{E}'_\sigma(\lambda_k;\nu)(z^2-\lambda_k)} \quad (1 \le n < m < +\infty). \tag{30}$$

Then, by use of Theorem 3.4-1, we obtain that

$$\left\{\|\Psi_{n,m}\|^+_{p,\omega}\right\}^p = \|\Psi_{n,m}(x^2)\|^p_{p,\kappa} \le B_1 \int_{-\infty}^{+\infty} \left|\Psi_{n,m}\left((x+i)^2\right)\right|^p |x+i|^\kappa \, dx$$

$$= B_1 \int_{-\infty}^{+\infty} \left|\mathcal{E}_\sigma\left((x+i)^2;\nu\right)\psi_{n,m}(x+i)\right|^p |x+i|^{\kappa+p} \, dx \quad (1 \le n,m < +\infty),$$

where $B_1 > 0$ is a constant independent of n and m(such constants will be denoted below by $B_j(j \ge 2)$). Thus, by (19),

$$\left\{\|\Psi_{n,m}\|^+_{p,\omega}\right\}^p \le B_2 \int_{-\infty}^{+\infty} |\psi_{n,m}(x+i)|^p |x+i|^{\kappa_3} \, dx, \tag{31}$$

where $\kappa_3 = \kappa + p(1-\nu)$, and $-1 < \kappa_3 < p-1$ by (26). But

$$\|\psi_{n,m}(x+i)\|_{p,\kappa_3,1} = \sup_{\|h\|_q \le 1} \left|\int_{-\infty}^{+\infty} \psi_{n,m}(x+i)h(x) |x+i|^{\kappa_3/p} \, dx\right|$$

by Han-Banach's theorem. Hence, there exists a function $h_0(x)$, such that $h_0(x)|x+i|^{-\kappa_3/p} \in L_q(-\infty,+\infty)$ $(q = p/(p-1))$ and

$$\|h_0\|_{q,\kappa_4,1} = \left\{\int_{-\infty}^{+\infty} |h_0(x)|^q |x+i|^{\kappa_4} \, dx\right\}^{1/q} \le 1,$$

where $\kappa_4 = -\kappa_3 q/p, -1 < \kappa_4 < q-1$ (by (26)), and

$$\|\psi_{n,m}(x+i)\|_{p,\kappa_3,1} \le B_3 \left|\int_{-\infty}^{+\infty} \psi_{n,m}(x+i)h_0(x)dx\right|. \tag{32}$$

Now consider the Cauchy type integral

$$\mathcal{N}_0^-(z) = \frac{1}{2\pi i} \int_{-\infty}^{+\infty} \frac{h_0(t)}{t-z}\,dt, \qquad z \in G_- = \{z : \operatorname{Im} z < 0\} \tag{33}$$

which, according to assertions $1°$ and $2°$ of Theorem 3.3-3, satisfies the conditions

$$(z-i)^{\kappa_4/q}\mathcal{N}_0^-(z) \in H_-^q, \quad \|(x-i)^{\kappa_4/q}\mathcal{N}_0^-(x)\|_q \le B_4\|h_0\|_{q,\kappa_4,1} \le B_4. \tag{34}$$

Denote $\lambda_k = x_k^2$ $(1 \le k < +\infty)$ and insert representation (30) of $\psi_{n,m}(z)$ into (32). Then, use of (33) gives

$$\|\psi_{n,m}(x+i)\|_{p,\kappa_3,1} \le B_3 \left| \sum_{k=n}^{m} \frac{b_k}{\lambda_k \mathcal{E}_\sigma'(\lambda_k;\nu)} \int_{-\infty}^{+\infty} \frac{(t+i)h_0(t)}{(t+i)^2 - x_k^2}\,dt \right|$$

$$\le \pi B_3 \left| \sum_{k=n}^{m} \frac{b_k}{\lambda_k \mathcal{E}_\sigma'(\lambda_k;\nu)} \left[\mathcal{N}_0^-(x_k - i) + \mathcal{N}_0^-(-x_k - i)\right] \right|. \tag{35}$$

But $\lambda_k \asymp (1+k)^2$ $(1 \le k < +\infty)$ and, according to Lemma 4.2-5, $|\mathcal{E}_\sigma'(\lambda_k;\nu)| \asymp (1+k)^{-1-\nu}$. Thus $\lambda_k|\mathcal{E}_\sigma'(\lambda_k;\nu)| \asymp (1+k)^{1-\nu}$ $(1 \le k < +\infty)$, and by use of Hölder's inequality we obtain from (35)

$$\|\psi_{n,m}(x+i)\|_{p,\kappa_3,1} \le B_5 \Bigg\{ \left(\sum_{k=1}^{\infty} |\mathcal{N}_0^-(x_k - i)|^q (1+k)^{\kappa_4}\right)^{1/q}$$
$$+ \left(\sum_{k=1}^{\infty} |\mathcal{N}_0^-(-x_k - i)|^q (1+k)^{\kappa_4}\right)^{1/q} \Bigg\} \|\{b_k\}_n^m\|_{p,\kappa} \tag{36}$$

for any $n, m(1 \le n < m < +\infty)$. Further, by Theorem 3.4-3 and (34),

$$\sum_{k=1}^{\infty} |\mathcal{N}_0^-(\pm x_k - i)|^q (1+k)^{\kappa_4} \le B_6 < +\infty.$$

Hence $\|\psi_{n,m}(x+i)\|_{p,\kappa_3,1} \le B_7\|\{b_k\}_n^m\|_{p,\kappa}$, $\|\Psi_{n,m}\|_{p,\omega}^+ \le B_8\|\{b_k\}_n^m\|_{p,\kappa}$, and, since $\{b_k\}_0^\infty \in l^{p,\kappa}$, it follows that the series

$$\Psi_{1,\infty}(z) = \sum_{k=1}^{\infty} b_k \frac{z\mathcal{E}_\sigma(z;\nu)}{\lambda_k \mathcal{E}_\sigma'(\lambda_k;\nu)(z - \lambda_k)}, \qquad z \in \mathbb{C}$$

is convergent, $\Psi_{1,\infty}(z) \in W_{1/2,\sigma}^{p,\omega}$ and

$$\|\Psi_{1,\infty}\|_{p,\omega}^+ \le B_8\|\{b_k\}_1^\infty\|_{p,\kappa}. \tag{37}$$

We also have

$$\Psi(z) \equiv \Psi_{0,\infty}(z) = b_0\Gamma(1+\nu)\mathcal{E}_\sigma(z;\nu) + \Psi_{1,\infty}(z).$$

Hence

$$\|\Psi\|^+_{p,\omega} \leq |b_0|\,\Gamma(1+\nu)\|\mathcal{E}_\sigma\|^+_{p,\omega} + \|\Psi_{1,\infty}\|^+_{p,\omega},$$

and, using assertion $2°$ of Lemma 4.2-4 (the case $k = 0$), we easily obtain

$$\|\Psi\|^+_{p,\omega} \leq B_9\|\{b_k\}^\infty_0\|_{p,\kappa}. \tag{38}$$

The interpolation data $\Psi(\lambda_k) = b_k (0 \leq k < +\infty)$ of the function $\Psi(z) \in W^{p,\omega}_{1/2,\sigma}$ was established earlier, in Lemma 4.3-1, and the inequality converse to (38) can be obtained by the technique used in the proof of $(25')$.

4.4 Interpolation series expansions

In this section we establish statements converse to Theorems 4.3-1 and 4.3-2 and prove some corollaries.

(a) The following is the converse of Theorem 4.3-1.

Theorem 4.4-1. *Let the parameter $\nu \in [0,2)$ satisfy the complementary condition*

$$\nu \in \Delta(\kappa,p) = \left(\frac{1+\kappa}{p} - 1, \frac{1+\kappa}{p}\right) \cap [0,2), \qquad -1 < \kappa < 2p-1. \tag{1}$$

Then any function $\Phi(z) \in W^{p,\omega}_{1/2,\sigma}$ $(1 < p < +\infty,\ \omega = (\kappa-1)/2$ can be expanded in the interpolation series

$$\Phi(z) = \sum_{k=1}^{\infty} \Phi(\lambda_k)\frac{\mathcal{E}_\sigma(z;\nu)}{\mathcal{E}'_\sigma(\lambda_k;\nu)(z - \lambda_k)}, \tag{2}$$

where $\{\lambda_k\}^\infty_1$ $(0 < \lambda_k < \lambda_{k+1}, 1 \leq k < +\infty)$ are the zeros of the function $\mathcal{E}_\sigma(z;\nu)$. This series converges uniformly in any compact $K \subset \mathbb{C}$ and it converges also in the norm $\|.\|^+_{p,\omega}$ of the space $W^{p,\omega}_{1/2,\sigma}$. Besides, the following two-sided inequality is true:

$$\|\Phi\|^+_{p,\omega} \asymp \left\{\sum_{k=1}^{\infty}|\Phi(\lambda_k)|^p(1+k)^\kappa\right\}^{1/p}, \tag{2'}$$

where the suitable constants do not depend on $\Phi(z)$.

Proof. It was established in the proof of Theorem 4.3-1 that the inequality 4.3(25), i.e.,

$$\|\{\Phi(\lambda_k)\}_1^\infty\|_{p,\kappa} \leq e_1\|\Phi\|_{p,\omega}^+ \qquad (\kappa = 1 + 2\omega), \tag{3}$$

is true for any function $\Phi(z) \in W_{1/2,\sigma}^{p,\omega}$ $(1 < p < +\infty, -1 < \omega < p - 1)$. Thus the series (2) converges in both senses, as it follows from Lemma 4.3-1 and Theorem 4.3-1. Hence the function

$$\Phi_*(z) = \Phi(z) - \sum_{k=1}^\infty \Phi(\lambda_k)\frac{\mathcal{E}_\sigma(z;\nu)}{\mathcal{E}_\sigma'(\lambda_k;\nu)(z - \lambda_k)}, \tag{4}$$

for which

$$\Phi_*(\lambda_k) = 0 (1 \leq k < +\infty), \tag{5}$$

is also of class $W_{1/2,\sigma}^{p,\omega}$. Further,

$$\Omega(z) = \frac{\Phi_*(z^2)}{\mathcal{E}_\sigma(z^2;\nu)} \tag{6}$$

is an entire function, since its only singularities $\{\pm\sqrt{\lambda_k}\}_1^\infty$ can be removed according to (5). Besides, by a well-known theorem (see, for example, Levin [1,§9]), $\Omega(z)$ is of exponential type $\leq \sigma$. To find an estimate for this function, remember that the two-sided estimates 4.2(9) are true by Lemma 4.2-2 in the sum $\Delta = \Delta_+ \cup \Delta_-$ of the corner domains $\Delta_\pm$ of openings equal to $\pi/2$. But, according to 4.2(2), $\mathcal{E}_\sigma(z^2;\nu) = z^{-1}S_\sigma(z;\nu)$. Hence the two-sided estimates

$$\|\mathcal{E}_\sigma(z^2;\nu)\| \asymp (1 + |z|)^{-\nu}e^{\sigma|\mathrm{Im}\, z|}, z \in \Delta, |z| \geq 1 \tag{7}$$

follow. On the other hand, $\Phi_*(z^2) \in W_\sigma^{p,\kappa}$ $(\kappa = 1 + 2\omega)$ since $\Phi_*(z) \in W_{1/2,\sigma}^{p,\omega}$. Thus by Lemma $3.4 - 2(1°)$

$$\left|\Phi_*(z^2)\right| \leq C_1(1 + |z|)^{-\kappa/p}e^{\sigma|\mathrm{Im}\, z|}, z \in \mathbb{C}, \tag{8}$$

where the constant $C_1 > 0$ does not depend on z. Now relations (6)-(8) and (1) give

$$|\Omega(z)| \leq C_2(1 + |z|)^{1/p}, \qquad z \in \Delta, \ |z| \geq 1, \tag{9}$$

where the constant $C_2 > 0$ is also independent of z. Observe that the function $\Omega_*(z) = [\Omega(z) - \Omega(0)]z^{-1}$ is also entire and of exponential type. In addition,

$$\Omega_*(z) = O(|z|^{1/p-1}), \qquad z \in \Delta, \ |z| \geq 1. \tag{10}$$

Thus it is bounded on the boundaries of the corner domains $\Delta_\pm$ and $\mathbb{C}/\Delta_\pm$ of openings equal to $\pi/2$ and it tends to zero as $|z| \to \infty$. Hence it follows from the Phragmen-Lindelöf principle that $\Omega_*(z) \equiv 0$, i.e. $\Omega(z) \equiv \Omega(0) = a_0, z \in \mathbb{C}$. Consequently,

$$\Phi_*(z) \equiv a_0\mathcal{E}_\sigma(z;\nu), z \in \mathbb{C}.$$

But $\Phi_*(z) \in W_{1/2,\sigma}^{p,\omega}$ and, on the other hand, if $\nu < (1 + \kappa)/p$, then $\mathcal{E}_\sigma(z;\nu) \notin W_{1/2,\sigma}^{p,\omega}$ according to Lemma 4.2-4 $(1°)$. Thus the last identity does not lead to a contradiction only in the case when $\Phi_*(z) \equiv a_0 = 0$, so the proof is now complete.

(b) The next theorem is the converse of Theorem 4.3-2.

Theorem 4.4-2. *Let the parameter $\nu \in [0,2)$ satisfy the complementary condition*

$$\nu \in \Delta^*(\kappa,p) = \left(\frac{1+\kappa}{p}, 1 + \frac{1+\kappa}{p} \right) \cap [0,2), \qquad -1 < \kappa < 2p - 1. \tag{11}$$

Then any function $\Phi(z) \in W_{1/2,\sigma}^{p,\omega}$ $(1 < p < +\infty, \omega = (\kappa - 1)/2)$ can be expanded in the interpolation series

$$\Phi(z) = \Phi(\lambda_0)\Gamma(1+\nu)\mathcal{E}_\sigma(z;\nu) + \sum_{k=1}^{\infty} \Phi(\lambda_k) \frac{z\mathcal{E}_\sigma(z;\nu)}{\lambda_k \mathcal{E}_\sigma'(\lambda_k;\nu)(z - \lambda_k)}, \tag{12}$$

where $\{\lambda_k\}_0^\infty$ $(\lambda_0 = 0 < \lambda_k < \lambda_{k+1}, 1 \leq k < +\infty)$ are the zeros of the function $z\mathcal{E}_\sigma(z;\nu)$. This series is uniformly convergent in any compact $\mathcal{K} \subset \mathbb{C}$ and it converges also in the norm $\|.\|_{p,\omega}^+$ of the space $W_{1/2,\sigma}^{p,\omega}$. In addition,

$$\|\Phi\|_{p,\omega}^+ \asymp \left\{ \sum_{k=0}^{\infty} |\Phi(\lambda_k)|^p (1+k)^\kappa \right\}^{1/p}, \tag{13}$$

where the suitable constants do not depend on $\Phi(z)$.

Proof. An inequality similar to (3) is true again. Thus the convergence of the series (12) in both desired senses follows from Lemma 4.3-1 and Theorem 4.3-2. Hence the function

$$\Phi_*(z) = \Phi(z) - \Phi(\lambda_0)\Gamma(1+\nu)\mathcal{E}_\sigma(z;\nu) - \sum_{k=1}^{\infty} \Phi(\lambda_k) \frac{z\mathcal{E}_\sigma(z;\nu)}{\lambda_k \mathcal{E}_\sigma'(\lambda_k;\nu)(z - \lambda_k)}, \tag{14}$$

for which

$$\Phi_*(\lambda_k) = 0 \qquad (0 \leq k < +\infty), \tag{15}$$

is of class $W_{1/2,\sigma}^{p,\omega}$. Now define the function

$$\tilde{\Omega}(z) = \frac{\Phi_*(z^2)}{z^2 \mathcal{E}_\sigma(z^2;\nu)} \tag{16}$$

which is obviously entire and of exponential type. From (7) it follows that

$$\left| z^2 \mathcal{E}_\sigma(z^2;\nu) \right| \asymp (1 + |z|)^{2-\nu} e^{\sigma|\mathrm{Im}\, z|}, \qquad z \in \tilde{\Delta}, \tag{17}$$

where

$$\tilde{\Delta} = \tilde{\Delta}_+ \cup \tilde{\Delta}_-, \qquad \tilde{\Delta}_\pm = \left\{ z : \left| \mathrm{Arg}\, z \mp \frac{\pi}{2} \right| < \frac{\pi}{4},\ |z| \geq 1 \right\}. \tag{18}$$

And, since estimate (8) is true for the function $\Phi_*(z^2) \in W_\sigma^{p,\kappa}$ ($\kappa = 1 + 2\omega$), it follows from (16) and (17) that

$$\left|\tilde{\Omega}(z)\right| \leq C_2(1 + |z|)^{\nu - 2 - \kappa/p}, \quad z \in \tilde{\Delta},$$

where the constant $C_2 > 0$ is independent of z and $\nu - 2 - \kappa/p < 0$. Therefore, the function $\tilde{\Omega}(z)$ is bounded on the boundaries of the corner domains $\Delta_\pm$ and $\mathbb{C}/\Delta_\pm$ of openings $\pi/2$, and it tends to zero as $|z| \to \infty$. Hence the Phragmén-Lindelöf principle gives $\tilde{\Omega}(z) \equiv 0, z \in \mathbb{C}$, and $\Phi_*(z) \equiv 0, z \in \mathbb{C}$, so (12) is true. Now the two-sided estimates (13) follow from the inequality

$$\|\Phi(\lambda_k)\}_0^\infty\|_{p,\kappa} \leq e_1\|\Phi\|_{p,\omega}^+$$

proved earlier and also from the inequality 4.3(38) established in the proof of Theorem 4.3-2. Thus the proof is complete.

(c) The following uniqueness theorem is an immediate consequence of the two preceding theorems.

Theorem 4.4-3. *Let* $\Phi(z) \in W_{1/2,\sigma}^{p,\omega}$ *($1 < p < +\infty, -1 < \omega < p - 1$) be an arbitrary function and, as everywhere, let* $\{\lambda_k\}_1^\infty$ *be the sequence of zeros of the function*

$$\mathcal{E}_\sigma(z; \nu) = E_{1/2}\left(-\sigma^2 z; 1 + \nu\right), \quad \nu \in [0, 2).$$

Then $\Phi(z) \equiv 0$ *in each of the following cases:*
1°. $\nu \in \Delta(\kappa, p)$ *and*

$$\Phi(\lambda_k) = 0 \quad (1 \leq k < +\infty). \tag{19}$$

2°. $\nu \in \Delta^*(\kappa, p)$ *and*

$$\Phi(\lambda_k) = 0 \quad (\lambda_0 = 0,\ 0 \leq k < +\infty). \tag{20}$$

The next general theorem follows from Theorems 4.3-1, 4.3-2 and 4.4-1, 4.4-2.

Theorem 4.4-4. *If* $\nu \in \Delta(\kappa, p)$ *(or* $\nu \in \Delta^*(\kappa, p)$*), then the series 4.3(3) (or 4.3(4)), which converges uniformly on compacts* $\mathcal{K} \subset \mathbb{C}$ *and also in the norm* $\|\cdot\|_{p,\omega}^+$ *of the space* $W_{1/2,\sigma}^{p,\omega}$, *represents a continuous one-to-one mapping of the space of sequences* $\{a_k\}_1^\infty \in l^{p,\kappa}$ *(or sequences* $\{b_k\}_0^\infty \in l^{p,\kappa}$*) ($1 < p < +\infty, -1 < \omega < p - 1, \kappa = 2\omega + 1$) onto the space* $W_{1/2,\sigma}^{p,\omega}$ *of entire functions* $\Phi(z)$ *(or* $\Psi(z)$*). These mappings correspondingly satisfy the conditions*

$$
\begin{aligned}
\Phi(\lambda_k) &= a_k \quad (1 \leq k < +\infty),\ \|\Phi\|_{p,\omega}^+ \asymp \|\{a_k\}_1^\infty\|_{p,\kappa}, \\
\Psi(\lambda_k) &= b_k \quad (\lambda_0 = 0,\ 0 \leq k < +\infty),\ \|\Psi\|_{p,\omega}^+ \asymp \|\{b_k\}_0^\infty\|_{p,\kappa}.
\end{aligned}
\tag{21}
$$

(d) Now we consider three particular choices of the parameter ν in which the sequence $\{\lambda_k\}_1^\infty$ and also the interpolation expansions of Theorems 4.4-1 and 4.4-2 are of the simplest types.

$1°$. If $-1 < \omega < p/2 - 1 (1 < p < +\infty)$ and $\nu = 0$, then it is easy to verify that $0 \in \Delta(\kappa, p)$, so Theorem 4.4-1 may be used. The sequence $\{\lambda_k\}_1^\infty = \{[\pi(k-1/2)/\sigma]^2\}_1^\infty$ $(1 \le k < +\infty)$ is the set of zeros of the function $\mathcal{E}_\sigma(z; 0) = E_{1/2}(-\sigma^2 z; 1) = \cos(\sigma\sqrt{z})$. Consequently, any function $\Phi(z) \in W^{p,\omega}_{1/2,\sigma}$ can be expanded in the series

$$\Phi(z) = \frac{2\pi}{\sigma^2} \sum_{k=1}^{\infty} (-1)^k \Phi\left(\left[\frac{\pi}{\sigma}\left(k - \frac{1}{2}\right)\right]^2\right) \frac{(k-1/2)\cos(\sigma\sqrt{z})}{z - [\pi(k-1/2)/\sigma]^2}, \qquad z \in \mathbb{C}. \quad (22)$$

$2°$. If $p/2 - 1 < \omega < p - 1$ and $\nu = 1$, then $1 \in \Delta(\kappa, p)$ and $\{\lambda_k\}_1^\infty = \{(\pi k/\sigma)^2\}_1^\infty$ is the sequence of zeros of the function

$$\mathcal{E}_\sigma(z; 1) = E_{1/2}\left(-\sigma^2 z; 2\right) = (\sigma\sqrt{z})^{-1} \sin(\sigma\sqrt{z}).$$

In this case Theorem 4.4-1 says: any function $\Phi(z) \in W^{p,\omega}_{1/2,\sigma}$ can be expanded in the series

$$\Phi(z) = 2 \left(\frac{\pi}{\sigma}\right)^2 \sum_{k=1}^{\infty} (-1)^k \Phi\left(\left(\frac{\pi}{\sigma}k\right)^2\right) \frac{k^2 \sin(\sigma\sqrt{z})}{\sigma\sqrt{z}[z - (\pi k/\sigma)^2]}, z \in \mathbb{C}. \quad (23)$$

$3°$. If $-1 < \omega < p/2 - 1$ and $\nu = 1$, then $1 \in \Delta^*(\kappa, p)$, and, according to Theorem 4.4-2, any function $\Phi(z) \in W^{p,\omega}_{1/2,\sigma}$ can be expanded in the series

$$\begin{aligned}
\Phi(z) =\ &\Phi(0)\frac{\sin(\sigma\sqrt{z})}{\sigma\sqrt{z}} \\
&+ \frac{2}{\sigma} \sum_{k=1}^{\infty} (-1)^k \Phi\left(\left(\frac{\pi}{\sigma}k\right)^2\right) \frac{\sqrt{z}\sin(\sigma\sqrt{z})}{z - (\pi k/\sigma)^2}, \qquad z \in \mathbb{C}.
\end{aligned} \quad (24)$$

Note finally, that the convergence of series (22) to (24), illustrated in Theorems 4.4-1 and 4.4-2, and also the inequalities of these theorems, certainly remain true in the special cases discussed above.

4.5 Notes

The results of this chapter are proved in M.M. Djrbashian-S.G. Raphaelian [3, §2]. The proofs of the auxiliary lemmas and the main theorems given here are essentially improved.

5 Fourier type basic systems in $L_2(0,\sigma)$

5.1 Introduction

As is well known, a function $\varphi(\tau) \in L_2(0,\sigma)$ $(0 < \sigma < +\infty)$ is the limit in mean of two different Fourier series constructed by the use of two different systems of trigonometric functions

$$\left\{\sqrt{\frac{2}{\sigma}}\sin\frac{\pi k}{\sigma}\tau\right\}_1^\infty \quad \text{and} \quad \left\{\frac{1}{\sqrt{\sigma}}, \left\{\sqrt{\frac{2}{\sigma}}\cos\frac{\pi k}{\sigma}\tau\right\}_1^\infty\right\}, \tag{1}$$

which are orthonormal in $L_2(0,\sigma)$. Plancherel's Theorem 1.7-2 relating to Fourier cos- and sin-transforms in $L_2(0,+\infty)$ is actually a continual analog of the mentioned expansions for the case when $\sigma = +\infty$. This theorem is contained in the general Theorem 1.7-1 as the special case when the parameter $\mu \in (1/2, 5/2)$ is equal to 1 or 2.

In this chapter the discrete analogs of the general Theorem 1.7-1 are established for arbitrary values of $\mu \in (1/2, 5/2)$. The results obtained appear to coincide in the cases $\mu = 1$ and $\mu = 2$ with the abovementioned classical statements of Fourier series theory in $L_2(0,\sigma)$. The proofs of this chapter are based on some of the main results of Chapter 4, namely, on the case $p = 2, -1 < \omega < 1$, of Theorems 4.3-1 and 4.3-2 relating to interpolation expansions of entire functions of classes $W_{1/2,\sigma}^{p,\omega}$. We pass from these interpolation theorems to the results of this chapter mainly using Theorem 2.2-1, which establishes the parametric representations of the classes $W_{1/2,\sigma}^{2,\omega}(-1 < \omega < 1)$ of entire functions. Remember that the last theorem was the simplest generalization of the classical Wiener-Paley Theorem 2.2-2 relating to parametric representations of the classes $W_{1/2,\sigma}^{2,-1/2}$ and $W_{1/2,\sigma}^{2,1/2}$ of entire functions. Finally, we arrive at some biorthogonal systems of Mittag-Leffler type functions forming Riesz bases of $L_2(0,\sigma)$ after suitable normalization.

It should be mentioned that the basic systems of this chapter, similar to trigonometric systems (1), appear to be the systems of eigenfunctions of definite boundary value problems on $(0,\sigma)$. But these new boundary value problems are absolutely non-ordinary, as they are formulated in terms of several model integro-differential operators of fractional orders. These results as well as more general ones of the same character are proved in Chapters 10, 11 and 12 of the book.

5.2 Biorthogonal systems of Mittag-Leffler type functions and their completeness in $L_2(0,\sigma)$.

It is now necessary to introduce short notations for the systems of entire functions contained in the interpolation expansions of Theorems 4.4-1 and 4.4-2.

Let $\{\lambda_k\}_1^\infty$ $(0 < \lambda_k < \lambda_{k+1}, 1 \leq k < +\infty)$ be, as in Chapters 1 and 4, the sequence of zeros of the entire function

$$\mathcal{E}_\sigma(z;\nu) = E_{1/2}\left(-\sigma^2 z; 1+\nu\right), \qquad \nu \in [0,2) \tag{1}$$

(remember that its zeros are simple). We introduce the following two sequences of entire functions of order $1/2$ and of type σ:

$$\{\omega_k(z)\}_1^\infty \text{ and } \{\omega_k^*(z)\}_0^\infty, \tag{2}$$

where

$$\omega_k(z) = \frac{\mathcal{E}_\sigma(z; \nu)}{\mathcal{E}_\sigma'(\lambda_k; \nu)(z - \lambda_k)} \qquad (1 \le k < +\infty), \tag{3}$$

$$\omega_0^*(z) = \Gamma(1 + \nu)\mathcal{E}_\sigma(z; \nu),$$

$$\omega_k^*(z) = \frac{z}{\lambda_k}\omega_k(z) \qquad (1 \le k < +\infty). \tag{4}$$

Note that according to Lemma 4.2-4 (the case $p = 2$)

$$\omega_k(z) \in W_{1/2,\sigma}^{2,\omega} \qquad (-1 < \omega < 1, 1 \le k < +\infty), \tag{5}$$

if $\nu \in [0, 2)$, and

$$\omega_k^*(z) \in W_{1/2,\sigma}^{2,\omega} \qquad (-1 < \omega < 1, 0 \le k < +\infty), \tag{6}$$

if $\nu \in (1 + \omega, 2) \subset [0, 2)$. Further, note that definitions (3) and (4) of systems (2) immediately give the interpolation data

$$\omega_k(\lambda_n) = \delta_{k,n} \qquad (1 \le k, n < +\infty), \tag{7}$$

$$\omega_k^*(\lambda_n) = \delta_{k,n} \qquad (0 \le k, n < +\infty), \tag{8}$$

where $\delta_{k,n}$ is the Kronecker's symbol.

Now, using the introduced notations, the last interpolation theorem of the preceding chapter, i.e. Theorem 4.4-4, may be formulated for $p = 2$ in the following way.

Theorem 5.2-1. *Let, as always, $\{\lambda_k\}_1^\infty$ and $\{\lambda_k\}_0^\infty$ be, respectively, the zeros of the functions*

$$\mathcal{E}_\sigma(z; \nu), \nu \in \Delta(\kappa, 2) = (\omega, \omega + 1) \cap [0, 2) = \left(\frac{\kappa - 1}{2}, \frac{\kappa + 1}{2}\right) \cap [0, 2) \tag{9}$$

and

$$z\mathcal{E}_\sigma(z; \nu), \ \nu \in \Delta^*(\kappa, 2) = (\omega + 1, \omega + 2) \cap [0, 2) = \left(\frac{\kappa + 1}{2}, 1 + \frac{\kappa + 1}{2}\right) \cap [0, 2), \tag{10}$$

where $\kappa = 2\omega + 1$ and $-1 < \omega < 1$. Then each of the sets of entire functions $\{\omega_k(z)\}_1^\infty$ and $\{\omega_k^*(z)\}_0^\infty$ is a basis of the space $W_{1/2,\sigma}^{2,\omega}$, and the two-sided inequalities 4.4(21) of Theorem 4.4-4 remain true.

(a) Remember that the points of interpolation $\{\lambda_k\}_1^\infty$ and $\{\lambda_k\}_0^\infty$ in Theorems 4.4-1 and 4.4-2 (relating to interpolation expansions of functions of the class $W_{1/2,\sigma}^{2,\omega}(-1 < \omega < 1, \kappa = 2\omega + 1))$ were the zeros,respectively, of the functions $\mathcal{E}_\sigma(z;\nu)$ and $z\mathcal{E}_\sigma(z;\nu)$. Remember also that the zeros of these functions are simple.

Let $-1 < \omega < 1$, and let $\nu \in [0,2)$ or $\nu \in (1 + \omega, 2) \subset [0,2)$ (so the assertions (5) or (6) are true correspondingly). Then, according to Theorem 2.2-1, the functions of the system $\{\omega_k(z)\}_1^\infty \subset W_{1/2,\sigma}^{2,\omega}$ or, correspondingly, the functions of the system $\{\omega_k^*(z)\}_0^\infty \subset W_{1/2,\sigma}^{2,\omega}$ are representable in the forms

$$\omega_k(z) = \int_0^\sigma E_{1/2}\left(-\tau^2 z; \mu\right) \tau^{\mu-1} \varphi_k(\tau) d\tau \qquad (1 \le k < +\infty), \qquad (11)$$

$$\omega_k^*(z) = \int_0^\sigma E_{1/2}\left(-\tau^2 z; \mu\right) \tau^{\mu-1} \varphi_k^*(\tau) d\tau \qquad (0 \le k < +\infty), \qquad (12)$$

where $\mu = 3/2 + \omega \in (1/2, 5/2)$, and where the functions $\varphi_k(\tau) \in L_2(0,\sigma)$ and $\varphi_k^*(\tau) \in L_2(0,\sigma)$ are unique. Note that it was tacitly supposed in the last representations that the parameters μ and ν satisfy the conditions $1/2 < \mu < 5/2, 0 \le \nu < 2$ and $1/2 < \mu < 5/2, \mu - 1/2 < \nu < 2$ respectively.

The systems of entire functions (2) are now associated respectively with the systems

$$\{\varphi_k(\tau)\}_1^\infty \text{ and } \{\varphi_k^*(\tau)\}_0^\infty \qquad (13)$$

of functions of $L_2(0,\sigma)$.

(b) The inversions of representations (11) and (12) can be obtained by use of the inversion formula 2.2(3) of Theorem 2.2-1. In this way the functions of systems (13) may be expressed as some improper integrals arising from formula 2.2(3), where $f(z)$ is replaced by $\omega_k(z)(1 \le k < +\infty)$ or $\omega_k^*(z)$ $(0 \le k < +\infty)$. The calculation of these integrals apparrently is far from being simple. Fortunately, it appears that there exists another way in which the functions (13) of representations (11) and (12) can be determined in explicit forms. Together with this, the following lemma permits the weakening of the assumptions relating to possible values of the parameters μ and ν.

Lemma 5.2-1. *1°. If*

$$0 \le \nu < 2 \text{ and } \qquad 0 \le \mu < 3 + \nu, \qquad (14)$$

then the functions of system $\{\omega_k(z)\}_1^\infty$ *are representable in the form of (11), where*

$$\varphi_k(\tau) = -\frac{\sigma^{-\nu}}{\mathcal{E}_\sigma'(\lambda_k;\nu)} E_{1/2}\left(-\lambda_k(\sigma - \tau)^2; 3 + \nu - \mu\right) (\sigma - \tau)^{\nu-\mu+2} \quad (1 \le k < +\infty)$$

$$(15)$$

when $\tau \in (0,\sigma)$.

$2°$. *If*

$$0 \leq \nu < 2 \qquad \text{and} \qquad 0 \leq \mu < 1 + \nu, \tag{16}$$

then the functions of system $\{\omega_k^*(z)\}_0^\infty$ are representable in the form of (12), where

$$\varphi_0^*(\tau) = \sigma^{-\nu} \frac{\Gamma(1+\nu)}{\Gamma(1+\nu-\mu)} (\sigma - \tau)^{\nu-\mu}, \qquad \varphi_k^*(\tau) = \frac{\sigma^{-\nu}}{\lambda_k \mathcal{E}_\sigma'(\lambda_k; \nu)}$$

$$\times E_{1/2}\left(-\lambda_k(\sigma - \tau)^2; 1 + \nu - \mu\right)(\sigma - \tau)^{\nu-\mu} \quad (1 \leq k < +\infty) \tag{17}$$

when $\tau \in (0, \sigma)$.

Proof. If the obvious equality $E_{1/2}(z; \mu) = 1/\Gamma(\mu) + z E_{1/2}(z; \mu + 2)$ is used, then the integral identity 1.2(10), where we assume $\rho = 1/2$, can be written down in both the following forms:

$$\mathcal{Y}_{\alpha,\beta}(z; \lambda) \equiv \int_0^\sigma E_{1/2}\left(-z\tau^2; \alpha\right) \tau^{\alpha-1} E_{1/2}\left(-\lambda(\sigma - \tau)^2; \beta\right)(\sigma - \tau)^{\beta-1} d\tau$$

$$= -\frac{E_{1/2}\left(-\sigma^2 z; \alpha + \beta - 2\right) - E_{1/2}\left(-\sigma^2 \lambda; \alpha + \beta - 2\right)}{z - \lambda} \sigma^{\alpha+\beta-3} \tag{18}$$

$$= \frac{z E_{1/2}\left(-\sigma^2 z; \alpha + \beta\right) - \lambda E_{1/2}\left(-\sigma^2 \lambda; \alpha + \beta\right)}{z - \lambda} \sigma^{\alpha+\beta-1},$$

where $z, \lambda \in \mathbb{C}$ and $0 \leq \alpha, \beta < +\infty$ are arbitrary numbers.
$1°$. If we put $\alpha = \mu$, $\beta = \nu - \mu + 3$ and $\lambda = \lambda_k$ $(1 \leq k < +\infty)$ (where λ_k are the zeros of $\mathcal{E}_\sigma(z; \nu)$) in the first of the identities (18), then it follows that

$$\int_0^\sigma E_{1/2}\left(-z\tau^2; \mu\right) \tau^{\mu-1} E_{1/2}\left(-\lambda_k(\sigma - \tau)^2; \nu - \mu + 3\right)(\sigma - \tau)^{\nu-\mu+2} d\tau$$

$$= -\sigma^\nu \frac{E_{1/2}\left(-\sigma^2 z; 1 + \nu\right)}{z - \lambda_k} \quad (1 \leq k < +\infty).$$

Hence, in view of definition (11) of the system $\{\varphi_k(\tau)\}_1^\infty$, we get formulas (15).
$2°$. We put $\alpha = \mu$, $\beta = \nu - \mu + 1$ in the second of the identities (18) and consider the cases $\lambda = \lambda_0 = 0$ and $\lambda = \lambda_k$ $(1 \leq k < +\infty)$. We arrive at the two following identities:

$$\int_0^\sigma E_{1/2}\left(-z\tau^2; \mu\right) \tau^{\mu-1}(\sigma - \tau)^{\nu-\mu} d\tau = \sigma^\nu \Gamma(1 + \nu - \mu) E_{1/2}\left(-\sigma^2 z; 1 + \nu\right),$$

$$\int_0^\sigma E_{1/2}\left(-z\tau^2; \mu\right) \tau^{\mu-1} E_{1/2}\left(-\lambda_k(\sigma - \tau)^2; 1 + \nu - \mu\right)(\sigma - \tau)^{\nu-\mu} d\tau$$

$$= \sigma^\nu \frac{z E_{1/2}(-\sigma^2 z; 1 + \nu)}{z - \lambda_k} \quad (1 \leq k < +\infty).$$

Hence, using definition (12) of the system $\{\varphi_k^*(\tau)\}_0^\infty$, we arrive at formulas (17).

(c) Now we shall consider, together with systems (15) and (17), the pair of systems

$$\{e_k(\tau;\mu)\}_1^\infty \qquad \text{and} \qquad \{e_k(\tau;\mu)\}_0^\infty, \tag{19}$$

where

$$e_0(\tau;\mu) = \frac{\tau^{\mu-1}}{\Gamma(\mu)}, \quad e_k(\tau;\mu) = E_{1/2}\left(-\lambda_k\tau^2;\mu\right)\tau^{\mu-1} \quad (1 \le k < +\infty). \tag{19'}$$

It is easy to see that all the functions of systems (19) are real and, in view of the identities

$$e_0(\tau;0) \equiv 0, e_k(\tau;0) = -\lambda_k E_{1/2}\left(-\lambda_k\tau^2;2\right)\tau \quad (1 \le k < +\infty),$$

they are all of $L_1(0,\sigma)$ for any $\mu \ge 0$. In addition, these functions may have integrable singularities only at $\tau = 0$, and they are of $L_2(0,\sigma)$ for any $\mu > 1/2$. Further, it follows from (15) and (17) that the functions of both those systems also are real, that they may have integrable singularities only at $\tau = \sigma$ and that they are of $L_1(0,\sigma)$ when ν and μ satisfy conditions (14) and (16) respectively. Thus, as a result, we obtain the pair of inclusions

$$e_k(\tau;\mu)\varphi_n(\tau) \in L_1(0,\sigma) \qquad (1 \le k, n < +\infty),$$
$$e_k(\tau;\mu)\varphi_n^*(\tau) \in L_1(0,\sigma) \qquad (0 \le k, n < +\infty)$$

which are true when the parameters μ and ν satisfy conditions (14) and (16) respectively.

Now we are ready to prove the following statement.

Lemma 5.2-2. *The left-hand system of (19) and system (15) are biorthogonal on $(0,\sigma)$. Also, the right-hand system of (19) and system (17) are biorthogonal on $(0,\sigma)$. In other words,*

$$\int_0^\sigma e_k(\tau;\mu)\varphi_n(\tau)d\tau = \delta_{k,n} \qquad (1 \le k, n < +\infty),$$
$$\int_0^\sigma e_k(\tau;\mu)\varphi_n^*(\tau)d\tau = \delta_{k,n} \qquad (0 \le k, n < +\infty), \tag{20}$$

where $\delta_{k,n}$ is Kronecker's symbol.

Proof. If we put $z = \lambda_k$ $(1 \le k < +\infty)$ and $z = \lambda_k$ $(0 \le k < +\infty)$ in representations (11) and (12) respectively, then the desired biorthogonality (20) immediately follows from (7) and (8).

(d) Now we shall prove a theorem on completeness of both systems of (19) in $L_2(0,\sigma)$.

Theorem 5.2-2. 1°. *If $\nu \in \Delta(\kappa,2)$ and $\mu = 3/2 + \omega$ $(-1 < \omega < 1)$, then the system of functions $\{e_k(\tau;\mu)\}_1^\infty$ is complete in $L_2(0,\sigma)$.*
2°. *If $\nu \in \Delta^*(\kappa,2)$ and $\mu = 3/2 + \omega(-1 < \omega < 1)$, then the system of functions $\{e_k(\tau;\mu)\}_0^\infty$ is complete in $L_2(0,\sigma)$.*

Proof. It is necessary to show that a function $\varphi(\tau) \in L_2(0,\sigma)$ vanishes almost everywhere in $(0,\sigma)$, if it satisfies any of the conditions

$$\int_0^\sigma \varphi(\tau)e_k(\tau;\mu)d\tau = 0 \qquad (1 \le k < +\infty), \tag{21}$$

$$\int_0^\sigma \varphi(\tau)e_k(\tau;\mu)d\tau = 0 \qquad (0 \le k < +\infty), \tag{22}$$

in the hypothesis that the parameter ν satisfies the suitable conditions. To this end we introduce the entire function

$$f_\varphi(z) = \int_0^\sigma E_{1/2}\left(-\tau^2 z;\mu\right)\tau^{\mu-1}\varphi(\tau)d\tau \tag{23}$$

which is of class $W_{1/2,\sigma}^{2,\omega}$, according to Theorem 2.2-1. Observe that (21) and (22) give respectively $f_\varphi(\lambda_k) = 0$ $(1 \le k < +\infty)$ and $f_\varphi(\lambda_k) = 0$ $(0 \le k < +\infty)$. Thus, by the uniqueness Theorem 4.4-3 proved for the classes $W_{1/2,\sigma}^{2,\omega}$, we have $f_\varphi(z) \equiv 0$. To complete the proof, it remains to use the inversion formula 2.2(3) of Theorem 2.2-1.

The following completeness theorem is also true.

Theorem 5.2-3. 1°. *If $\nu \in \Delta(\kappa,2)$ and $\mu = 3/2+\omega$ $(0 < \omega < 1)$, then the system of functions $\{\varphi_k(\tau)\}_1^\infty$ is complete in $L_2(0,\sigma)$.*
2°. *If $\nu \in \Delta^*(\kappa,2)$ and $\mu = 3/2 + \omega$ $(-1 < \omega < 0)$, then the system of functions $\{\varphi_k^*(\tau)\}_0^\infty$ is complete in $L_2(0,\sigma)$.*

Proof. We apply arguments based mainly on the preceding theorem.
1°. If we put $3 - \mu + \nu = \tilde{\mu} = 3/2 + \tilde{\omega}$ in representation (15) of functions $\{\varphi_k(\tau)\}_1^\infty$, then, obviously we obtain $\nu = \omega + \tilde{\omega}$. Hence $\tilde{\omega} \in (0,1)$, since $\nu \in \Delta(\kappa,2) = (\omega, \omega+1) \cap [0,2)$. It is easy to verify that

$$\varphi_k(\tau) = -\frac{\sigma^{-\nu}}{\mathcal{E}_\sigma'(\lambda_k;\nu)}e_k(\sigma - \tau;\tilde{\mu}) \qquad (1 \le k < +\infty).$$

Thus the functions $\{\varphi_k(\tau)\}_1^\infty$ differ from $\{e_k(\sigma-\tau;\tilde{\mu})\}_1^\infty$ only by constant multipliers, and the last system is complete in $L_2(0,\sigma)$ by Theorem 5.2-2(1°). Indeed, if we take $\tilde{\kappa} = 1+2\tilde{\omega}$, then it remains only to observe that $\nu \in \Delta(\tilde{\kappa},2) = (\tilde{\omega},\tilde{\omega}+1)\cap[0,2)$ since $0 < \omega < 1$.
2°. If we put $1 + \nu - \mu = \tilde{\mu} = 3/2 + \tilde{\omega}$ in the representation (17) of functions $\{\varphi_k^*(\tau)\}_0^\infty$, then we obtain $\nu - 2 = \omega + \tilde{\omega}$. Hence $\tilde{\omega} \in (-1,0)$ since $\nu \in \Delta^*(\kappa,2) =$

$(\omega + 1, \omega + 2) \cap [0, 2)$. It is also easy to verify that

$$\varphi_0^*(\tau) = \sigma^{-\nu}\Gamma(1 + \nu)e_0(\sigma - \tau; \tilde{\mu}),$$

$$\varphi_k^*(\tau) = \frac{\sigma^{-\nu}}{\lambda_k \mathcal{E}_\sigma'(\lambda_k; \nu)} e_k(\sigma - \tau; \tilde{\mu}) \qquad (1 \le k < +\infty).$$

Therefore, the desired assertion follows from Theorem 5.2-2($2°$). Indeed, if we put $\tilde{\kappa} = 1 + 2\tilde{\omega}$, then $\nu \in \Delta^*(\tilde{\kappa}, 2) = (\tilde{\omega} + 1, \tilde{\omega} + 2) \cap [0, 2)$ since $-1 < \omega < 0$.

Remark. It turns out that the preceding theorem remains true if the restrictions on the parameter ω are omitted. This fact is an obvious consequence of Theorems 5.3-1 and 5.3-2 of the next chapter.

5.3 Fourier series type biorthogonal expansions in $L_2(0,\sigma)$

Here we shall prove the main theorems relating to expansions of functions of $L_2(0,\sigma)$ in terms of the pairs of biorthogonal systems

$$\{e_k(\tau; \mu)\}_1^\infty, \qquad \{\varphi_k(\tau)\}_1^\infty \tag{1}$$

and

$$\{e_k(\tau; \mu)\}_0^\infty, \qquad \{\varphi_k^*(\tau)\}_0^\infty \tag{2}$$

which were introduced in Section 5.2.

(a) Theorem 5.3-1. *Let $\{\lambda_k\}_1^\infty$ be, as always, the sequence of zeros of the function*

$$\mathcal{E}_\sigma(z; \nu) = E_{1/2}\left(-\sigma^2 z; 1 + \nu\right), \qquad \nu \in [0, 2),$$

and let the parameters ν, ω and μ satisfy the conditions

$$\nu \in \Delta(\kappa, 2) = (\omega, \omega+1) \cap [0, 2), \quad -1 < \omega < 1 \text{ and } \mu = 3/2 + \omega \quad (\kappa = 1 + 2\omega). \tag{3}$$

Then:

$1°$. The series of the form

$$\varphi(\tau) = \sum_{k=1}^\infty a_k \varphi_k(\tau) \tag{4}$$

converge in the norm of $L_2(0,\sigma)$ and present a continuous one-to-one mapping of the space $l^{2,\kappa}$ ($\kappa = 1 + 2\omega$) of sequences $\{a_k\}_1^\infty$ for which

$$\|\{a_k\}_1^\infty\|_{2,\kappa} = \left\{\sum_{k=1}^\infty |a_k|^2 (1 + k)^\kappa\right\}^{1/2} < +\infty, \tag{5}$$

onto the space $L_2(0,\sigma)$ of functions $\varphi(\tau)$. And the coefficients of expansion (4) of any function $\varphi(\tau) \in L_2(0,\sigma)$ can be determined by the formula

$$a_k = \int_0^\sigma \varphi(t)e_k(t; \mu)dt \qquad (1 \le k < +\infty). \tag{6}$$

In addition, the the following two-sided inequalities are true:

$$\|\varphi\|_2 = \left\{ \int_0^\sigma |\varphi(\tau)|^2 \, d\tau \right\}^{1/2} \asymp \|\{a_k\}_1^\infty\|_{2,\kappa}. \tag{7}$$

$2°$. The series of the form

$$\varphi(\tau) = \sum_{k=1}^\infty b_k e_k(\tau; \mu) \tag{8}$$

converge in the norm of $L_2(0, \sigma)$ and present a continuous one-to-one mapping of the space $l^{2,-\kappa}$ of sequences $\{b_k\}_1^\infty$ onto the space $L_2(0, \sigma)$ of functions $\varphi(\tau)$. And the coefficients of the expansion (8) of any function $\varphi(\tau) \in L_2(0, \sigma)$ can be determined by the formula

$$b_k = \int_0^\sigma \varphi(t)\varphi_k(t)dt \qquad (1 \leq k < +\infty). \tag{9}$$

In addition, the following two-sided inequalities are true:

$$\|\varphi\|_2 \asymp \|\{b_k\}_1^\infty\|_{2,-\kappa}. \tag{10}$$

Proof. $1°$. According to Theorem 4.4-4 (the case $p = 2$), the series

$$\Phi(z) = \sum_{k=1}^\infty a_k \omega_k(z), \tag{11}$$

where $\{\omega_k(z)\}_1^\infty$ is the first of the systems 5.2(2) of entire functions of $W_{1/2,\sigma}^{2,\omega}$, presents a continuous, one-to-one mapping of $l^{2,\kappa}$ onto $W_{1/2,\sigma}^{2,\omega}$. By the same theorem,

$$\|\Phi\|_{2,\omega}^+ \asymp \|\{a_k\}_1^\infty\|_{2,\kappa}, \quad \Phi(\lambda_k) = a_k (1 \leq k < +\infty). \tag{12}$$

On the other hand, according to Theorem 2.2-1, the formula

$$\Phi(z) = \int_0^\sigma E_{1/2}\left(-z\tau^{2}; \mu\right) \tau^{\mu-1}\varphi(\tau)d\tau, \qquad \mu = 3/2 + \omega \tag{13}$$

gives a continuous one-to-one mapping of $L_2(0, \sigma)$ onto $W_{1/2,\sigma}^{2,\omega}$ and

$$\|\Phi\|_{2,\omega}^+ \asymp \|\varphi\|_2. \tag{14}$$

In addition, 5.2(11) implies

$$\omega_k(z) = \int_0^\sigma E_{1/2}\left(-\tau^2 z; \mu\right) \tau^{\mu-1}\varphi_k(\tau)d\tau \qquad (1 \leq k < +\infty).$$

The two-sided inequalities (7) follow from (12) and (14). Thus the spaces $l^{2,\kappa}$ and $L_2(0,\sigma)$ are homeomorphic, and the proof of $1°$ will be complete, if we observe that (12) and (13) together with the notations 5.2(19′) give

$$\Phi(\lambda_k) = a_k = \int_0^\sigma \varphi(t)e_k(t;\mu)dt \qquad (1 \le k < +\infty).$$

$2°$. We pass from the systems (1) to the systems

$$\{\tilde{e}_k(\tau;\mu)\}_1^\infty, \qquad \{\tilde{\varphi}_k(\tau)\}_1^\infty, \tag{1′}$$

where

$$\tilde{e}_k(\tau;\mu) = (1+k)^{\kappa/2}e_k(\tau;\mu), \quad \tilde{\varphi}_k(\tau) = (1+k)^{-\kappa/2}\varphi_k(\tau) \qquad (1 \le k < +\infty). \tag{15}$$

It is obvious that these new systems also are biorthogonal on $(0,\sigma)$. Besides, they are complete in $L_2(0,\sigma)$, according to Theorem 5.2-2 and assertion $1°$, which is already proved.

Evidently, assertion $1°$ can be reformulated now using the systems (1′). Then, as a result we obtain that the series of the form

$$\varphi(\tau) = \sum_{k=1}^\infty \tilde{a}_k\tilde{\varphi}_k(\tau) \tag{4′}$$

presents a continuous one-to-one mapping of the space l^2 of sequences for which

$$\|\{\tilde{a}_k\}_1^\infty\|_2 = \left\{\sum_{k=1}^\infty |\tilde{a}_k|^2\right\}^{1/2} < +\infty$$

onto $L_2(0,\sigma)$ and, instead of formula (6) and inequalities (7), we have

$$\tilde{a}_k = \int_0^\sigma \varphi(t)\tilde{e}_k(t;\mu)dt \qquad (1 \le k < +\infty), \qquad \|\varphi\|_2 \asymp \|\{\tilde{a}_k\}_1^\infty\|_2.$$

So, the system $\{\tilde{\varphi}_k(\tau)\}_1^\infty$, which is complete in $L_2(0,\sigma)$, is a Riesz basis. Hence, according to a well-known result, the system $\{\tilde{e}_k(\tau;\mu)\}_1^\infty$, which is biorthogonal to $\{\tilde{\varphi}_k(\tau)\}_1^\infty$ and is complete in $L_2(0,\sigma)$, is also a Riesz basis. Using this and returning to the initial system $\{e_k(\tau;\mu\}_1^\infty$, we complete the proof.

Theorem 5.3-2. *All the statements of Theorem 5.3-1 remain true also for the systems (2) which are biorthogonal on $(0,\sigma)$, if only the parameters ν, ω and μ satisfy the conditions*

$$\nu \in \Delta^*(\kappa,2) = (\omega+1,\omega+2)\cap[0,2), \quad -1 < \omega < 1, \ \mu = 3/2+\omega \ (\kappa = 1+2\omega). \tag{16}$$

Proof. It suffices just to repeat the arguments used in the proof of Theorem 5.3-1, but it is also necessary to apply the second parallel assertion of Theorem 4.4-4, again in the case $p = 2$. In this way we shall arrive at the biorthogonal expansions

$$
\varphi(\tau) = \sum_{k=0}^{\infty} b_k e_k(\tau;\mu), \quad b_k = \int_0^{\sigma} \varphi(t)\varphi_k^*(t)dt \quad (0 \le k < +\infty),
$$
$$
\varphi(\tau) = \sum_{k=0}^{\infty} a_k \varphi_k^*(\tau), \quad a_k = \int_0^{\sigma} \varphi(t)e_k(t;\mu)dt \quad (0 \le k < +\infty),
\tag{17}
$$

and we shall arrive also at the following inequalities between the norms of a function $\varphi(\tau) \in L_2(0,\sigma)$ and the sequences of coefficients $\{a_k\}_0^{\infty}$, $\{b_k\}_0^{\infty}$:

$$
\|\varphi\|_2 \asymp \|\{a_k\}_0^{\infty}\|_{2,\kappa}, \qquad \|\varphi\|_2 \asymp \|\{b_k\}_0^{\infty}\|_{2,-\kappa}.
\tag{18}
$$

(b) Now we shall state some particular cases of the proved biorthogonal expansions, as they are of independent interest.

Theorem 5.3-3. *Let $\{\lambda_k\}_0^{\infty}$ $(0 = \lambda_0 < \lambda_k < \lambda_{k+1}, 1 \le k < +\infty)$ be the zeros of the function $z\mathcal{E}_\sigma(z;\nu)$, where $\nu \in (1/2,3/2)$. Then, after suitable normalization, each system of biorthogonal pairs*

$$
\left\{\frac{\sin(\sqrt{\lambda_k}\tau)}{\sqrt{\lambda_k}}\right\}_1^{\infty}, \quad \left\{-\frac{2\sigma^{-\nu}\lambda_k(\sigma-\tau)^\nu}{E_{1/2}(-\sigma^2\lambda_k;\nu)} E_{1/2}\left(-\lambda_k(\sigma-\tau)^2;1+\nu\right)\right\}_1^{\infty}
\tag{19}
$$

and

$$
\left\{\cos(\sqrt{\lambda_k}\tau)\right\}_0^{\infty},
$$
$$
\left\{\nu\sigma^{-\nu}(\sigma-\tau)^{\nu-1}, \left\{\frac{2\sigma^{-\nu}(\sigma-\tau)^{\nu-1}}{E_{1/2}(-\sigma^2\lambda_k;\nu)} E_{1/2}\left(-\lambda_k(\sigma-\tau)^2;\nu\right)\right\}_1^{\infty}\right\}
\tag{20}
$$

forms a Riesz basis of $L_2(0,\sigma)$.

Proof. Observe that, if we take $\mu = 2$ or $\mu = 1$, then we obtain correspondingly

$$
E_{1/2}(z;2) = \frac{\sinh\sqrt{z}}{\sqrt{z}} \quad \text{and} \quad E_{1/2}(z;1) = \cosh\sqrt{z}.
$$

In addition, note that hypothesis (3) of Theorem 5.3-1 and hypothesis (16) of Theorem 5.3-2 are satisfied in the cases $\mu = 2$ and $\mu = 1$ respectively, if $\nu \in (1/2,3/2) \subset [0,2)$. Therefore, the desired statements relating to systems (19) and (20) follow correspondingly from the same theorems, if we use definitions 5.2((15),(17),(19$'$)) of biorthogonal systems (1) and (2) and also identity 4.2(35).

Remark. Obviously, the zeros of the function

$$E_{1/2}\left(-\sigma^2 z; 2\right) = \frac{\sin(\sigma\sqrt{z})}{\sqrt{z}}$$

coincide when $\nu = 1$ with the set of numbers $\lambda_k = (\pi k/\sigma)^2 (1 \le k < +\infty)$. Besides, it is easy to see that, in the case considered in the previous theorem, the suitable normalization transforms the systems (19) and (20) into the well-known trigonometric systems 5.1(1) which are orthonormal and closed in $L_2(0, \sigma)$. Thus the aim of this chapter is attained completely.

(c) The concluding theorem of this section is a corollary of those statements of Theorems 5.3-1 and 5.3-2 which relate to systems $\{e_k(\tau; \mu)\}_1^\infty$ and $\{e_k(\tau; \mu)\}_0^\infty$.

Theorem 5.3-4. *Let $\{\lambda_k\}_1^\infty$ be the zeros of the function $E_{1/2}(-\sigma^2 z; 1+\nu)$. Then the following statements are true:*
$1°$. *If $\nu \in \Delta(\kappa, 2) = (\omega, 1+\omega) \cap [0, 2)$, $-1 < \omega < 1$, then, after suitable normalization, the system*
$$\left\{E_{1/2}\left(-\lambda_k\tau^2; \mu\right)\right\}_1^\infty, \qquad \mu = 3/2 + \omega \tag{21}$$
forms a Riesz basis of the space $L_{2,\kappa}(0, \sigma)(\kappa = 1+2\omega)$ of functions $\varphi(\tau)$ for which

$$\|\varphi\|_{2,\kappa} = \left\{\int_0^\sigma |\varphi(\tau)|^2 \tau^\kappa d\tau\right\}^{1/2} < +\infty. \tag{22}$$

$2°$. *If $\nu \in \Delta^*(\kappa, 2) = (1 + \omega, 2 + \omega) \cap [0, 2)$, $-1 < \omega < 1$, then, after suitable normalization, the system*

$$\left\{E_{1/2}\left(-\lambda_k\tau^2; \mu\right)\right\}_0^\infty \qquad (\lambda_0 = 0), \ \mu = 3/2 + \omega \tag{23}$$

forms a Riesz basis of the space $L_{2,\kappa}(0, \sigma)$.

5.4 Notes

The results of this chapter were established in M.M. Djrbashian-S.G. Raphaelian [3, §3] in somewhat different forms. Here we use only one of the well-known equivalent definitions of the Riesz bawe use only one of the well-known equivalent definitions of the Riesz basis. In the proof of Theorem 5.3-1 we used the following well-known assertion: *if one of two biorthogonal systems is a Riesz basis of a Hilbert space, then the other one is also a Riesz basis.* A detailed account of the theory of Riesz bases is given, for example, in the monographs of I. Gohberg-M.G. Krein [1, Chapter 7, §1,2] and S. Kaczmarz-H. Steinhaus [1, "Review", §9]. The mentioned well-known assertion will also be used in later chapters of this book.

It should be mentioned that some systems of linear combinations of Mittag-Leffler type functions $E_\rho(z; \mu)$ $(\rho \ge 1/2)$, which are biorthogonal on the segment $[0, l]$ of the real axis, were constructed in the early paper of M.M. Djrbashian-A.B.

Nersesian [1]. Later it was established by M.M. Djrbashian-A.B. Nersesian [2,3] that the expansions in terms of these systems are equiconvergent with the ordinary Fourier series of any function $\varphi(\tau) \in L_1(0,l)$. This was proved by applying the well-known contour integration method ascending to Cauchy. It must be noted, however, that these methods and results have nothing in common with the contents of Chapters 4 and 5 of this book.

6 Interpolation series expansions in spaces $W^{p,\omega}_{s+1/2,\sigma}$ of entire functions

6.1 Introduction

In this chapter we establish some theorems relating to interpolation series expansions of entire functions $\Phi(z)$ of order $s+1/2$ (where $s \geq 1$ is an arbitrary natural number) and of type $\leq \sigma$, which are of spaces $W^{p,\omega}_{s+1/2,\sigma}$, i.e. for which

$$\left\{ \sum_{j=-s}^{s} \int_0^{+\infty} \left| \Phi\left(re^{i\pi j/(s+1/2)} \right) \right|^p r^\omega dr \right\}^{1/p} < +\infty. \tag{1}$$

It will be supposed, as always, that

$$1 < p < +\infty \text{ and } -1 < \omega < p-1. \tag{2}$$

The expansions of this chapter are established on the base of the concluding Theorems 4.4-1 and 4.4-2 of Chapter 4, which relate to interpolation expansions of functions of the simplest spaces $W^{p,\omega}_{1/2,\sigma}$ of the mentioned type. But, in contrast to Chapter 4, here we prefer to give the necessary notations and to formulate the main theorems on interpolation in classes $W^{p,\omega}_{s+1/2,\sigma}$ right in the beginning of the chapter, in spite of the fact that these theorems are proved only in its concluding section, by use of some auxiliary results which we prove in its earlier sections.

6.2 The formulation of the main theorems

First we have to introduce some necessary notations often used later.

(a) Denote by

$$\Gamma_{2s+1} = \bigcup_{j=-s}^{s} \Gamma_{1,j} (s \geq 0) \tag{1}$$

the sum of $2s+1$ rays

$$\Gamma_{1,j} = \left\{ z : z = r \exp\left\{ i\frac{\pi j}{s+1/2} \right\}, \quad 0 \leq r < +\infty \right\}. \tag{2}$$

Evidently, these rays turn into the semi-axis $[0, +\infty)$ when $s = 0$. The restriction 6.1(1), defining the Banach space $W^{p,\omega}_{s+1/2,\sigma}$ of entire functions $\Phi(z)$ of order $s + 1/2(s \geq 1)$ and of type $\leq \sigma$(which is more general than $W^{p,\omega}_{1/2,\sigma}$), may now be written down as follows:

$$\|\Phi; \Gamma_{2s+1}\|_{p,\omega} \equiv \left\{ \int_{\Gamma_{2s+1}} |\Phi(z)|^p |z|^\omega |dz| \right\}^{1/p}$$

$$= \left\{ \sum_{j=-s}^{s} \int_0^{+\infty} \left| \Phi\left(re^{i\frac{\pi j}{s+1/2}} \right) \right|^p r^\omega dr \right\}^{1/p} < +\infty. \tag{3}$$

(b) Using the function $\mathcal{E}_\sigma(z;\nu)$ defined earlier and assuming, as always, that $\nu \in [0,2)$, we introduce the entire function

$$\mathcal{E}_{s+1/2,\sigma}(z;\nu) \equiv \mathcal{E}_\sigma(z^{2s+1};\nu) = E_{1/2}(-\sigma^2 z^{2s+1}; 1+\nu) \tag{4}$$

which is of order $s + 1/2$ and of type σ. The zeros $\{\lambda_k\}_1^\infty$ $(0 < \lambda_k < \lambda_{k+1}$, $1 \le k < +\infty)$ of $\mathcal{E}_\sigma(z;\nu)$ are simple and positive by Theorem 1.4-3. Hence zeros of $\mathcal{E}_{s+1/2,\sigma}(z;\nu)$ are also simple and are situated on the sum of rays Γ_{2s+1}. And, if we denote

$$\alpha_s = \exp\left\{i\frac{\pi}{s+1/2}\right\}, \tag{5}$$

then, evidently,

$$\{\mu_{j,k}\}_1^\infty = \left\{\alpha_s^j \lambda_k^{1/(2s+1)}\right\}_1^\infty \subset \Gamma_{1,j}(-s \le j \le s) \tag{6}$$

is the set of zeros of $\mathcal{E}_{s+1/2,\sigma}(z;\nu)$ which are all situated on the rays $\Gamma_{1,j}$ and

$$\mu_{j,k}^{2s+1} = \lambda_k \qquad (-s \le j \le s, 1 \le k < +\infty). \tag{7}$$

Obviously, we give a uniform numeration of all zeros $\{\mu_n\}_1^\infty \subset \Gamma_{2s+1}$ of $\mathcal{E}_{s+1/2,\sigma}(z;\nu)$, if we put

$$\mu_{(2s+1)k+j-s} = \mu_{j,k} \qquad (-s \le j \le s, 1 \le k < +\infty). \tag{8}$$

Now note that $\lambda_k \asymp (1+k)^2(1 \le k < +\infty)$ by Theorem 1.4-3. Thus (6)-(8) imply the two-sided inequalities

$$|\mu_{j,k}| \asymp (1+k)^{2/(2s+1)} \qquad (-s \le j \le s, 1 \le k < +\infty),$$
$$|\mu_n| \asymp (1+n)^{2/(2s+1)}(1 \le n < +\infty). \tag{9}$$

(c) Let $\{\Phi_n\}_{-r}^\infty$ $(0 \le r \le s)$ be an arbitrary sequence of complex numbers, such that the quantity

$$\|\{\Phi_n\}_{-r}^\infty\|_{p,\kappa_{-s}} \equiv \left\{\sum_{n=0}^r \left|\frac{\Phi_{-n}}{n!}\right|^p + \sum_{n=1}^\infty |\Phi_n|^p (1+n)^{\kappa_{-s}}\right\}^{1/p}, \tag{10}$$

where

$$\kappa_{-s} = \frac{1+2(\omega-s)}{2s+1}, \tag{11}$$

is finite. Then we shall call this quantity *the norm of the sequence* $\{\Phi_n\}_{-r}^\infty$ and write

$$\{\Phi_n\}_{-r}^\infty \in L^{(r)}_{p,\kappa_{-s}}. \tag{12}$$

One can easily be convinced that, in the considered case $1 < p < +\infty$, the class $L_{p,\kappa_{-s}}^{(r)}$, defined in this way, is a Banach space. Finally, we put

$$\gamma = \frac{2(1+\omega)}{p(2s+1)} \quad (s \geq 1) \tag{13}$$

and introduce the pair of intervals

$$\Delta_s(1^0) = \left(\gamma + \frac{2s-1}{2s+1}, \gamma + \frac{2s}{2s+1}\right) \subset [0,2), \tag{14}$$

$$\Delta_s(2^0) = \left(\gamma + \frac{2s}{2s+1}, \gamma + 1\right) \subset [0,2). \tag{15}$$

It can easily be verified that these intervals have no common points and that they both lie in $[0,2)$. Thus, the function $\mathcal{E}_{s+1/2,\sigma}(z;\nu)$ has in both cases

$$\nu \in \Delta_s(1^0) \text{ and } \nu \in \Delta_s(2^0) \tag{16}$$

an infinite set of simple zeros $\{\mu_n\}_1^\infty$ situated on the sum of rays Γ_{2s+1}.

(d) Now we pass to formulations of the main interpolation theorems of this chapter, but we shall be ready to prove these theorems only at the end of the chapter, after proving some necessary lemmas and propositions.

Theorem 6.2-1. *If $\nu \in \Delta_s(1^\circ)$, then the series*

$$\Phi(z) = \Gamma(1+\nu) \left(\sum_{k=0}^{s-1} \frac{\Phi_{-k}}{k!} z^k\right) \mathcal{E}_{s+1/2,\sigma}(z;\nu)$$

$$+ \sum_{n=1}^{\infty} \Phi_n \frac{z^s \mathcal{E}_{s+1/2,\sigma}(z;\nu)}{\mu_n^s \mathcal{E}'_{s+1/2,\sigma}(\mu_n;\nu)(z-\mu_n)} \tag{17}$$

represents a continuous one-to-one mapping of the space $L_{p,\kappa_{-s}}^{(s-1)}$ of sequences $\{\Phi_n\}_{-(s-1)}^{\infty}$ onto the space $W_{s+1/2,\sigma}^{p,\omega}$ of entire functions $\Phi(z)$. In addition, the following assertions are true:

1°. A series (17) converges to its sum in the norm $\|\,.\,;\Gamma_{2s+1}\|_{p,\omega}$ of the space $W_{s+1/2,\sigma}^{p,\omega}$ and it also converges to the same limit uniformly in any disk $|z| \leq R < +\infty$.

2°. The following two-sided estimates are true:

$$\|\Phi;\Gamma_{2s+1}\|_{p,\omega} \asymp \|\{\Phi_n\}_{-(s-1)}^{\infty}\|_{p,\kappa_{-s}}. \tag{18}$$

Here, as everywhere, the suitable constants are independent of both estimated elements of Banach spaces.

3°. The function $\Phi(z)$ of (17) has the interpolation data

$$\Phi^{(k)}(0) = \Phi_{-k} \quad (0 \leq k \leq s-1), \quad \Phi(\mu_n) = \Phi_n \quad (1 \leq n < +\infty). \tag{19}$$

Theorem 6.2-2. *If $\nu \in \Delta_s(2^\circ)$, then the series*

$$
\Phi(z) = \Gamma(1+\nu)\left(\sum_{k=0}^{s} \frac{\Phi_{-k}}{k!}z^k\right)\mathcal{E}_{s+1/2,\sigma}(z;\nu)
$$
$$
+ \sum_{n=1}^{\infty}\Phi_n \frac{z^{s+1}\mathcal{E}_{s+1/2,\sigma}(z;\nu)}{\mu_n^{s+1}\mathcal{E}'_{s+1/2,\sigma}(\mu_n;\nu)(z-\mu_n)}
\tag{20}
$$

represents a continuous one-to-one mapping of the space $L^{(s)}_{p,\kappa_{-s}}$ of sequences $\{\Phi_n\}^{\infty}_{-s}$ onto the space $W^{p,\omega}_{s+1/2,\sigma}$ of entire functions $\Phi(z)$. In addition, the following assertions are true:
1°. A series (20) converges to its sum in the norm $\|.;\Gamma_{2s+1}\|_{p,\omega}$ of the space $W^{p,\omega}_{s+1/2,\sigma}$, and it also converges to the same limit uniformly in any disk $|z| \leq R < +\infty$.
2°. The following two-sided inequalities are true:

$$
\|\Phi;\Gamma_{2s+1}\|_{p,\omega} \asymp \|\{\Phi_n\}^{\infty}_{-s}\|_{p,\kappa_{-s}}.
\tag{21}
$$

3°. The function $\Phi(z)$ of (20) has the interpolation data

$$
\Phi^{(k)}(0) = \Phi_{-k} \quad (0 \leq k \leq s), \quad \Phi(\mu_n) = \Phi_n \quad (1 \leq n < +\infty).
\tag{22}
$$

(e) The following uniqueness theorem is an immediate consequence of the two preceding theorems.

Theorem 6.2-3. *If $\Phi(z) \in W^{p,\omega}_{s+1/2,\sigma}$, and $\{\mu_n\}^{\infty}_1$ is the sequence of zeros of the function*
$$
\mathcal{E}_{s+1/2,\sigma}(z;\nu) = E_{1/2}\left(-\sigma^2 z^{2s+1};1+\nu\right), \qquad \nu \in [0,2),
$$
then in both the following cases
1° $\nu \in \Delta_s(1^\circ)$ and

$$
\Phi^{(k)}(0) = 0 \quad (0 \leq k \leq s-1), \quad \Phi(\mu_n) = 0 \quad (1 \leq n < +\infty),
\tag{23}
$$

2° $\nu \in \Delta_s(2^\circ)$ and

$$
\Phi^{(k)}(0) = 0 \quad (0 \leq k \leq s), \quad \Phi(\mu_n) = 0 \quad (1 \leq n < +\infty)
\tag{24}
$$

we have
$$
\Phi(z) \equiv 0.
\tag{25}
$$

The identity (25) follows in both cases 1° and 2° from the interpolation data (19) and (22) of functions of $W^{p,\omega}_{s+1/2,\sigma}$ which are representable in the forms (17) and (20) when $\nu \in \Delta_s(1^\circ)$ and $\nu \in \Delta_s(2^\circ)$ respectively.

6.3 Auxiliary relations and lemmas

(a) If the power expansion

$$\Phi(z) = \sum_{n=0}^{\infty} c_n z^n \tag{1}$$

represents an entire function of a finite order, then, as is well known, the following relations are true for the order ρ and for the type σ:

$$\rho = \limsup_{n\to\infty} \frac{n \ln n}{\ln 1/|c_n|}, \quad \sigma = \frac{1}{e\rho} \limsup_{n\to\infty} \sup n\,|c_n|^{\rho/n}. \tag{2}$$

Thus, if $\Phi(z)$ is an entire function of order $\rho = s + 1/2$ (where $s \geq 0$ is a given integer) and of type $\sigma(0 < \sigma < +\infty)$, then the relations (2) pass to

$$\limsup_{n\to\infty} \frac{n \ln n}{\ln 1/|c_n|} = s + \frac{1}{2}, \quad \limsup_{n\to\infty} n\,|c_n|^{\frac{s+1/2}{n}} = e\sigma\left(s + \frac{1}{2}\right). \tag{3}$$

We associate with such a function $\Phi(z)$ a set $\{\varphi_j(w)\}_{-s}^{s}$ of $2s + 1$ entire functions which have the following power expansions:

$$\varphi_j(w) = \sum_{n=0}^{\infty} c_{n_j} w^n, \quad n_j = (2s + 1)n + s + j \quad (-s \leq j \leq s, 0 \leq n < +\infty). \tag{4}$$

If we now denote by ρ_j and σ_j the order and the type of $\varphi_j(w)$ $(-s \leq j \leq s)$, then relations (2)-(4) give $0 \leq \rho_j \leq 1/2$ and $0 \leq \sigma_j \leq \sigma$, if $\rho_j = 1/2$ exactly.

(b) If we apply to the simple equalities 1.2(11)-(12), i.e., to

$$\alpha_s = \exp\left\{i\frac{\pi}{s + 1/2}\right\}, \quad \sum_{h=-s}^{s} \alpha_s^{kh} = \begin{cases} 2s + 1 & \text{when } k \equiv 0(\mod 2s + 1), \\ 0 & \text{when } k \not\equiv 0(\mod 2s + 1), \end{cases} \tag{5}$$

then the relation between the functions $\Phi(z)$ and $\{\varphi_j(w)\}_{-s}^{s}$ can be given in the following way.

Lemma 6.3-1. *$1°$. The representations*

$$z^{s+j}\varphi_j\left(z^{2s+1}\right) = \frac{1}{2s + 1} \sum_{h=-s}^{s} \alpha_s^{-(s+j)h} \Phi\left(\alpha_s^h z\right) \quad (-s \leq j \leq s) \tag{6}$$

and their inversions

$$\Phi(z) = \sum_{j=-s}^{s} z^{s+j}\varphi_j\left(z^{2s+1}\right) \tag{7}$$

are true.
$2°$. The sequence of entire functions $\{\varphi_j(w)\}_{-s}^{s}$ contains at least one function of order $1/2$ and of type σ.

Proof. $1°$. Representations (6) and (7) follow directly from expansions (1) and (4) of the functions $\Phi(z)$ and $\{\varphi_j(w)\}^s_{-s}$, if formulas (5) are used.

$2°$. Contrary to our assertion, suppose $\rho_j < 1/2$ $(-s \leq j \leq s)$ or, if $\rho_j = 1/2$ for some j, then $\sigma_j < \sigma$. Then it follows from representation (7) that the order ρ of $\Phi(z)$ is less than or equal to $s + 1/2$, and if $\rho = s + 1/2$, then its type is less than σ.

Note that, if we put $\Phi(z) = E_{s+1/2}(z; \mu)$, then formulas (6) and (7) give the identities 1.2(14), but only for the case when $\rho = 1/2$. Hence the following lemma is true.

Lemma 6.3-2. *If $s \geq 0$ is any integer, then*

$$z^{s+j} E_{1/2}\left(z^{2s+1}; \mu + \frac{s+j}{s+1/2}\right)$$

$$= \frac{1}{2s+1} \sum_{h=-s}^{s} \alpha_s^{-(s+j)h} E_{s+1/2}\left(\alpha_s^h z; \mu\right) \quad (-s \leq j \leq s), \tag{8}$$

$$E_{s+1/2}(z; \mu) = \sum_{j=-s}^{s} z^{s+j} E_{1/2}\left(z^{2s+1}; \mu + \frac{s+j}{s+1/2}\right). \tag{9}$$

(c) In accordance with the notations 6.2(3) and 4.1(1) denote

$$\|\varphi; \Gamma_1\|_{p,\omega} \equiv \|\varphi\|^+_{p,\omega} \equiv \left\{\int_0^{+\infty} |\varphi(r)|^p r^\omega dr\right\}^{1/p} \tag{10}$$

as the norm of a function $\varphi(w) \in W^{p,\omega}_{1/2,\sigma}$. Further, along with parameters $1 < p < +\infty$ and $-1 < \omega < p - 1$ we consider some new ones:

$$\omega_j = \frac{\omega - 2s + p(s+j)}{2s+1}, \quad -1 < \omega_j < p - 1 \quad (-s \leq j \leq s) \tag{11}$$

and prove the following lemma.

Lemma 6.3-3. $1°$. *The class $W^{p,\omega}_{s+1/2,\sigma}$ coincides with the set of those entire functions $\Phi(z)$ which can be represented by a sum*

$$\Phi(z) = \sum_{j=-s}^{s} z^{s+j} \varphi_j\left(z^{2s+1}\right), \tag{12}$$

where

$$\varphi_j(w) \in W^{p,\omega_j}_{1/2,\sigma} \text{ and } \varphi_j\left(z^{2s+1}\right) = \frac{1}{2s+1} \sum_{h=-s}^{s} \left(\alpha_s^h z\right)^{-(s+j)} \Phi\left(\alpha_s^h z\right). \tag{13}$$

$2°$. *The two-sided inequalities*

$$\|\Phi; \Gamma_{2s+1}\|^\delta_{p,\omega} \asymp \sum_{j=-s}^{s} \|\varphi_j; \Gamma_1\|^\delta_{p,\omega_j} = \sum_{j=-s}^{s} \left\{\|\varphi_j\|^+_{p,\omega_j}\right\}^\delta \tag{14}$$

are true in both cases $\delta = 1$ and $\delta = p$.

Proof. $1°$ First note that a simple change of variable leads to the equalities

$$\int_0^{+\infty} \left| r^{s+j} \varphi_j \left(r^{2s+1} \right) \right|^p r^\omega dr$$

$$= (2s+1)^{-1} \int_0^{+\infty} |\varphi_j(r)|^p r^{\omega_j} dr = (2s+1)^{-1} \|\varphi_j; \Gamma_1\|_{p,\omega_j}^p \quad (-s \le j \le s),$$

if the notation (11) is taken into account. Next, observe that these equalities may be written down also in the form

$$\int_{\Gamma_{2s+1}} \left| z^{s+j} \varphi_j(z^{2s+1}) \right|^p |z|^\omega |dz| = \|\varphi_j; \Gamma_1\|_{p,\omega_j}^p \quad (-s \le j \le s), \tag{15}$$

if we take into account that $z^{2s+1} = r^{2s+1}$ when $z \in \Gamma_{2s+1}$, i.e. when $z = r\alpha_s^j$ $(0 \le r < +\infty, -s \le j \le s)$. Now let the inclusions of (13) be true and let the function $\Phi(z)$ be representable in the form (12). Then, by Minkowski's inequality and (15),

$$\|\Phi; \Gamma_{2s+1}\|_{p,\omega} \le \sum_{j=-s}^s \|\varphi_j; \Gamma_1\|_{p,\omega_j}. \tag{16}$$

Hence, it follows in particular that $\Phi(z) \in W_{s+1/2,\sigma}^{p,\omega}$. Conversely, if $\Phi(z) \in W_{s+1/2,\sigma}^{p,\omega}$, then we determine the set of functions $\{\varphi_j(w)\}_{-s}^s$ by (6), the inversion of which is (12). Therefore, by (15) and Minkowski's inequality,

$$\|\varphi_j; \Gamma_1\|_{p,\omega_j} = (2s+1)^{-1} \left\{ \int_{\Gamma_{2s+1}} \left| \sum_{h=-s}^s \alpha_s^{-(s+j)h} \Phi \left(\alpha_s^h z \right) \right|^p |z|^\omega |dz| \right\}^{1/p}$$

$$\le (2s+1)^{-1} \sum_{h=-s}^s \left\{ \int_{\Gamma_{2s+1}} \left| \Phi \left(\alpha_s^h z \right) \right|^p |z|^\omega |dz| \right\}^{1/p} \quad (-s \le j \le s).$$

But $\alpha_s^h z \in \Gamma_{2s+1}$ for any $z \in \Gamma_{2s+1}$ and $h(-s \le h \le s)$. Hence

$$\|\varphi_j; \Gamma_1\|_{p,\omega_j}^\delta \le \|\Phi; \Gamma_{2s+1}\|_{p,\omega}^\delta \quad (-s \le j \le s)$$

in both cases $\delta = 1$ and $\delta = p$. Thus $\varphi_j(w) \in W_{1/2,\sigma}^{p,\omega_j}$ $(-s \le j \le s)$ and

$$\|\Phi; \Gamma_{2s+1}\|_{p,\omega}^\delta \ge (2s+1)^{-1} \sum_{j=-s}^s \|\varphi_j; \Gamma_1\|_{p,\omega_j}^\delta \tag{17}$$

in both cases $\delta = 1$ and $\delta = p$.

2°. The inequalities (16) and (17) imply (14) in both cases $\delta = 1$ and $\delta = p$, as (16) implies the inequality

$$\|\Phi;\Gamma_{2s+1}\|^p_{p,\omega} \leq (2s+1)^p \sum_{j=-s}^{s} \|\varphi_j;\Gamma_1\|^p_{p,\omega_j}.$$

Remark 1. One may prove that, if any of the functions $\varphi_j(w)$ $(-s \leq j \leq s)$ of the representation (12) of $\Phi(z) \in W^{p,\omega}_{s+1/2,\sigma}$ is of order $\rho_j < 1/2$, then $\varphi_j(w) \equiv 0$.

Remark 2. It may be noted in connection with functions $\varphi_j \in W^{p,\omega_j}_{1/2,\sigma}$, that (11) implies

$$\frac{p(s+j)}{2s+1} - 1 < \omega_j < \frac{p(s+j+1)}{2s+1} - 1 \qquad (-s \leq j \leq s). \tag{18}$$

Thus the intervals, in which $\omega_j(-s \leq j \leq s)$ vary, have no common points and they completely cover the interval $(-1, p-1)$ (where ω vary), excluding their endpoints $p(s+j)/(2s+1) - 1$ $(-s < j \leq s)$.

(d) If we put $z = \lambda_k^{1/(2s+1)}$ and $z = \mu_{h,k} = \alpha_s^h \lambda_k^{1/(2s+1)}$ correspondingly in expansions (6) and (7) of Lemma 6.3-1, then we obtain the following pair of formulas, which give a connection between the sequences of numbers $\{\Phi(\mu_{h,k})\}_1^{\infty}$ $(-s \leq h \leq s)$ and $\{\varphi_j(\lambda_k)\}_1^{\infty}$ $(-s \leq j \leq s)$:

$$\varphi_j(\lambda_k) = \frac{1}{2s+1} \sum_{h=-s}^{s} \mu_{h,k}^{-(s+j)} \Phi(\mu_{h,k}) \tag{19}$$

$$\Phi(\mu_{h,k}) = \sum_{j=-s}^{s} \mu_{h,k}^{s+j} \varphi_j(\lambda_k). \tag{20}$$

The following lemma establishes similar invertible relations between arbitrary sequences of complex numbers.

Lemma 6.3-4. *Let*

$$\{\Phi_{h,k}\}_1^{\infty}(-s \leq h \leq s) \text{ and } \{\varphi_{j,k}\}_1^{\infty} \qquad (-s \leq j \leq s) \tag{21}$$

be any sequences of complex numbers, such that one of the equalities

$$\varphi_{j,k} = \frac{1}{2s+1} \sum_{h=-s}^{s} \mu_{h,k}^{-(s+j)} \Phi_{h,k} \quad (-s \leq j \leq s, 1 \leq k < +\infty), \tag{22}$$

$$\Phi_{h,k} = \sum_{j=-s}^{s} \mu_{h,k}^{s+j} \varphi_{j,k} \quad (-s \leq h \leq s, 1 \leq k < \infty) \tag{23}$$

is true. Then the other equality is also true.

Proof. Obviously

$$\sum_{j=-s}^{s}\left(\frac{\mu_{h,k}}{\mu_{\nu,k}}\right)^{s+j} = \sum_{j=-s}^{s}\alpha_s^{(h-\nu)(s+j)} = \begin{cases} 2s+1 & \text{when } \nu = h, \\ 0 & \text{when } \nu \neq h. \end{cases}$$

Thus, if the equality (22) is true, then

$$\sum_{j=-s}^{s}\mu_{h,k}^{s+j}\varphi_{j,k} = \frac{1}{2s+1}\sum_{j=-s}^{s}\mu_{h,k}^{s+j}\sum_{\nu=-s}^{s}\mu_{\nu,k}^{-(s+j)}\Phi_{\nu,k}$$

$$= \frac{1}{2s+1}\sum_{\nu=-s}^{s}\Phi_{\nu,k}\sum_{j=-s}^{s}\left(\frac{\mu_{h,k}}{\mu_{\nu,k}}\right)^{s+j} = \Phi_{h,k}.$$

Further, obviously

$$\sum_{h=-s}^{s}\mu_{h,k}^{\nu-j} = \lambda_k^{\frac{\nu-j}{2s+1}}\sum_{h=-s}^{s}\alpha_s^{(\nu-j)h} = \begin{cases} 2s+1 & \text{when } \nu = j, \\ 0 & \text{when } \nu \neq j. \end{cases}$$

Therefore, equality (23) gives

$$\sum_{h=-s}^{s}\mu_{h,k}^{-(s+j)}\Phi_{h,k} = \sum_{h=-s}^{s}\mu_{h,k}^{-(s+j)}\sum_{\nu=-s}^{s}\mu_{h,k}^{s+\nu}\varphi_{\nu,k}$$

$$= \sum_{\nu=-s}^{s}\varphi_{\nu,k}\sum_{h=-s}^{s}\mu_{h,k}^{\nu-j} = (2s+1)\varphi_{j,k}.$$

(e) To prove another lemma relating to sequences of complex numbers, first observe that the equalities

$$\Phi_n = \Phi_{h,k} \qquad (n = (2s+1)k + h - s, -s \leq h \leq s, 1 \leq k < +\infty) \qquad (24)$$

identify an arbitrary sequence of form $\{\Phi_{h,k}\}_1^\infty(-s \leq h \leq s)$ with a sequence of form $\{\Phi_n\}_1^\infty$. And, if the equalities (24) are true, then

$$\{\Phi_n\}_1^\infty = \bigcup_{k=1}^{\infty}\left\{\bigcup_{h=-s}^{s}\Phi_{h,k}\right\}. \qquad (25)$$

Next, note that Lemma 6.3-4 establishes a one-to-one correspondence between some classes of sequences $\{\Phi_{h,k}\}_1^\infty$ and $\{\varphi_{j,k}\}_1^\infty$ $(-s \leq h, j \leq s)$. Finally, remember that the space $l^{p,\kappa}(1 < p < +\infty, \kappa \in (-1, +\infty))$ was defined in 4.3(a) as the set of sequences $\{c_n\}_k^\infty(k = 0, 1)$, for which

$$\|\{c_n\}_k^\infty\|_{p,\kappa} = \left\{\sum_{n=k}^{\infty}|c_n|^p(1+n)^\kappa\right\}^{1/p} < +\infty \quad (k = 0, 1). \qquad (26)$$

Obviously, this definition can be extended on the case of any $\kappa \in (-\infty, +\infty)$. We are ready to prove the following lemma.

Lemma 6.3-5. *If the sequences $\{\Phi_{h,k}\}_1^\infty$ and $\{\varphi_{j,k}\}_1^\infty$ $(-s \le h, j \le s)$ are connected by formulas (22) and (23), and $\{\Phi_n\}_1^\infty$ is the sequence (24), then the inclusions*

$$\{\Phi_n\}_1^\infty \in l^{p,\kappa_{-s}} \text{ and } \{\varphi_{j,k}\}_1^\infty \in l^{p,\kappa_j} \qquad (-s \le j \le s), \tag{27}$$

where

$$\kappa_{-s} = \frac{2(\omega - s) + 1}{2s + 1}, \kappa_j = \frac{2(\omega - s) + 2p(s + j) + 1}{2s + 1} \qquad (-s \le j \le s) \tag{28}$$

are equivalent. And, if they are true, then

$$\|\{\Phi_n\}_1^\infty\|_{p,\kappa_{-s}}^\delta \asymp \sum_{j=-s}^s \|\{\varphi_{j,k}\}_1^\infty\|_{p,\kappa_j}^\delta \quad (\delta = 1, p). \tag{29}$$

Proof. It follows from representations (22) and (23) and from the two-sided inequalities 6.2(9), that

$$|\varphi_{j,k}|^p \le C_1 \sum_{h=-s}^s |\Phi_{h,k}|^p (1 + k)^{-\frac{2p(s+j)}{2s+1}},$$

$$|\Phi_{h,k}|^p \le C_2 \sum_{j=-s}^s |\varphi_{j,k}|^p (1 + k)^{\frac{2p(s+j)}{2s+1}}, \tag{30}$$

where C_1 and C_2 are independent of $\{\Phi_{h,k}\}_1^\infty$ and $\{\varphi_{j,k}\}_1^\infty$. The constants C_m $(m = 3, 4, \ldots)$, appearing further in the proof, will be assumed to be of the same kind. Observe now that (29) implies

$$\kappa_j = \kappa_{-s} + \frac{2p(s + j)}{2s + 1} \qquad (-s \le j \le s).$$

It follows from (30) that

$$|\varphi_{j,k}|^p (1 + k)^{\kappa_j} \le C_1 \sum_{h=-s}^s |\Phi_{h,k}|^p (1 + k)^{\kappa_{-s}} \quad (-s \le j \le s, 1 \le k < +\infty) \tag{31_1}$$

and

$$|\Phi_{h,k}|^p (1 + k)^{\kappa_{-s}} \le C_2 \sum_{j=-s}^s |\varphi_{j,k}|^p (1 + k)^{\kappa_j} \quad (-s \le h \le s, 1 \le k < +\infty). \tag{31_2}$$

But, obviously,

$$1 + (2s + 1)k + h - s \asymp 1 + k \quad (-s \le h \le s, 1 \le k < +\infty), \tag{32}$$

thus, if we take (25) into account, then the summation of inequalities (31_1) gives

$$\sum_{k=1}^{\infty} |\varphi_{j,k}|^p (1+k)^{\kappa_j} = \|\{\varphi_{j,k}\}_1^\infty\|_{p,\kappa_j}^p \le C_1 \sum_{k=1}^{\infty} \sum_{h=-s}^{s} |\Phi_{h,k}|^p (1+k)^{\kappa-s}$$

$$\le C_3 \sum_{k=1}^{\infty} \sum_{h=-s}^{s} |\Phi_{h,k}|^p [1+(2s+1)k+h-s]^{\kappa-s}$$

$$= C_3 \sum_{n=1}^{\infty} |\Phi_n|^p (1+n)^{\kappa-s} \quad (-s \le j \le s). \tag{33}$$

Finally, we sum the last inequalities over j and get

$$\sum_{j=-s}^{s} \|\{\varphi_{j,k}\}_1^\infty\|_{p,\kappa_j}^\delta \le C_4 \|\{\Phi_n\}_1^\infty\|_{p,\kappa-s}^\delta \quad (\delta = 1,p). \tag{29_1}$$

To prove the converse inequality, we sum (31_2) over k and obtain

$$\sum_{k=1}^{\infty} |\Phi_{h,k}|^p (1+k)^{\kappa-s} \le C_2 \sum_{j=-s}^{s} \sum_{k=1}^{\infty} |\varphi_{j,k}|^p (1+k)^{\kappa_j} = C_2 \sum_{j=-s}^{s} \|\{\varphi_{j,k}\}_1^\infty\|_{p,\kappa_j}^p$$

for any h $(-s \le h \le s)$. Hence, by (32),

$$\sum_{n=1}^{\infty} |\Phi_n|^p (1+n)^{\kappa-s} = \sum_{h=-s}^{s} \sum_{k=1}^{\infty} |\Phi_{h,k}|^p [1+(2s+1)k+h-s]^{\kappa-s}$$

$$\le C_5 \sum_{h=-s}^{s} \sum_{k=1}^{\infty} |\Phi_{h,k}|^p (1+k)^{\kappa-s} \le C_6 \sum_{j=-s}^{s} \|\{\varphi_{j,k}\}_1^\infty\|_{p,\kappa_j}^p,$$

i.e.,

$$\|\{\Phi_n\}_1^\infty\|_{p,\kappa-s}^p \le C_6 \sum_{j=-s}^{s} \|\{\varphi_{j,k}\}_1^\infty\|_{p,\kappa_j}^p.$$

Hence it follows that

$$\|\{\Phi_n\}_1^\infty\|_{p,\kappa-s}^\delta \le C_7 \sum_{j=-s}^{s} \|\{\varphi_{j,k}\}_1^\infty\|_{p,\kappa_j}^\delta \quad (\delta = 1,p), \tag{29_2}$$

and the proof is complete.

Remark. The obvious estimate

$$\left(\sum_{j=-s}^{s} c_j \right)^p \le (2s+1)^p \sum_{j=-s}^{s} c_j^p \quad (0 < p < +\infty, c_j \ge 0)$$

was used in the proofs of Lemmas 6.3-3 and 6.3-5.

(f) We pass to the last lemma of this section.

Lemma 6.3-6. *If* $\Phi(z) \in W^{p,\omega}_{s+1/2,\sigma}$, *and the functions* $\{\varphi_j(w)\}^s_{-s}$ *are defined by* (6), *then*

$$\varphi_{j,0} \equiv \varphi_j(0) = \frac{\Phi^{(s+j)}(0)}{(s+j)!} \quad (-s \le j \le s). \tag{34}$$

Proof. It is easy to see that

$$\Omega_{s,j}(z) \equiv \frac{d^{s+j}}{dz^{s+j}} \left\{ z^{s+j}\varphi_j\left(z^{2s+1}\right) \right\}$$

$$= \sum_{k=0}^{s+j} C^k_{s+j}(s+j)\ldots(s+j-k+1)z^{s+j-k} \frac{d^{s+j-k}}{dz^{s+j-k}}\varphi_j\left(z^{2s+1}\right).$$

Hence $\Omega_{s,j}(0) = (s+j)!\varphi_j(0)$ $(-s \le j \le s)$, but it follows from (6) that

$$\Omega_{s,j}(z) = \frac{1}{2s+1} \sum_{h=-s}^{s} \Phi^{(s+j)}\left(\alpha_s^h z\right).$$

Hence $\Omega_{s,j}(0) = \Phi^{(s+j)}(0)$ $(-s \le j \le s)$. The comparison of the representations obtained for $\Omega_{s,j}(0)$ gives formula (34).

Later it will be necessary to consider the set of numbers

$$\Phi_{-k} \equiv \Phi^{(k)}(0) \quad (k = 0, 1, 2, \ldots) \tag{35}$$

along with the set of numbers $\{\varphi_{j,0}\}^s_{-s}$ connected with the function $\Phi(z) \in W^{p,\omega}_{s+1/2,\sigma}$. It is obvious that formulas (34) and (35) give

$$P_{s-1}(z;\Phi) \equiv \sum_{j=-s}^{-1} \varphi_{j,0}z^{s+j} = \sum_{k=0}^{s-1} \Phi_{-k}\frac{z^k}{k!},$$

$$P_s(z;\Phi) \equiv \sum_{j=-s}^{0} \varphi_{j,0}z^{s+j} = \sum_{k=0}^{s} \Phi_{-k}\frac{z^k}{k!}. \tag{36}$$

6.4 Further auxiliary results

(a) In this section we frequently use the notations of the previous ones, but we begin with some new notations which will also be necessary. First we introduce for a given natural $s \ge 1$ the following two sets of intervals:

$$\Delta_j = \left(\frac{2(1+\omega_j)}{p} - 1, \frac{2(1+\omega_j)}{p}\right)$$

$$\Delta^*_j = \left(\frac{2(1+\omega_j)}{p}, 1 + \frac{2(1+\omega_j)}{p}\right) \quad (-s \le j \le s). \tag{1}$$

Further, if we use the definitions 6.3$((11), (29))$ of the numbers ω_j and $\kappa_j = 1+2\omega_j$ $(-s \le j \le s)$, then these intervals can be expressed as follows:

$$\Delta_j = \left(\frac{1+\kappa_j}{p} - 1, \frac{1+\kappa_j}{p}\right) = \left(\gamma + \frac{2j-1}{2s+1}, \gamma + \frac{2(s+j)}{2s+1}\right),$$
$$\Delta_j^* = \left(\frac{1+\kappa_j}{p}, 1 + \frac{1+\kappa_j}{p}\right) = \left(\gamma + \frac{2(s+j)}{2s+1}, \gamma + \frac{2(2s+j)+1}{2s+1}\right),$$

$$(2)$$

where

$$\gamma = \frac{2(1+\omega)}{p(2s+1)}, \qquad 0 < \gamma < \frac{2}{2s+1}. \tag{3}$$

Besides, the notations of Chapter 4 lead to the equalities

$$\Delta_j \cap [0,2) = \Delta(\kappa_j, p), \quad \Delta_j^* \cap [0,2) = \Delta^*(\kappa_j, p) \quad (-s \le j \le s). \tag{1'}$$

Now we suppose

$$J \subset \{j\}_{-s}^{s} \tag{4}$$

to be arbitrary set of indices and put

$$\Delta_J = \bigcap_{j \in J} \Delta_j \text{ and } \Delta_J^* = \bigcap_{j \in J} \Delta_j^*. \tag{5}$$

Further, we denote by

$$J^* = \{j\}_{-s}^{s} \setminus J \tag{6}$$

the set of remaining indices and prove the following lemma.

Lemma 6.4-1. *The intersection of the sets of intervals $\Delta_{J^*}^*$ and Δ_J is not empty, i.e.,*

$$\Delta_{J^*}^* \bigcap \Delta_J \neq \emptyset, \tag{7}$$

only in the following two cases:

$$1°. \qquad J^* = \{j\}_{-s}^{-1}, \qquad J = \{j\}_0^s. \tag{8}$$
$$2°. \qquad J^* = \{j\}_{-s}^{0}, \qquad J = \{j\}_1^s. \tag{9}$$

And if (7) is true, then

$$\Delta_s(1°) \equiv \left\{\bigcap_{j=-s}^{-1} \Delta_j^*\right\} \cap \left\{\bigcap_{j=0}^{s} \Delta_j\right\} = \left(\gamma + \frac{2s-1}{2s+1}, \gamma + \frac{2s}{2s+1}\right) \subset [0,2), \tag{10}$$

$$\Delta_s(2°) \equiv \left\{\bigcap_{j=-s}^{0} \Delta_j^*\right\} \cap \left\{\bigcap_{j=1}^{s} \Delta_j\right\} = \left(\gamma + \frac{2s}{2s+1}, \gamma + 1\right) \subset [0,2). \tag{11}$$

Proof. First observe that if $J = (j_1, j_2, \ldots, j_r)$, where $-s \leq j_1 < j_2 < \ldots < j_r \leq s$, then, in view of (2), $\Delta^*_J = \Delta_J = \emptyset$ when $j_r - j_1 > s$.

Particularly, it is so when $r > s+1$. Therefore, if (7) is true, then, necessarily, $J = (j_1, \ldots, j_r)$, $J^* = (j^*_1, \ldots j^*_\tau)$, where $r = s+1$ and $\tau = s$, or $r = s$ and $\tau = s+1$, but in these cases we inevitably have $j_{k+1} = j_k + 1$ $(1 \leq k < r)$ and $j^*_{k+1} = j^*_k + 1$ $(1 \leq k < \tau)$. Observe also that $\Delta^*_j \cap \Delta_{-s} = \emptyset$ and $\Delta^*_s \cap \Delta_j = \emptyset$ $(-s \leq j \leq s)$. Hence we conclude that, if (7) is true, then one of the representations (8) and (9) is valid. As to representations (10) and (11), their validity is easy to verify, and we omit their proofs. One can also easily verify the inclusions $\Delta_s(1^\circ) \subset [0,2)$ and $\Delta_s(2^\circ) \subset [0,2)$.

(b) It was established in the theorems of Chapter 4 that any entire function of class $W^{p,\omega}_{1/2,\sigma}$ can be expanded in two different interpolation series generated respectively by the points of interpolation $\{\lambda_k\}^\infty_1$ and $\{\lambda_k\}^\infty_0$ $(\lambda_0 = 0)$ which were the (simple) zeros of the functions $\mathcal{E}_\sigma(z; \nu)$ and $z\mathcal{E}_\sigma(z; \nu)$. It was also supposed that $\nu \in \Delta(\kappa, p)$ or $\nu \in \Delta^*(\kappa, p)$, respectively, where $\Delta(\kappa, p) \subset [0,2)$ and $\Delta^*(\kappa, p) \subset [0,2)$ were definite intervals. Now we formulate two theorems which are similar to each other, but have also an essential difference. These theorems easily follow from the concluding Theorem 4.4-4 of Chapter 4, Lemma 6.4-1, definitions (10) and (11) of intervals $\Delta_s(1^\circ)$ and $\Delta_s(2^\circ)$ and from equalities $(1')$. Therefore we omit their proofs.

Theorem 6.4-1. *If $\nu \in \Delta_s(1^\circ)$, then the series*

$$\varphi_j(w) = \varphi_{j,0}\Gamma(1 + \nu)\mathcal{E}_\sigma(w; \nu) + \sum_{k=1}^{\infty} \varphi_{j,k}\frac{w\mathcal{E}_\sigma(w; \nu)}{\lambda_k\mathcal{E}'_\sigma(\lambda_k; \nu)(w - \lambda_k)} \qquad (12)$$
$$(-s \leq j \leq -1),$$

$$\varphi_j(w) = \sum_{k=1}^{\infty} \varphi_{j,k}\frac{\mathcal{E}_\sigma(w; \nu)}{\mathcal{E}'_\sigma(\lambda_k; \nu)(w - \lambda_k)} \qquad (0 \leq j \leq s) \qquad (13)$$

represent correspondingly continuous one-to-one mappings of the spaces of sequences

$$\{\varphi_{j,k}\}^\infty_0 \in l^{p,\kappa_j} \quad (-s \leq j \leq -1) \quad \text{and} \quad \{\varphi_{j,k}\}^\infty_1 \in l^{p,\kappa_j} \quad (0 \leq j \leq s) \qquad (14)$$

with finite norms $\|.\|_{p,\kappa_j}$ $(\kappa_j = 1+2\omega_j)$ onto the space $W^{p,\omega_j}_{1/2,\sigma}$ $(-s \leq j \leq s)$ of entire functions. The series (12) and (13) converge to their sums $\varphi_j(w)$ $(-s \leq j \leq s)$ in the norms of the spaces $W^{p,\omega_j}_{1/2,\sigma}$, and they converge to the same limits uniformly in any disk $|z| \leq R < +\infty$. In addition, the functions $\varphi_j(w)$ $(-s \leq j \leq s)$ have the interpolation data

$$\begin{aligned} \varphi_j(\lambda_k) &= \varphi_{j,k} \qquad (-s \leq j \leq -1, 0 \leq k < +\infty), \\ \varphi_j(\lambda_k) &= \varphi_{j,k} \qquad (0 \leq j \leq s, 1 \leq k < +\infty), \end{aligned} \qquad (15)$$

and the following two-sided inequalities are true:

$$\|\varphi_j\|_{p,\omega_j}^+ \asymp \|\{\varphi_{j,k}\}_0^\infty\|_{p,\kappa_j} = \left\{\sum_{k=0}^\infty |\varphi_{j,k}|^p (1+k)^{\kappa_j}\right\}^{1/p} \quad (-s \le j \le -1),$$

$$\|\varphi_j\|_{p,\omega_j}^+ \asymp \|\{\varphi_{j,k}\}_1^\infty\|_{p,\kappa_j} = \left\{\sum_{k=1}^\infty |\varphi_{j,k}|^p (1+k)^{\kappa_j}\right\}^{1/p} \quad (0 \le j \le s). \tag{16}$$

Theorem 6.4-2. *Let $\nu \in \Delta_s(2^\circ)$. Then the assertions of Theorem 6.4-1 remain true, if we take $-s \le j \le 0$ everywhere instead of $-s \le j \le -1$ and $1 \le j \le s$ instead of $0 \le j \le s$.*

The statements of Lemma 6.3-3 can be complemented on the basis of the last two theorems. Namely, the following lemma can be proved.

Lemma 6.4-2. *Let a function $\Phi(z) \in W_{s+1/2,\sigma}^{p,\omega}$ be representable in the form*

$$\Phi(z) = \sum_{j=-s}^s z^{s+j} \varphi_j\left(z^{2s+1}\right), \quad \varphi_j(w) \in W_{1/2,\sigma}^{p,\omega_j} \quad (-s \le j \le s)$$

and let $\delta = 1$ or $\delta = p$. Then, in any case:
1° When $\nu \in \Delta_s(1^\circ)$,

$$\|\Phi; \Gamma_{2s+1}\|_{p,\omega}^\delta \asymp \sum_{j=-s}^{-1} \|\{\varphi_j(\lambda_k)\}_0^\infty\|_{p,\kappa_j}^\delta + \sum_{j=0}^s \|\{\varphi_j(\lambda_k)\}_1^\infty\|_{p,\kappa_j}^\delta. \tag{17}$$

2°. When $\nu \in \Delta_s(2^\circ)$,

$$\|\Phi; \Gamma_{2s+1}\|_{p,\omega}^\delta \asymp \sum_{j=-s}^{\circ} \|\{\varphi_j(\lambda_k)\}_0^\infty\|_{p,\kappa_j}^\delta + \sum_{j=1}^s \|\{\varphi_j(\lambda_k)\}_1^\infty\|_{p,\kappa_j}^\delta. \tag{18}$$

Proof. 1°. The two-sided inequalities (17) immediately follow from Lemma 6.3 − 3(2°) and from relations (15) and (16) of Theorem 6.4 − 1. 2°. The inequalities (18) follow from the same lemma and relations in a similar way.

(c) Let $\{\varphi_{j,k}\}_1^\infty$ $(-s \le j \le s)$ be an arbitrary set of complex numbers. We shall connect with such a set, as before, another set of numbers $\{\Phi_{h,k}\}_1^\infty$ defined by the formula

$$\Phi_{h,k} = \sum_{j=-s}^s \mu_{h,k}^{s+j} \varphi_{j,k} \quad (-s \le h \le s, 1 \le k < +\infty), \tag{19}$$

where, as always, $\mu_{h,k} = \alpha_s^h \lambda_k^{1/(2s+1)}$. Conversely, if initially a set of numbers $\{\Phi_{h,k}\}_1^\infty$ $(-s \le h \le s)$ is given, then we define the corresponding set $\{\varphi_{j,k}\}_1^\infty$ by the formula

$$\varphi_{j,k} = \frac{1}{2s+1} \sum_{h=-s}^{s} \mu_{h,k}^{-(s+j)} \Phi_{h,k} \quad (-s \le j \le s, 1 \le k < +\infty). \qquad (20)$$

Note that formulas (19) and (20) are inversions of each other, as was established in Lemma 6.3-4. Now we shall prove a lemma concerning the sums

$$R_{s,k}(z) = \sum_{j=-s}^{-1} z^{3s+j+1} \varphi_{j,k} + \lambda_k \sum_{j=0}^{s} z^{s+j} \varphi_{j,k} \quad (1 \le k < +\infty), \qquad (21)$$

$$R_{s,k}^*(z) = \sum_{j=-s}^{0} z^{3s+j+1} \varphi_{j,k} + \lambda_k \sum_{j=1}^{s} z^{s+j} \varphi_{j,k} \quad (1 \le k < +\infty). \qquad (22)$$

Lemma 6.4-3. *The following identities are true for any* $k(1 \le k < +\infty)$:

$$R_{s,k}(z) = \frac{z^s(z^{2s+1} - \lambda_k)}{2s+1} \sum_{h=-s}^{s} \frac{\Phi_{h,k}}{\mu_{h,k}^{s-1}(z - \mu_{h,k})}, \qquad (21')$$

$$R_{s,k}^*(z) = \frac{z^{s+1}(z^{2s+1} - \lambda_k)}{2s+1} \sum_{h=-s}^{s} \frac{\Phi_{h,k}}{\mu_{h,k}^{s}(z - \mu_{h,k})}. \qquad (22')$$

Proof. Inserting the representations (20) of the quantities $\varphi_{j,k}$ into (21) and inverting the order of summation, we obtain

$$R_{s,k}(z) = \frac{z^s}{2s+1}\Big\{ \sum_{h=-s}^{s} \Phi_{h,k} z^{s+1} \sum_{j=-s}^{-1} \left(\frac{z}{\mu_{h,k}}\right)^{s+j}$$

$$+ \lambda_k \sum_{h=-s}^{s} \Phi_{h,k} \mu_{h,k}^{-s} \sum_{j=0}^{s} \left(\frac{z}{\mu_{h,k}}\right)^{j}\Big\}.$$

Hence (21') follows, when we calculate the sums over j and use the equalities $\mu_{h,k}^{2s+1} = \lambda_k$ $(-s \le h \le s)$. To prove (22'), we insert the representations (20) into (22) and again invert the order of summation. The calculation of the sum over j gives (22').

(d) Remember that formulas 6.3(24), 6.3(25) and the linear transformations 6.3(22), 6.3(23) of Lemma 6.3-4, which are inverse to each other, can be used to establish a one-to-one correspondence between the following three sequences:

$$\{\Phi_n\}_1^\infty, \ \{\Phi_{h,k}\}_1^\infty \ (-s \le h \le s) \text{ and } \{\varphi_{j,k}\}_1^\infty \ (-s \le j \le s). \qquad (23)$$

Besides, the two-sided inequalities

$$\sum_{n=1}^{\infty} |\Phi_n|^p \, (1+n)^{\kappa-s} \asymp \sum_{h=-s}^{s} \sum_{k=1}^{\infty} |\Phi_{h,k}|^p \, (1+k)^{\kappa-s}$$

$$\asymp \sum_{j=-s}^{s} \sum_{k=1}^{\infty} |\varphi_{j,k}|^p \, (1+k)^{\kappa_j} \tag{24}$$

were established in Lemma 6.3-5. Now we add to a given sequence $\{\varphi_{j,k}\}_1^{\infty}$ $(-s \leq j \leq s)$ an arbitrary set of numbers $\{\varphi_{j,0}\}$ $(-s \leq j \leq s)$. As a result, we obtain a sequence of form $\{\varphi_{j,k}\}_0^{\infty}$ $(-s \leq j \leq s)$, and it is obvious that the norms

$$\|\{\varphi_{j,k}\}_1^{\infty}\|_{p,\kappa_j} = \left\{ \sum_{k=1}^{\infty} |\varphi_{j,k}|^p \, (1+k)^{\kappa_j} \right\}^{1/p}$$
$$\|\{\varphi_{j,k}\}_0^{\infty}\|_{p,\kappa_j} = \left\{ \sum_{k=0}^{\infty} |\varphi_{j,k}|^p \, (1+k)^{\kappa_j} \right\}^{1/p} \quad (-s \leq j \leq s) \tag{25}$$

can be finite only simultaneously. Also, we add to the first two sequences of (23), i.e. to $\{\Phi_n\}_1^{\infty} \equiv \bigcup_{h=-s}^{s} \{\Phi_{h,k}\}_1^{\infty}$, an arbitrary set of numbers $\{\Phi_n\}_{-r}^{\circ} (r \geq 0)$ and form the sequence

$$\{\Phi_n\}_{-r}^{\infty} \equiv \{\Phi_n\}_{-r}^{\circ} \bigcup \{\Phi_n\}_1^{\infty} (r \geq 0). \tag{26}$$

Then, obviously, the norms

$$\|\{\Phi_n\}_1^{\infty}\|_{p,\kappa-s} = \left\{ \sum_{n=1}^{\infty} |\Phi_n|^p \, (1+n)^{\kappa-s} \right\}^{1/p},$$

$$\|\{\Phi_n\}_{-r}^{\infty}\|_{p,\kappa-s} = \left\{ \sum_{n=0}^{r} \left| \frac{\Phi_{-n}}{n!} \right|^p + \sum_{n=1}^{\infty} |\Phi_n|^p \, (1+n)^{\kappa-s} \right\}^{1/p}, \tag{27}$$

also can be finite only simultaneously. Finally, we suppose that the additional sets $\{\varphi_{j,0}\}$ $(-s \leq j \leq 0)$ and $\{\Phi_n\}_{-s}^{\circ}$ are connected as follows:

$$\varphi_{j,0} = \Phi_{-(s+j)}/(s+j)! \quad (-s \leq j \leq 0). \tag{28}$$

And we suppose also that these additions are done in accordance with the following two cases:

1°. If $\nu \in \Delta_s(1°)$, then we consider the sequences

$$\{\Phi_n\}_{-(s-1)}^{\infty} \text{ and } \{\varphi_{j,0}\}_{-s}^{-1} \bigcup \left\{ \bigcup_{j=-s}^{s} \{\varphi_{j,k}\}_1^{\infty} \right\}. \tag{29}$$

$2°$. If $\nu \in \Delta_s(2°)$, then we consider the sequences

$$\{\Phi_n\}^\infty_{-s} \text{ and } \{\varphi_{j,0}\}^\circ_{-s} \bigcup \left\{ \bigcup_{j=-s}^{s} \{\varphi_{j,k}\}^\infty_1 \right\}. \tag{30}$$

Of course, we have to suppose that in both these cases formulas (28) are true. Now we are ready to prove the following theorem which is based on the previous lemmas.

Theorem 6.4-3. *Let* $\Phi(z) \in W^{p,\omega}_{s+1/2,\sigma}$ *and let*

$$\Phi_{-n} = \Phi^{(n)}(0) \ (n = 0, 1, 2, \dots), \qquad \Phi_n = \Phi(\mu_n) \ (n = 1, 2, \dots). \tag{31}$$

Then the following assertions are true:
$1°$. If $\nu \in \Delta_s(1°)$, *then*

$$\|\Phi; \Gamma_{2s+1}\|_{p,\omega} \asymp \|\{\Phi_n\}^\infty_{-(s-1)}\|_{p,\kappa_{-s}}. \tag{32_1}$$

$2°$. If $\nu \in \Delta_s(2°)$, *then*

$$\|\Phi; \Gamma_{2s+1}\|_{p,\omega} \asymp \|\{\Phi_n\}^\infty_{-s}\|_{p,\kappa_{-s}}. \tag{32_2}$$

Proof. According to (17) and (18) (where we take $\delta = p$),

$$\int_{\Gamma_{2s+1}} |\Phi(z)|^p |z|^\omega |dz|$$
$$\asymp \sum_{j=-s}^{-1} |\varphi_j(\lambda_0)|^p + \sum_{j=-s}^{s} \sum_{k=1}^{\infty} |\varphi_j(\lambda_k)|^p (1+k)^{\kappa_j} \tag{33_1}$$

in the case when $\nu \in \Delta_s(1°)$, and

$$\int_{\Gamma_{2s+1}} |\Phi(z)|^p |z|^\omega |dz|$$
$$\asymp \sum_{j=-s}^{\circ} |\varphi_j(\lambda_0)|^p + \sum_{j=-s}^{s} \sum_{k=1}^{\infty} |\varphi_j(\lambda_k)|^p (1+k)^{\kappa_j} \tag{33_2}$$

in the case when $\nu \in \Delta_s(2°)$. But, by 6.3(34),

$$\varphi_j(\lambda_0) = \varphi_j(0) = \varphi_{j,0} = \frac{\Phi^{(s+j)}(0)}{(s+j)!} \quad (-s \le j \le s).$$

Therefore, in view of the notations (31), the first sums of the right-hand sides of (33_1) and (33_2) can be written in the form

$$\sum_{j=-s}^{-r} |\varphi_j(\lambda_0)|^p = \sum_{n=0}^{s-r} \left| \frac{\Phi_{-n}}{n!} \right|^p \quad (r = 1, 0). \tag{34_1}$$

Now observe that, by formulas 6.3(19)-(20) and by Lemma 6.3-5, the two-sided inequalities (24) are true particularly for the sequences $\{\Phi(\mu_n)\}_1^\infty$, $\{\Phi(\mu_{h,k})\}_1^\infty$ $(-s \le h \le s)$ and $\{\varphi_j(\lambda_k)\}_1^\infty$ $(-s \le j \le s)$. Therefore

$$\sum_{j=-s}^{s} \sum_{k=1}^{\infty} |\varphi_j(\lambda_k)|^p (1+k)^{\kappa_j} \asymp \sum_{n=1}^{\infty} |\Phi(\mu_n)|^p (1+n)^{\kappa-s}, \tag{34_2}$$

and it remains to see that the two-sided inequalities (32_1) and (32_2) follow from (33_1), (33_2) and (34_1), (34_2).

Remark. The uniqueness Theorem 6.2-3 follows immediately from Theorem 6.4-3, without using expansion Theorems 6.2-1 and 6.2-2.

6.5 Proofs of the main theorems

The aim of this section is to prove the main Theorems 6.2-1 and 6.2-2 relating to expansions of entire functions of spaces $W^{p,\omega}_{s+1/2,\sigma}$ $(s = 1, 2, \dots)$ in terms of interpolation series. To this end it is necessary to prove first two expansion theorems.

(a) Theorem 6.5-1. *If $\nu \in \Delta_s(1^\circ)$, then any function $\Phi(z) \in W^{p,\omega}_{s+1/2,\sigma}$ can be expanded in the interpolation series*

$$\Phi(z) = \Gamma(1+\nu) P_{s-1}(z;\Phi)\mathcal{E}_{s+1/2,\sigma}(z;\nu)$$
$$+ \sum_{n=1}^{\infty} \Phi(\mu_n) \frac{z^s \mathcal{E}_{s+1/2,\sigma}(z;\nu)}{\mu_n^s \mathcal{E}'_{s+1/2,\sigma}(\mu_n;\nu)(z-\mu_n)} \tag{1}$$

which converges to its sum $\Phi(z)$ in the norm $\|.;\Gamma_{2s+1}\|_{p,\omega}$ of the space $W^{p,\omega}_{s+1/2,\sigma}$ and which converges to the same limit uniformly in any disk $|\le z| R < +\infty$. Here

$$P_{s-1}(z;\Phi) = \sum_{k=0}^{s-1} \frac{\Phi^{(k)}(0)}{k!} z^k \tag{2}$$

and

$$\|\Phi;\Gamma_{2s+1}\|_{p,\omega} \asymp \|\{\Phi_n\}_{-(s-1)}^{\infty}\|_{p,\kappa-s}, \tag{3}$$

where

$$\Phi_{-n} = \Phi^{(n)}(0) \quad (0 \le n \le s-1), \quad \Phi_n = \Phi(\mu_n) \quad (1 \le n < +\infty) \tag{4}$$

and

$$\kappa_{-s} = \frac{2(\omega - s) + 1}{2s + 1}. \tag{5}$$

Proof. According to Lemma 6.3-3,

$$\Phi(z) = \sum_{j=-s}^{s} z^{s+j}\varphi_j(z^{2s+1}), \tag{6}$$

where $\varphi_j(w) \in W^{p,\omega_j}_{1/2,\sigma}$ and $\omega_j = [\omega - 2s + p(s+j)](2s+1)^{-1}$. Besides, it is easy to verify that

$$\int_{\Gamma_{2s+1}} \left| z^{s+j}\varphi_j(z^{2s+1}) \right|^p |z|^\omega \, |dz| = \int_0^{+\infty} |\varphi_j(r)|^p \, r^{\omega_j} \, dr < +\infty$$

for any $j(-s \le j \le s)$. Hence

$$z^{s+j}\varphi_j(z^{2s+1}) \in W^{p,\omega}_{s+1/2,\sigma} \quad (-s \le j \le s). \tag{7}$$

Now the use of Theorem 6.4-1 gives the expansions

$$z^{s+j}\varphi_j(z^{2s+1}) = \Gamma(1+\nu)\varphi_j(0)z^{s+j}\mathcal{E}_\sigma(z^{2s+1};\nu)$$
$$+ \sum_{k=1}^\infty \varphi_j(\lambda_k)\frac{z^{3s+j+1}\mathcal{E}_\sigma(z^{2s+1};\nu)}{\lambda_k\mathcal{E}'_\sigma(\lambda_k;\nu)(z^{2s+1}-\lambda_k)} \quad (-s \le j \le -1), \tag{8}$$

$$z^{s+j}\varphi_j(z^{2s+1}) = \sum_{k=1}^\infty \varphi_j(\lambda_k)\frac{z^{s+j}\mathcal{E}_\sigma(z^{2s+1};\nu)}{\mathcal{E}'_\sigma(\lambda_k;\nu)(z^{2s+1}-\lambda_k)} \quad (0 \le j \le s) \tag{9}$$

which converge to their sums in the norm of $W^{p,\omega}_{s+1/2,\sigma}$ and which converge uniformly in any disk $|z| \le R < +\infty$. Substituting these expansions in the right-hand side of (6) and inverting the order of summations over j and k, we obtain

$$\Phi(z) = \Gamma(1+\nu)\Big\{ \sum_{j=-s}^{-1} \varphi_j(0)z^{s+j} \Big\}\mathcal{E}_\sigma(z^{2s+1};\nu)$$
$$+ \sum_{k=1}^\infty \frac{\mathcal{E}_\sigma(z^{2s+1};\nu)\tilde{R}_{s,k}(z)}{\lambda_k\mathcal{E}'_\sigma(\lambda_k;\nu)(z^{2s+1}-\lambda_k)}, \tag{10}$$

where

$$\tilde{R}_{s,k}(z) = \sum_{j=-s}^{-1} z^{3s+j+1}\varphi_j(\lambda_k) + \lambda_k \sum_{j=0}^{s} z^{s+j}\varphi_j(\lambda_k). \tag{11}$$

But, in the considered case, formula 6.4(21)–(21′) gives

$$\tilde{R}_{s,k}(z) = \frac{z^s(z^{2s+1}-\lambda_k)}{2s+1} \sum_{h=-s}^{s} \frac{\Phi(\mu_{h,k})}{\mu_{h,k}^{s-1}(z-\mu_{h,k})}, \tag{11′}$$

since

$$\varphi_j(\lambda_k) = \frac{1}{2s+1} \sum_{h=-s}^{s} \mu_{h,k}^{-(s+j)} \Phi(\mu_{h,k}),$$

according to relations 6.3(19)-(20). Thus, by representations (10) and (11'),

$$\Phi(z) = \Gamma(1+\nu)P_{s-1}(z;\Phi)\mathcal{E}_\sigma(z^{2s+1};\nu)$$
$$+ \sum_{k=1}^{\infty} \frac{z^s \mathcal{E}_\sigma(z^{2s+1};\nu)}{(2s+1)\lambda_k \mathcal{E}_\sigma'(\lambda_k;\nu)} \sum_{h=-s}^{s} \frac{\Phi(\mu_{h,k})}{\mu_{h,k}^{s-1}(z-\mu_{h,k})}. \tag{10'}$$

Now observe that, according to accepted definition,

$$\mathcal{E}_{s+1/2,\sigma}(z;\nu) = \mathcal{E}_\sigma(z^{2s+1};\nu),$$
$$\mathcal{E}_{s+1/2,\sigma}'(z;\nu) = (2s+1)z^{2s}\mathcal{E}_\sigma'(z^{2s+1};\nu), \tag{12}$$

and also remember that

$$\{\mu_n\}_1^\infty = \bigcup_{h=-s}^{s} \{\mu_{h,k}\}_1^\infty \subset \Gamma_{2s+1} \tag{13}$$

is the set of zeros of the function $\mathcal{E}_{s+1/2,\sigma}(z;\nu)$. So, if we take $z = \mu_{h,k} = \alpha_s^h \lambda_k^{1/(2s+1)}$, then the second identity of (12) gives

$$(2s+1)\lambda_k \mathcal{E}_\sigma'(\lambda_k;\nu) = \mu_{h,k}\mathcal{E}_{s+1/2,\sigma}'(\mu_{h,k};\nu) \quad (-s \le h \le s, 1 \le k < +\infty). \tag{14}$$

Hence, if we substitute the quantity $(2s+1)\lambda_k \mathcal{E}_\sigma'(\lambda_k;\nu)$ in the denominator of the inner sum of (10'), then (10') takes the form

$$\Phi(z) = \Gamma(1+\nu)P_{s-1}(z;\Phi)\mathcal{E}_{s+1/2,\sigma}(z;\nu)$$
$$+ \sum_{k=1}^{\infty} z^s \mathcal{E}_{s+1/2,\sigma}(z;\nu) \sum_{h=-s}^{s} \frac{\Phi(\mu_{h,k})}{\mu_{h,k}^s \mathcal{E}_{s+1/2,\sigma}'(\mu_{h,k};\nu)(z-\mu_{h,k})}. \tag{10''}$$

Hence the expansion (1) of the theorem follows, if we take into account definition (13) of the sequence $\{\mu_n\}_1^\infty$ and prove that, for any fixed h $(-s \le h \le s)$, the common term of the last series or, which is the same, the common term of series (10') tends to zero in the norm of $W_{s+1/2,\sigma}^{p,\omega}$ as $k \to +\infty$. To this end we put

$$\omega_r^* = \frac{\omega + p(s-r-1) - 2s}{2s+1}, \quad \kappa_r^* = 1 + 2\omega_r^* \quad (0 \le r \le s-1),$$
$$\omega_r^* = \frac{\omega + p(3s-r) - 2s}{2s+1}, \quad \kappa_r^* = 1 + 2\omega_r^* \quad (s \le r \le 2s)$$

and observe that $-1 < \omega^*_r < p - 1$ for any $r(0 \le r \le 2s)$. Further, we introduce the sequences

$$\{a^{j,h}_{k,r}\}^\infty_{k=1} = \{\mu^{j+r+1}_{h,k}\varphi_j(\lambda_k)\}^\infty_{k=1} \quad (0 \le r \le s - 1, -s \le j, h \le s),$$

$$\{a^{j,h}_{k,r}\}^\infty_{k=1} = \{\lambda^{-1}_k\mu^{j+r+1}_{h,k}\varphi_j(\lambda_k)\}^\infty_{k=1} \quad (s \le r \le 2s, -s \le j, h \le s)$$

and note that it can be easily verified that $\{a^{j,h}_{k,r}\}^\infty_{k=1} \in l^{p,\kappa^*_r}$ $(0 \le r \le 2s, -s \le j, h \le s)$, if the inclusions $\{\varphi_j(\lambda_k)\}^\infty_1 \in l^{p,\kappa_j}$ $(-s \le j \le s)$ (which follow from Theorem 6.4-1) and the equalities 6.2(9) are taken into account. Now observe that 6.3(20) gives

$$\frac{z^s\mathcal{E}_\sigma(z^{2s+1};\nu)\Phi(\mu_{h,k})}{(2s+1)\lambda_k\mathcal{E}'_\sigma(\lambda_k;\nu)\mu^{s-1}_{h,k}(z-\mu_{h,k})}$$

$$= \sum^s_{j=-s} \frac{z^s\mathcal{E}_\sigma(z^{2s+1};\nu)\mu^{j+1}_{h,k}\varphi_j(\lambda_k)}{(2s+1)\lambda_k\mathcal{E}'_\sigma(\lambda_k;\nu)(z-\mu_{h,k})} \quad (-s \le h \le s).$$

Hence it follows that

$$\frac{z^s\mathcal{E}_\sigma(z^{2s+1};\nu)\Phi(\mu_{h,k})}{(2s+1)\lambda_k\mathcal{E}'_\sigma(\lambda_k;\nu)\mu^{s-1}_{h,k}(z-\mu_{h,k})}$$

$$= \sum^s_{j=-s} \frac{z^s\mathcal{E}_\sigma(z^{2s+1};\nu)\mu^{j+1}_{h,k}\varphi_j(\lambda_k)}{(2s+1)\lambda_k\mathcal{E}'_\sigma(\lambda_k;\nu)(z^{2s+1}-\mu^{2s+1}_{h,k})} \sum^{2s}_{r=0}z^{2s-r}\mu^r_{h,k}$$

$$= \sum^{2s}_{r=0}\sum^s_{j=-s} \frac{z^{3s-r}\mathcal{E}_\sigma(z^{2s+1};\nu)\mu^{j+r+1}_{h,k}\varphi_j(\lambda_k)}{(2s+1)\lambda_k\mathcal{E}'_\sigma(\lambda_k;\nu)(z^{2s+1}-\lambda_k)} \quad (-s \le h \le s),$$

since

$$(z - \mu_{h,k})\sum^{2s}_{r=0}z^{2s-r}\mu^r_{h,k} = z^{2s+1} - \mu^{2s+1}_{h,k}$$

and $\mu^{2s+1}_{h,k} = \lambda_k$. Thus it suffices to prove only that

$$\lim_{k\to+\infty}\int_{\Gamma_{2s+1}}\left|\frac{z^{3s-r}\mathcal{E}_\sigma(z^{2s+1};\nu)\mu^{j+r+1}_{h,k}\varphi_j(\lambda_k)}{\lambda_k\mathcal{E}'_\sigma(\lambda_k;\nu)(z^{2s+1}-\lambda_k)}\right|^p |z|^\omega |dz| = 0$$

for any $r(0 \le r \le 2s)$ and any $j, h(-s \le j, h \le s)$. But, since

$$I = \int_{\Gamma_{2s+1}}\left|\frac{z^{3s-r}\mathcal{E}_\sigma(z^{2s+1};\nu)\mu^{j+r+1}_{h,k}\varphi_j(\lambda_k)}{\lambda_k\mathcal{E}'_\sigma(\lambda_k;\nu)(z^{2s+1}-\lambda_k)}\right|^p |z|^\omega |dz|$$

$$= \int^{+\infty}_0\left|a^{j,h}_{k,r}\frac{x\mathcal{E}_\sigma(x;\nu)}{\lambda_k\mathcal{E}'_\sigma(\lambda_k;\nu)(x-\lambda_k)}\right|^p x^{\omega^*_r}dx$$

when $0 \le r \le s - 1$, and

$$I = \int_0^{+\infty} \left| a_{k,r}^{j,h} \frac{\mathcal{E}_\sigma(x;\nu)}{\mathcal{E}_\sigma'(\lambda_k;\nu)(x - \lambda_k)} \right|^p x^{\omega_r^*} dx$$

when $s \le r \le 2s$, the last limit is zero, if the series

$$\Psi_r^{j,h}(z) = \sum_{k=1}^\infty a_{k,r}^{j,h} \frac{z\mathcal{E}_\sigma(z;\nu)}{\lambda_k \mathcal{E}_\sigma'(\lambda_k;\nu)(z - \lambda_k)} \quad (0 \le r \le s-1, -s \le j, h \le s),$$

$$\Psi_r^{j,h}(z) = \sum_{k=1}^\infty a_{k,r}^{j,h} \frac{\mathcal{E}_\sigma(z;\nu)}{\mathcal{E}_\sigma'(\lambda_k;\nu)(z - \lambda_k)} \quad (s \le r \le 2s, -s \le j, h \le s)$$

are convergent in the norm of $W_{1/2,\sigma}^{p,\omega_r^*}$. The last fact follows from Theorems 4.3-1 and 4.3-2. To this end, it is necessary to take into account the inclusions

$$\{a_{k,r}^{j,h}\}_{k=1}^\infty \in l^{p,\kappa_r^*} \quad (0 \le r \le 2s, -s \le j, h \le s)$$

and to observe that $\Delta_s(1^\circ) \subset \Delta^*(\kappa_r^*, p)(0 \le r \le s - 1)$ and $\Delta_s(1^\circ) \subset \Delta(\kappa_r^*, p)$ $(s \le r \le 2s)$, as is easy to verify. So the validity of the expansion (1) is proved, and, since the two-sided inequalities (3) were proved earlier in Theorem $6.4-3(1^\circ)$, the proof of Theorem 6.5-1 is complete.

(b) We shall call expansion (1) of preceding theorem *a first type expansion*. Similarly, we shall call the expansion of the following theorem *a second type expansion* of a function of $W_{s+1/2,\sigma}^{p,\omega}$.

Theorem 6.5-2. *If* $\nu \in \Delta_s(2^\circ)$, *then any function* $\Phi(z) \in W_{s+1/2,\sigma}^{p,\omega}$ *can be expanded in the interpolation series*

$$\Phi(z) = \Gamma(1 + \nu) P_s(z; \Phi) \mathcal{E}_{s+1/2,\sigma}(z;\nu)$$
$$+ \sum_{n=1}^\infty \Phi(\mu_n) \frac{z^{s+1} \mathcal{E}_{s+1/2,\sigma}(z;\nu)}{\mu_n^{s+1} \mathcal{E}_{s+1/2,\sigma}'(\mu_n;\nu)(z - \mu_n)} \tag{15}$$

which converges to its sum in the norm $\|.; \Gamma_{2s+1}\|_{p,\omega}$ *of the space* $W_{s+1/2,\sigma}^{p,\omega}$ *and also converges to the same limit uniformly in any disk* $|z| \le R < +\infty$. *Here*

$$P_s(z; \Phi) = \sum_{k=0}^s \frac{\Phi^{(k)}(0)}{k!} z^k \tag{16}$$

and

$$\|\Phi; \Gamma_{2s+1}\|_{p,\omega} \asymp \|\{\Phi_n\}_{-s}^\infty\|_{p,\kappa_{-s}}, \tag{17}$$

where

$$\Phi_{-n} = \Phi^{(n)}(0) \quad (0 \le n \le s), \qquad \Phi_n = \Phi(\mu_n) \quad (1 \le n < +\infty). \tag{18}$$

Proof. The arguments we use are similar to the ones used to prove Theorem 6.5-1, so, we shall note only the differences. If $\nu \in \Delta_s(2^\circ)$, then, according to Theorem 6.4-2, expansions (8) and (9), which have the same character of convergence, remain true when $-s \leq j \leq 0$ and $1 \leq j \leq s$ respectively. Therefore, substituting them into representation (6) of $\Phi(z)$ and inverting the order of summations over j and k we obtain

$$
\begin{aligned}
\Phi(z) = \Gamma(1+\nu)\{ \sum_{j=-s}^{\overset{o}{}} \varphi_j(0)z^{s+j} \}\mathcal{E}_\sigma(z^{2s+1};\nu) \\
+ \sum_{k=1}^{\infty} \frac{\mathcal{E}_\sigma(z^{2s+1};\nu)\tilde{R}^*_{s,k}(z)}{\lambda_k \mathcal{E}'_\sigma(\lambda_k;\nu)(z^{2s+1}-\lambda_k)},
\end{aligned}
\tag{19}
$$

where

$$
\tilde{R}^*_{s,k}(z) = \sum_{j=-s}^{\overset{o}{}} z^{3s+j+1}\varphi_j(\lambda_k) + \lambda_k \sum_{j=1}^{s} z^{s+j}\varphi_j(\lambda_k).
$$

Further, by formulas 6.4(22) $-$ (22$'$),

$$
\tilde{R}^*_{s,k}(z) = \frac{z^{s+1}(z^{2s+1}-\lambda_k)}{2s+1} \sum_{h=-s}^{s} \frac{\Phi(\mu_{h,k})}{\mu^s_{h,k}(z-\mu_{h,k})}.
$$

Hence, by formula (19),

$$
\begin{aligned}
\Phi(z) = \Gamma(1+\nu)P_s(z;\Phi)\mathcal{E}_\sigma(z^{2s+1};\nu) \\
+ \sum_{k=1}^{\infty} \frac{z^{s+1}\mathcal{E}_\sigma(z^{2s+1};\nu)}{(2s+1)\lambda_k \mathcal{E}'_\sigma(\lambda_k;\nu)} \sum_{h=-s}^{s} \frac{\Phi(\mu_{h,k})}{\mu^s_{h,k}(z-\mu_{h,k})}.
\end{aligned}
\tag{19$'$}
$$

A literal repetition of the remaining arguments of the proof of Theorem 6.5-1 proves all the desired assertions.

(c) Finally, we can pass to the proofs of the main interpolation theorems of this chapter which were formulated in Section 6.2.

Proof of Theorem 6.2-1. According to Theorem 6.5-1, any function $\Phi(z) \in W^{p,\omega}_{s+1/2,\sigma}$ is representable by a series of the form 6.2(17) and the assertions 1°, 2°, 3° of Theorem 6.2-1 are true. Therefore, it only remains to prove the following: if $\{\Phi_n\}^{\infty}_{-(s-1)} \in L^{(s-1)}_{p,\kappa_{-s}}$ is an arbitrary sequence of numbers, then there exists an entire function $\Phi(z) \in W^{p,\omega}_{s+1/2,\sigma}$, such that

$$
\Phi^n(0) = \Phi_{-n} \quad (0 \leq n \leq s-1), \qquad \Phi(\mu_n) = \Phi_n \quad (1 \leq n < +\infty).
\tag{20}
$$

To this end, suppose $\{\Phi_n\}^{\infty}_{-(s-1)} \equiv \{\Phi_n\}^{\overset{o}{}}_{-(s-1)} \bigcup \{\Phi_n\}^{\infty}_1 \in L^{(s-1)}_{p,\kappa_{-s}}$ is an arbitrary sequence. Then $\{\Phi_n\}^{\infty}_1 \equiv \bigcup^{s}_{h=-s}\{\Phi_{h,k}\}^{\infty}_1$ according to 6.3(24)-(25). Further, using formulas 6.3(22) and 6.3(23), we put $\{\Phi_{h,k}\}_1^\infty$ $(-s \leq h \leq s)$ into correspondence with a new set of sequences $\{\varphi_{j,k}\}_1^\infty$ $(-s \leq j \leq s)$. Then, by 6.4(24),

$$\sum_{j=-s}^{s}\sum_{k=1}^{\infty}|\varphi_{j,k}|^p (1+k)^{\kappa_j} \asymp \sum_{n=1}^{\infty}|\Phi_n|^p (1+n)^{\kappa-s}. \tag{21}$$

Now we add to $\{\varphi_{j,k}\}_1^\infty$ $(-s \leq j \leq s)$ the set of numbers

$$\varphi_{j,0} = \frac{\Phi_{-(s+j)}}{(s+j)!} \quad (-s \leq j \leq -1). \tag{22}$$

So, we put $\{\Phi_n\}_{-(s-1)}^\infty$ into correspondence with the set of numbers

$$\{\varphi_{j,0}\}_{-s}^{-1}\bigcup\{\bigcup_{j=-s}^{s} \{\varphi_{j,k}\}_1^\infty\},$$

as it was done in 6.4(29). Then, by (21),

$$\sum_{j=-s}^{-1} \|\{\varphi_{j,k}\}_0^\infty\|_{p,\kappa_j} < +\infty \text{ and } \sum_{j=0}^{s} \|\{\varphi_{j,k}\}_1^\infty\|_{p,\kappa_j} < +\infty$$

simultaneously, since $\{\Phi_n\}_{-(s-1)}^\infty \in L_{p,\kappa_{-s}}^{(s-1)}$. Using Theorem 6.4-1 we construct the entire functions $\varphi_j(w) \in W_{1/2,\sigma}^{p,\omega_j}$ $(-s \leq j \leq s)$ by means of series 6.4(12)-(13). Then, obviously, $z^{s+j}\varphi_j(z^{2s+1})$ $(-s \leq j \leq s)$ are entire functions of class $W_{s+1/2,\sigma}^{p,\omega}$. Hence

$$\Phi(z) \equiv \sum_{j=-s}^{s} z^{s+j}\varphi_j(z^{2s+1}) \in W_{s+1/2,\sigma}^{p,\omega}. \tag{23}$$

We have put into correspondence any sequence $\{\Phi_n\}_{-(s-1)}^\infty$ of $L_{p,\kappa_{-s}}^{(s-1)}$ with a definite function $\Phi(z) \in W_{s+1/2,\sigma}^{p,\omega}$. Thus, it remains only to verify the validity of relations (20). To this end the suitable steps of the proof of Theorem 6.5-1 must be repeated. First we obtain that the constructed function $\Phi(z)$ is representable in the form (10)-(11), if only the quantities $\{\varphi_j(\lambda_k)\}_1^\infty$ $(-s \leq j \leq s)$ and $\varphi_j(0)$ $(-s \leq j \leq -1)$ of expansions (10)-(11) are replaced correspondingly by $\{\varphi_{j,k}\}_1^\infty$ $(-s \leq j \leq s)$ and $\varphi_{j,0}$ $(-s \leq j \leq -1)$. Next, we use formulas 6.4(21) $-$ (21′). Then the suitable simplifications lead to the expansion $(10'')$ of the function $\Phi(z)$, where $\Phi(\mu_{h,k})$ is replaced by $\Phi_{h,k}$ $(1 \leq k < +\infty, -s \leq h \leq s)$ and $\Phi^{(k)}(0)$ - by Φ_{-k} $(0 \leq k \leq s-1)$, i.e.

$$\Phi(z) = \Gamma(1+\nu)(\sum_{k=0}^{s-1} \frac{\Phi_{-k}}{k!} z^k)\mathcal{E}_{s+1/2,\sigma}(z;\nu)$$

$$+ \sum_{k=1}^{\infty} z^s \mathcal{E}_{s+1/2,\sigma}(z;\nu) \sum_{h=-s}^{s} \frac{\Phi_{h,k}}{\mu_{h,k}^s \mathcal{E}'_{s+1/2,\sigma}(\mu_{h,k};\nu)(z-\mu_{h,k})},$$

where $\{\mu_n\}_1^\infty = \bigcup_{h=-s}^s \{\mu_{h,k}\}_1^\infty$ are zeros of the function $\mathcal{E}_{s+1/2,\sigma}(z;\nu)$. Hence the relations (20) can be easily deduced. Indeed, taking $z = \mu_n$ $(1 \le n < +\infty)$, we immediately obtain $\Phi(\mu_n) = \Phi_n$ $(1 \le n < +\infty)$, and since

$$\mathcal{E}_{s+1/2,\sigma}(0;\nu) = \frac{1}{\Gamma(1+\nu)}, \quad \frac{d^h}{dz^h}\left\{\sum_{k=0}^{s-1} \frac{\Phi_{-k}}{k!} z^k\right\}\Bigg|_{z=0} = \Phi_{-h} (0 \le h \le s-1)$$

and also

$$\frac{d^h}{dz^h}\left\{\frac{z^s \mathcal{E}_{s+1/2,\sigma}(z;\nu)}{z-\mu_n}\right\}\Bigg|_{z=0} = 0 \quad (0 \le h \le s-1, 1 \le n < +\infty),$$

we arrive at the equalities $\Phi^h(0) = \Phi_{-h}$ $(0 \le h \le s-1)$.

Proof of Theorem 6.2-2. As was stated in Theorem 6.5-2, any function $\Phi(z) \in W^{p,\omega}_{s+1/2,\sigma}$ is representable by a series 6.2(20) and assertions $1°$, $2°$, $3°$ are true. The rest of the proof is similar to the proof of Theorem 6.2-1. Namely, we put into correspondence a sequence $\{\Phi_n\}_{-s}^\infty \in L^{(s)}_{p,\kappa_{-s}}$ with a sequence

$$\{\varphi_{j,0}\}_{-s}^0 \bigcup \left\{ \bigcup_{j=-s}^s \{\varphi_{j,k}\}_1^\infty \right\},$$

as was done in 6.4(30). Then

$$\sum_{j=-s}^0 \|\{\varphi_{j,k}\}_0^\infty\|_{p,\kappa_j} < +\infty \text{ and } \sum_{j=1}^s \|\{\varphi_{j,k}\}_1^\infty\|_{p,\kappa_j} < +\infty.$$

Further, by use of Theorem 6.4-2, we obtain the functions $\varphi_j(w) \in W^{p,\omega_j}_{1/2,\sigma}$ $(-s \le j \le s)$. It is obvious that the functions $\{z^{s+j}\varphi_j(z^{2s+1})\}_{-s}^s$ and their sum $\Phi(z)$ are of class $W^{p,\omega}_{s+1/2,\sigma}$. To complete the proof, we have to repeat the final arguments of the proof of Theorem 6.2-1 and obtain the equalities $\Phi^n(0) = \Phi_{-n}(0 \le n \le s)$ and $\Phi(\mu_n) = \Phi_n(1 \le n < +\infty)$.

6.6 Notes

The results of this chapter were obtained by M.M. Djrbashian-S.G. Raphaelian [4, §1–3] in the particular case $s = 1$, when the considered problems and their solutions are essentially simpler.

7 Basic Fourier type systems in L_2 spaces of odd-dimensional vector functions

7.1 Introduction

In this chapter we pass from the interpolation theorems of Chapter 6 to theorems on the basis property of several Fourier type biorthogonal systems of odd-dimensional vector functions. The first step to this is the construction of the mentioned systems in explicit form. This is achieved by generalization of the methods developed in Chapter 5. The second step is the proof of the completeness and of the basis property in the Riesz sense of the constructed systems of $2s + 1$-dimensional vector functions in the space L_2 of vector functions. This is established by use of the case $p = 2$ of interpolation Theorems 6.2-1 and 6.2-2 relating to classes $W^{2,\omega}_{s+1/2,\sigma}$ of entire functions and also by essential use of Theorem 2.4-1 on parametric representations of these classes. As will be shown in Chapter 11, some simple reformulations of these results lead to an explicit and completed apparatus of Fourier type systems of entire functions. These systems are the bases of the space L_2 considered over $2s + 1 (s \geq 1)$ segments of equal length with a common endpoint at the origin and forming equal angles of opening $2\pi/(2s + 1)$ in the complex plane. In this chapter the main notations of Section 6.2 of Chapter 6 are frequently used without any special indication.

7.2 Some identities

(a) Note that, if $\rho = s + 1/2$, where $s \geq 1$ is any integer, then the first of the identities 1.2(10) becomes

$$\mathcal{Y}_s(z; \lambda; \eta_1; \eta_2)$$

$$\equiv \int_0^\sigma E_{s+1/2}\left(z\tau^{\frac{2}{2s+1}}; \eta_1\right) \tau^{\eta_1-1} E_{s+1/2}\left(\lambda(\sigma - \tau)^{\frac{2}{2s+1}}; \eta_2\right)(\sigma - \tau)^{\eta_2-1} d\tau \tag{1}$$

$$= \frac{E_{s+1/2}\left(\sigma^{\frac{2}{2s+1}} z; \eta_3\right) - E_{s+1/2}\left(\sigma^{\frac{2}{2s+1}} \lambda; \eta_3\right)}{z - \lambda} \sigma^{\eta_4} \quad (\eta_1 \geq 0, \ \eta_2 \geq 0),$$

where

$$\eta_3 = \eta_1 + \eta_2 - \frac{2}{2s + 1}, \quad \eta_4 = \eta_3 - 1. \tag{2}$$

Next note that, according to formula 6.3(8), the following identities are true for all $j(-s \leq j \leq s)$:

$$z^{s+j} E_{1/2}\left(z^{2s+1}; \mu + \frac{2(s + j)}{2s + 1}\right) = \frac{1}{2s + 1} \sum_{h=-s}^{s} \alpha_s^{-(s+j)h} E_{s+1/2}(\alpha_s^h z; \mu), \tag{3}$$

where

$$\alpha_s = \exp\left\{i\frac{2\pi}{2s + 1}\right\}. \tag{4}$$

Suppose $j(-s \leq j \leq s)$ is any fixed integer and put in (1)

$$\eta_1 = \eta_{1,j} = \mu, \qquad \eta_2 = \eta_{2,j} = \frac{3 - 2j}{2s + 1} + \nu - \mu \tag{5_1}$$

noting that in this case

$$\eta_3 = \eta_{3,j} = \nu + \frac{1 - 2j}{2s + 1}, \qquad \eta_4 = \eta_{4,j} = \eta_{3,j} - 1 = \nu - \frac{2(s + j)}{2s + 1}. \tag{5_2}$$

Then, replacing z and λ respectively by $\alpha_s^{h+1/2} z$ and $\alpha_s^{h+1/2} \lambda$, we arrive at the identities

$$\begin{aligned}
U_{s,h,j}(z; \lambda) &= \mathcal{Y}_s \left(\alpha_s^{h+1/2} z; \alpha_s^{h+1/2} \lambda; \mu; \eta_{2,j} \right) \\
&= \int_0^\sigma E_{s+1/2} \left(\alpha_s^{h+1/2} z \tau^{\frac{2}{2s+1}}; \mu \right) \tau^{\mu-1} \\
&\quad \times E_{s+1/2} \left(\alpha_s^{h+1/2} \lambda (\sigma - \tau)^{\frac{2}{2s+1}}; \eta_{2,j} \right) (\sigma - \tau)^{\eta_{2,j}-1} d\tau \\
&= \frac{E_{s+1/2} \left(\sigma^{\frac{2}{2s+1}} \alpha_s^{h+1/2} z; \eta_{3,j} \right) - E_{s+1/2} \left(\sigma^{\frac{2}{2s+1}} \alpha_s^{h+1/2} \lambda; \eta_{3,j} \right)}{z - \lambda} \\
&\quad \times \alpha_s^{-(h+1/2)} \sigma^{\eta_{4,j}} \qquad (-s \leq j \leq s).
\end{aligned} \tag{6}$$

Further, replacing z by $\alpha_s^{1/2} \sigma^{2/(2s+1)} z$ and μ – by $\eta_{3,j}$ in (3), for any $j(-s \leq j \leq s)$ we obtain

$$\begin{aligned}
&z^{s+j} E_{1/2} \left(-\sigma^2 z^{2s+1}; 1 + \nu \right) \\
&= \frac{\sigma^{-\frac{2(s+j)}{2s+1}}}{2s + 1} \sum_{h=-s}^{s} \alpha_s^{-(h+1/2)(s+j)} E_{s+1/2} \left(\sigma^{\frac{2}{2s+1}} \alpha_s^{h+1/2} z; \nu + \frac{1 - 2j}{2s + 1} \right).
\end{aligned} \tag{7}$$

(b) Further, we shall use only the cases $j = 0$ and $j = 1$ of the identities (6) and (7) assuming that the suitable parameters $\eta_{i,j}$ are defined by $(5_1) - (5_2)$. In the mentioned cases these identities pass to the following ones:

1°. If $j = 0$, then (6) becomes

$$\begin{aligned}
U_{s,h}^{(1)}(z; \lambda) &\equiv U_{s,h,0}(z; \lambda) = \int_0^\sigma E_{s+1/2} \left(\alpha_s^{h+1/2} z \tau^{\frac{2}{2s+1}}; \mu \right) \tau^{\mu-1} \\
&\quad \times E_{s+1/2} \left(\alpha_s^{h+1/2} \lambda (\sigma - \tau)^{\frac{2}{2s+1}}; \eta_{2,0} \right) (\sigma - \tau)^{\eta_{2,0}-1} d\tau \\
&= \frac{E_{s+1/2} \left(\sigma^{\frac{2}{2s+1}} \alpha_s^{h+1/2} z; \eta_{3,0} \right) - E_{s+1/2} \left(\sigma^{\frac{2}{2s+1}} \alpha_s^{h+1/2} \lambda; \eta_{3,0} \right)}{z - \lambda} \\
&\quad \times \alpha_s^{-(h+1/2)} \sigma^{\eta_{4,0}},
\end{aligned} \tag{6_1}$$

where

$$\eta_{2,0} = \frac{3}{2s+1} + \nu - \mu, \quad \eta_{3,0} = \nu + \frac{1}{2s+1}, \quad \eta_{4,0} = \nu - \frac{2s}{2s+1}, \tag{8_1}$$

and (7) becomes

$$(1 + 2s)z^s E_{1/2}\left(-\sigma^2 z^{2s+1}; 1 + \nu\right)$$
$$= \sigma^{-\frac{2s}{2s+1}} \sum_{h=-s}^{s} \alpha_s^{-(h+1/2)s} E_{s+1/2}\left(\sigma^{\frac{2}{2s+1}} \alpha_s^{h+1/2} z; \nu + \frac{1}{2s+1}\right). \tag{7_1}$$

$2°$. If $j = 1$, then (6) becomes

$$U_{s,h}^{(2)}(z; \lambda) \equiv U_{s,h,1}(z; \lambda) = \int_0^\sigma E_{s+1/2}\left(\alpha_s^{h+1/2} z \tau^{\frac{2}{2s+1}}; \mu\right) \tau^{\mu-1}$$
$$\times E_{s+1/2}\left(\alpha_s^{h+1/2}\lambda(\sigma - \tau)^{\frac{2}{2s+1}}; \eta_{2,1}\right)(\sigma - \tau)^{\eta_{2,1}-1} d\tau$$
$$= \frac{E_{s+1/2}\left(\sigma^{\frac{2}{2s+1}}\alpha_s^{h+1/2} z; \eta_{3,1}\right) - E_{s+1/2}\left(\sigma^{\frac{2}{2s+1}}\alpha_s^{h+1/2}\lambda; \eta_{3,1}\right)}{z - \lambda} \tag{6_2}$$
$$\times \alpha_s^{-(h+1/2)}\sigma^{\eta_{4,1}},$$

where

$$\eta_{2,1} = \frac{1}{2s+1} + \nu - \mu, \quad \eta_{3,1} = \nu - \frac{1}{2s+1}, \quad \eta_{4,1} = \nu - \frac{2(s+1)}{2s+1}, \tag{8_2}$$

and (7) becomes

$$(1 + 2s)z^{s+1} E_{1/2}\left(-\sigma^2 z^{2s+1}; 1 + \nu\right)$$
$$= \sigma^{-\frac{2(s+1)}{2s+1}} \sum_{h=-s}^{s} \alpha_s^{-(h+1/2)(s+1)} E_{s+1/2}\left(\sigma^{\frac{2}{2s+1}} \alpha_s^{h+1/2} z; \nu - \frac{1}{2s+1}\right). \tag{7_2}$$

(c) It is easy to see that the identities (6_1) and (6_2) are true for any $\mu \geq 0$ and $\eta_{2,0} \geq 0, \eta_{2,1} \geq 0$. But it is necessary to obtain explicitly what conditions μ and ν must satisfy in order to apply these identities correctly when proving the main results of this chapter. To this end, first note that two intervals in which the parameter $\nu \in [0, 2)$ varies were defined by means of formulas 6.2(13)-(15). These intervals can be written down as follows:

$$\Delta_s(1°) = \left(\frac{2s + \omega}{2s + 1}, \frac{2s + \omega + 1}{2s + 1}\right), \quad \Delta_s(2°) = \left(\frac{2s + \omega + 1}{2s + 1}, \frac{2s + \omega + 2}{2s + 1}\right) \tag{9}$$

in the case when $p = 2$ and $-1 < \omega < 1$. Next note that we shall suppose everywhere

$$\mu = \frac{3/2 + s + \omega}{2s + 1} \qquad (s \geq 1, -1 < \omega < 1) \tag{10}$$

and, consequently, we shall have

$$\frac{1}{2} < \mu < \frac{1}{2} + \frac{1}{s + 1/2}. \tag{11}$$

Therefore, the multiplier $\tau^{\mu-1}$ in the integrals $U_{s,h}^{(1)}(z; \lambda)$ and $U_{s,h}^{(2)}(z; \lambda)$ is of class $L_2(0, \sigma)$ for any $\sigma \in (0, +\infty)$. And, taking into account the values $(8_1), (8_2)$ and (10) of parameters $\eta_{2,0}, \eta_{2,1}$ and μ, we conclude that

$1°$. if $\nu \in \Delta_s(1°)$, then

$$\eta_{2,0} = \frac{3/2 - s - \omega}{2s + 1} + \nu > \frac{3/2 + s}{2s + 1} > \frac{1}{2}$$

$2°$. if $\nu \in \Delta_s(2°)$, then

$$\eta_{2,1} = \frac{-1/2 - s - \omega}{2s + 1} + \nu > \frac{1/2 + s}{2s + 1} = \frac{1}{2}.$$

Therefore, for any $\sigma > 0$

$$(\sigma - \tau)^{\eta_{2,0}-1} \in L_2(0, \sigma), \qquad \nu \in \Delta_s(1°), \tag{12_1}$$

$$(\sigma - \tau)^{\eta_{2,1}-1} \in L_2(0, \sigma), \qquad \nu \in \Delta_s(2°) \tag{12_2}$$

in the integrals $U_{s,h}^{(1)}(z; \lambda)$ and $U_{s,h}^{(2)}(z; \lambda)$.

According to the previous arguments, in the considered cases $\nu \in \Delta_s(1°)$ and $\nu \in \Delta_s(2°)$, (6_1) and (6_2) are both integrals of products of two functions of $L_2(0, \sigma)$ which are continuous at least in $(0, \sigma)$. Besides, the singularities of these functions are at different points — $\tau = 0$ and $\tau = \sigma$. Consequently, the functions integrated in (6_1) and (6_2) are both of $L_2(0, \sigma) \subset L_1(0, \sigma)$.

(d) For brevity we introduce the following notations:

$1°$.
$$U_s^{(1)}(z; \lambda) \equiv \sum_{h=-s}^{s} \alpha_s^{-(h+1/2)(s-1)} U_{s,h}^{(1)}(z; \lambda)$$

$$= \sum_{h=-s}^{s} \alpha_s^{-(h+1/2)(s-1)} \int_0^{\sigma} E_{s+1/2}\left(\alpha_s^{h+1/2} z \tau^{\frac{2}{2s+1}}; \mu\right) \tau^{\mu-1}$$

$$\times E_{s+1/2}\left(\alpha_s^{h+1/2} \lambda (\sigma - \tau)^{\frac{2}{2s+1}}; \eta_{2,0}\right) (\sigma - \tau)^{\eta_{2,0}-1} d\tau, \qquad \nu \in \Delta_s(1°), \tag{13_1}$$

$$\eta_{2,0} = \nu + \frac{3/2 - s - \omega}{2s + 1},$$

$2°.$ $U_s^{(2)}(z;\lambda) \equiv \sum_{h=-s}^{s} \alpha_s^{-(h+1/2)s} U_{s,h}^{(2)}(z;\lambda)$

$$= \sum_{h=-s}^{s} \alpha_s^{-(h+1/2)s} \int_0^{\sigma} E_{s+1/2}\left(\alpha_s^{h+1/2} z\tau^{\frac{2}{2s+1}};\mu\right) \tau^{\mu-1}$$

$$\times E_{s+1/2}\left(\alpha_s^{h+1/2}\lambda(\sigma-\tau)^{\frac{2}{2s+1}};\eta_{2,1}\right)(\sigma-\tau)^{\eta_{2,1}-1}d\tau, \qquad \nu \in \Delta_s(2°), \tag{13_2}$$

$$\eta_{2,1} = \nu - \frac{1/2 + s + \omega}{2s + 1}.$$

Now we prove the following lemma, where all previous assertions concerning the functions integrated in (13_1) and (13_2) obviously remain true.

Lemma 7.2-1. *If $z, \lambda \in \mathbb{C}$ are arbitrary numbers, then the following identities are true:*
$1°.$ *When $\nu \in \Delta_s(1°)$*

$$U_s^{(1)}(z;\lambda) = \frac{2s+1}{z-\lambda}\sigma^{\nu}\Big\{z^s E_{1/2}\left(-\sigma^2 z^{2s+1};1+\nu\right)$$
$$- \lambda^s E_{1/2}\left(-\sigma^2\lambda^{2s+1};1+\nu\right)\Big\}. \tag{14_1}$$

$2°.$ *When $\nu \in \Delta_s(2°)$*

$$U_s^{(2)}(z;\lambda) = \frac{2s+1}{z-\lambda}\sigma^{\nu}\Big\{z^{s+1} E_{1/2}\left(-\sigma^2 z^{2s+1};1+\nu\right)$$
$$- \lambda^{s+1} E_{1/2}\left(-\sigma^2\lambda^{2s+1};1+\nu\right)\Big\}. \tag{14_2}$$

Before proving these identities, observe that the notation

$$\mathcal{E}_{s+1/2,\sigma}(z;\nu) = E_{1/2}\left(-\sigma^2 z^{2s+1};1+\nu\right) \tag{15}$$

may be used in both of them.

Proof. $1°.$ We multiply identity (6_1) by $\alpha_s^{(h+1/2)(1-s)}$ and sum up both sides over $h(-s \leq h \leq s)$. Then formula (7_1) leads to identity (14_1).
$2°.$ Similarly, if we multiply both sides of the identity (6_2) by $\alpha_s^{-(h+1/2)s}$ and sum over $h(-s \leq h \leq s)$, then, using the formula (7_2), we obtain (14_2).

(e) Remember finally some facts relating to zeros of the function $\mathcal{E}_{s+1/2,\sigma}(z;\nu)$, $\nu \in [0,2)$. As was mentioned in Section 6.2, all zeros $\{\mu_n\}_1^\infty$ of this function are simple and are situated on the set of rays

$$\Gamma_{2s+1} \equiv \bigcup_{h=-s}^{s} \{z = r\alpha_s^h : 0 \leq r < +\infty\}. \tag{16}$$

A universal numeration of these zeros was introduced there by the equalities

$$\{\mu_{h,k}\}_1^\infty = \left\{\alpha_s^h \lambda_k^{\frac{1}{2s+1}}\right\}_1^\infty \qquad (-s \leq h \leq s),$$
$$\mu_{(2s+1)k+h-s} = \mu_{h,k} \qquad (-s \leq h \leq s,\ 1 \leq k < +\infty), \tag{17}$$

where $\{\lambda_k\}_1^\infty$ $(0 < \lambda_k < \lambda_{k+1}, k \geq 1)$ is the set of zeros of the function

$$\mathcal{E}_\sigma(z;\nu) = E_{1/2}\left(-\sigma^2 z; 1+\nu\right), \qquad \nu \in [0,2), \tag{18}$$

which are also simple.

7.3 Biorthogonal systems of odd-dimensional vector functions.

7.3 Biorthogonal systems of odd-dimensional vector functions.7.3 Biorthogonal systems**(a)** First we introduce two sequences of functions on $(0,\sigma)$. For any $\nu \in \Delta_s(1^\circ)$ we denote

$$\Omega_{m,h}^{(1)}(\tau;s) \equiv \frac{\alpha_s^{-(h+1/2)(s-1)}\sigma^{-\nu}}{(2s+1)\mu_m^s \mathcal{E}'_{s+1/2,\sigma}(\mu_m;\nu)}$$
$$\times E_{s+1/2}\left(\alpha_s^{h+1/2}\mu_m(\sigma-\tau)^{\frac{2}{2s+1}};\eta_{2,0}\right)(\sigma-\tau)^{\eta_{2,0}-1} \quad \begin{array}{c}(1 \leq m < +\infty,\\ -s \leq h \leq s)\end{array} \tag{1_1}$$

where

$$\eta_{2,0} = \nu + \frac{3/2 - s - \omega}{2s+1}. \tag{2_1}$$

For any $\nu \in \Delta_s(2^\circ)$ we denote

$$\Omega_{m,h}^{(2)}(\tau;s) \equiv \frac{\alpha_s^{-(h+1/2)s}\sigma^{-\nu}}{(2s+1)\mu_m^{s+1}\mathcal{E}'_{s+1/2,\sigma}(\mu_m;\nu)}$$
$$\times E_{s+1/2}\left(\alpha_s^{h+1/2}\mu_m(\sigma-\tau)^{\frac{2}{2s+1}};\eta_{2,1}\right)(\sigma-\tau)^{\eta_{2,1}-1} \quad \begin{array}{c}(1 \leq m < +\infty,\\ -s \leq h \leq s),\end{array} \tag{1_2}$$

where

$$\eta_{2,1} = \nu - \frac{1/2 + s + \omega}{2s+1}. \tag{2_2}$$

Recalling that

$$\mu = \frac{3/2 + s + \omega}{2s+1} \qquad (-1 < \omega < 1), \tag{3}$$

we shall now prove two lemmas.

Lemma 7.3-1. *The following identities are true in the whole z-plane:*
if $\nu \in \Delta_s(1^\circ)$, *then*

$$\sum_{h=-s}^{s} \int_0^\sigma E_{s+1/2}\left(\alpha_s^{h+1/2} z \tau^{\frac{2}{2s+1}};\mu\right) \tau^{\mu-1}\Omega_{m,h}^{(1)}(\tau;s)d\tau$$

$$= \frac{z^s \mathcal{E}_{s+1/2,\sigma}(z;\nu)}{\mu_m^s \mathcal{E}'_{s+1/2,\sigma}(\mu_m;\nu)(z-\mu_m)} \qquad (1 \le m < +\infty); \tag{4_1}$$

if $\nu \in \Delta_s(2^\circ)$, *then*

$$\sum_{h=-s}^{s} \int_0^\sigma E_{s+1/2}\left(\alpha_s^{h+1/2} z \tau^{\frac{2}{2s+1}};\mu\right) \tau^{\mu-1}\Omega_{m,h}^{(2)}(\tau;s)d\tau$$

$$= \frac{z^{s+1} \mathcal{E}_{s+1/2,\sigma}(z;\nu)}{\mu_m^{s+1} \mathcal{E}'_{s+1/2,\sigma}(\mu_m;\nu)(z-\mu_m)} \qquad (1 \le m < +\infty). \tag{4_2}$$

Proof. If $\nu \in \Delta_s(1^\circ)$, then using the representations $7.2(13_1)$ and $7.2(14_1)$ of the function $U_s^{(1)}(z;\mu_m)$ $(1 \le m < +\infty)$, we arrive at the identities

$$\sum_{h=-s}^{s} \alpha_s^{-(h+1/2)(s-1)} \int_0^\sigma E_{s+1/2}\left(\alpha_s^{h+1/2} z \tau^{\frac{2}{2s+1}};\mu\right) \tau^{\mu-1}$$

$$\times E_{s+1/2}\left(\alpha_s^{h+1/2}\mu_m(\sigma-\tau)^{\frac{2}{2s+1}};\eta_{2,0}\right)(\sigma-\tau)^{\eta_{2,0}-1}d\tau \tag{5_1}$$

$$= (2s+1)\sigma^\nu \frac{z^s E_{1/2}\left(-\sigma^2 z^{2s+1};1+\nu\right)}{z-\mu_m} \qquad (1 \le m < +\infty).$$

Hence we obtain identity (4_1), if we use definition (1_1) of the function $\Omega_{m,h}^{(1)}(\tau;s)$. If $\nu \in \Delta_s(2^\circ)$, then, similarly, using representations $7.2(13_2)$ and $7.2(14_2)$ of the function $U_s^{(2)}(z;\mu_m)$ $(1 \le m < +\infty)$, we arrive at the identity

$$\sum_{h=-s}^{s} \alpha_s^{-(h+1/2)s} \int_0^\sigma E_{s+1/2}\left(\alpha_s^{h+1/2} z \tau^{\frac{2}{2s+1}};\mu\right) \tau^{\mu-1}$$

$$\times E_{s+1/2}\left(\alpha_s^{h+1/2}\mu_m(\sigma-\tau)^{\frac{2}{2s+1}};\eta_{2,1}\right)(\sigma-\tau)^{\eta_{2,1}-1}d\tau \tag{5_2}$$

$$= (2s+1)\sigma^\nu \frac{z^{s+1} E_{1/2}\left(-\sigma^2 z^{2s+1};1+\nu\right)}{z-\mu_m} \qquad (1 \le m < +\infty).$$

The use of definition (1_2) of the function $\Omega_{m,h}^{(2)}(\tau;s)$ gives identities (4_2), so the lemma is proved.

Now we add finite sets of some new functions to systems (1_1) and (1_2). We define these new functions as follows:

if $\nu \in \Delta_s(1^\circ)$, then we denote

$$\Omega_{m,h}^{(1)}(\tau;s) \equiv \alpha_s^{(h+1/2)m} \frac{\sigma^{-\nu}\Gamma(1+\nu)}{(2s+1)\Gamma(1-m)}$$
$$\times \frac{(\sigma-\tau)^{\frac{2(s+m-1)}{2s+1}+\eta_{2,0}-1}}{\Gamma\left(\eta_{2,0}+\frac{2(s+m-1)}{2s+1}\right)} \qquad (-(s-1)\le m \le 0, -s \le h \le s);$$

$$(6_1)$$

if $\nu \in \Delta_s(2^\circ)$, then

$$\Omega_{m,h}^{(2)}(\tau;s) \equiv \alpha_s^{(h+1/2)m} \frac{\sigma^{-\nu}\Gamma(1+\nu)}{(2s+1)\Gamma(1-m)}$$
$$\times \frac{(\sigma-\tau)^{\frac{2(s+m)}{2s+1}+\eta_{2,1}-1}}{\Gamma\left(\eta_{2,1}+\frac{2(s+m)}{2s+1}\right)} \qquad (-s\le m \le 0, -s \le h \le s).$$

$$(6_2)$$

Lemma 7.3-2. *The following identities are true in the whole z-plane:*
  if $\nu \in \Delta_s(1^\circ)$, then

$$\sum_{h=-s}^{s} \int_0^\sigma E_{s+1/2}\left(\alpha_s^{h+1/2} z\tau^{\frac{2}{2s+1}};\mu\right)\tau^{\mu-1}\Omega_{m,h}^{(1)}(\tau;s)d\tau$$
$$(7_1)$$
$$= \frac{\Gamma(1+\nu)}{\Gamma(1-m)} z^{-m} E_{1/2}\left(-\sigma^2 z^{2s+1};1+\nu\right) \qquad (-(s-1)\le m \le 0);$$

if $\nu \in \Delta_s(2^\circ)$, then

$$\sum_{h=-s}^{s} \int_0^\sigma E_{s+1/2}\left(\alpha_s^{h+1/2} z\tau^{\frac{2}{2s+1}};\mu\right)\tau^{\mu-1}\Omega_{m,h}^{(2)}(\tau;s)d\tau$$
$$(7_2)$$
$$= \frac{\Gamma(1+\nu)}{\Gamma(1-m)} z^{-m} E_{1/2}\left(-\sigma^2 z^{2s+1};1+\nu\right) \qquad (-s\le m \le 0).$$

Proof. Using representations $7.2(13_1)-(14_1)$ of the function $U_s^{(1)}(z;\lambda)$, we obtain

$$\frac{d^{s+m-1}}{d\lambda^{s+m-1}}U_s^{(1)}(z;\lambda)\bigg|_{\lambda=0} = \sum_{h=-s}^{s} \int_0^\sigma E_{s+1/2}\left(\alpha_s^{h+1/2} z\tau^{\frac{2}{2s+1}};\mu\right)\tau^{\mu-1}$$
$$\times \left\{\alpha_s^{(h+1/2)m}\Gamma(s+m)\frac{(\sigma-\tau)^{\frac{2(s+m-1)}{2s+1}+\eta_{2,0}-1}}{\Gamma\left(\eta_{2,0}+\frac{2(s+m-1)}{2s+1}\right)}\right\}d\tau \qquad (-(s-1)\le m \le 0).$$

$$(8_1)$$

The second term of the right-hand side of 7.2(14_1) may be considered to be a function depending on λ and having (if $z \neq 0$) a zero of order s at the point $\lambda = 0$. Therefore,

$$\frac{d^{s+m-1}}{d\lambda^{s+m-1}} U_s^{(1)}(z;\lambda)\Big|_{\lambda=0} = \frac{(2s+1)\sigma^\nu \Gamma(s+m)}{z^m} E_{1/2}\left(-\sigma^2 z^{2s+1}; 1+\nu\right) \qquad (9_1)$$

for any $m(-(s-1) \leq m \leq 0)$. Identity (7_1) follows from the representations (8_1), (9_1) and from (6_1). To prove identity (7_2), we use representations 7.2(13_2) — (14_2) of the function $U_s^{(2)}(z;\lambda)$ and also (6_2). Then the desired identity follows in a similar way.

(b) Now we move to definitions of some systems of vector functions which are biorthogonal on $(0,\sigma)$. We shall say that a $2s+1$-dimensional vector function is of class $L_2^{2s+1}(0,\sigma)$, if all its components are functions of class $L_2(0,\sigma)$. Further, we define the inner product of any two vector functions

$$y(\tau) = \{y_j(\tau)\}_{-s}^s \quad \text{and} \quad z(\tau) = \{z_j(\tau)\}_{-s}^s \qquad (s \geq 1) \qquad (10)$$

of class $L_2^{2s+1}(0,\sigma)$ as follows:

$$\{y(\tau), z(\tau)\} \equiv \{y, z\} \equiv \sum_{j=-s}^s \int_0^\sigma y_j(\tau)\overline{z_j(\tau)}d\tau. \qquad (11)$$

Then, evidently, the norm of a vector function $y(\tau) \in L_2^{2s+1}(0,\sigma)$ is the quantity

$$\|y\| \equiv \{y,y\}^{1/2} = \left\{ \sum_{j=-s}^s \int_0^\sigma |y_j(\tau)|^2 d\tau \right\}^{1/2} \geq 0. \qquad (12)$$

And it is obvious that $\|y\| = 0$ if and only if $y_j(\tau) = 0$ $(-s \leq j \leq s)$ almost everywhere in $(0,\sigma)$.

Note that by formulas (1_1), (6_1) and (1_2), (6_2) we actually have defined two systems of $2s+1$-dimensional vector functions of $L_2^{2s+1}(0,\sigma)$. Now, if $\nu \in \Delta_s(1^\circ)$, we also introduce the system

$$\left\{ \omega_{n,h}^{(1)}(\tau) \right\}_{-(s-1)}^\infty = \left\{ \overline{\Omega_{n,h}^{(1)}(\tau;s)} \right\}_{-(s-1)}^\infty \qquad (-s \leq h \leq s) \qquad (13_1)$$

and, if $\nu \in \Delta_s(2^\circ)$, the system

$$\left\{ \omega_{n,h}^{(2)}(\tau) \right\}_{-s}^\infty = \left\{ \overline{\Omega_{n,h}^{(2)}(\tau;s)} \right\}_{-s}^\infty \qquad (-s \leq h \leq s). \qquad (13_2)$$

Further, we put

$$\kappa_{n,h}(\tau) \equiv \alpha_s^{-(h+1/2)n} \frac{\Gamma(1-n)}{\Gamma\left(\mu - \frac{2n}{2s+1}\right)} \tau^{-\frac{2n}{2s+1}+\mu-1} \tag{14_1}$$

$$(-\infty < n \leq 0, -s \leq h \leq s),$$

$$\kappa_{n,h}(\tau) \equiv E_{s+1/2}\left(\alpha_s^{h+1/2}\mu_n \tau^{\frac{2}{2s+1}}; \mu\right)\tau^{\mu-1} \tag{14_2}$$

$$(1 \leq n < +\infty, -s \leq h \leq s)$$

and introduce the following systems of $2s+1$-dimensional vector functions of $L_2^{2s+1}(0,\sigma)$:

if $\nu \in \Delta_s(1^\circ)$, we introduce the system

$$\{\kappa_{n,h}(\tau)\}^\infty_{-(s-1)} \qquad (-s \leq h \leq s) \tag{15_1}$$

and, if $\nu \in \Delta_s(2^\circ)$, the system

$$\{\kappa_{n,h}(\tau)\}^\infty_{-s} \qquad (-s \leq h \leq s). \tag{15_2}$$

It is useful to mention that the functions of systems (15_1) and (15_2) are of class $L_2^{2s+1}(0,\sigma)$, and they may have singularities only at the point $\tau = 0$. This is in contrast to the functions of systems (13_1) and (13_2) which are also of $L_2^{2s+1}(0,\sigma)$ and which may have singularities only at the point $\tau = \sigma$. Consequently,

if $\nu \in \Delta_s(1^\circ)$, then, for any $j, h(-s \leq j, h \leq s)$

$$\kappa_{n,h}(\tau)\overline{\omega_{m,j}^{(1)}(\tau)} \in L_2(0,\sigma) \subset L_1(0,\sigma) \qquad (-(s-1) \leq n, m < +\infty),$$

and, if $\nu \in \Delta_s(2^\circ)$, then, for any $j, h(-s \leq j, h \leq s)$

$$\kappa_{n,h}(\tau)\overline{\omega_{m,j}^{(2)}(\tau)} \in L_2(0,\sigma) \subset L_1(0,\sigma) \qquad (-s \leq n, m < +\infty).$$

Now we can prove the main theorem of this section.

Theorem 7.3-1. *1°. If $\nu \in \Delta_s(1^\circ)$, then the systems of vector functions (13_1) and (15_1) are biorthogonal in the sense of their inner product (11). In other words, if we denote*

$$\kappa_n(\tau) = \{\kappa_{n,h}(\tau)\}^s_{-s} \text{ and } \omega_n^{(1)}(\tau) = \{\omega_{n,h}^{(1)}(\tau)\}^s_{-s} \qquad (-(s-1) \leq n < +\infty), \tag{16_1}$$

then

$$\{\kappa_n, \omega_m^{(1)}\} = \delta_{n,m} \qquad (-(s-1) \leq n, m < +\infty), \tag{17_1}$$

where $\delta_{n,m}$ is the Kronecker's symbol.
2°. If $\nu \in \Delta_s(2^\circ)$, then the systems of vector functions (13_2) and (15_2) are biothogonal in the same sense. In other words, if we denote

$$\kappa_n(\tau) = \{\kappa_{n,h}(\tau)\}^s_{-s} \quad \text{and} \quad \omega_n^{(2)}(\tau) = \{\omega_{n,h}^{(2)}(\tau)\}^s_{-s} \qquad (-s \leq n < +\infty), \tag{16_2}$$

then

$$\{\kappa_n, \omega_m^{(2)}\} = \delta_{n,m} \qquad (-s \leq n, m < +\infty). \tag{17_2}$$

Proof. Note that the zeros of the function $\mathcal{E}_{s+1/2,\sigma}(z;\nu)$ in both cases $\nu \in \Delta_s(1^\circ)$ and $\nu \in \Delta_s(2^\circ)$ are simple and coincide with the sequence $\{\mu_n\}_1^\infty$. Besides, it is obvious that

$$\left.\frac{z^s \mathcal{E}_{s+1/2,\sigma}(z;\nu)}{\mu_m^s \mathcal{E}'_{s+1/2,\sigma}(\mu_m;\nu)(z-\mu_m)}\right|_{z=\mu_n} = \delta_{n,m} \qquad (1 \le n,m < +\infty), \qquad (18_1)$$

$$\left.\frac{z^{s+1} \mathcal{E}_{s+1/2,\sigma}(z;\nu)}{\mu_m^{s+1} \mathcal{E}'_{s+1/2,\sigma}(\mu_m;\nu)(z-\mu_m)}\right|_{z=\mu_n} = \delta_{n,m} \qquad (1 \le n,m < +\infty). \qquad (18_2)$$

Now, if we put $z = \mu_n(1 \le n < +\infty)$ in identities (4_1) and (4_2) of Lemma 7.3-1, then, in view of definitions $(13_1) - (13_2), (15_1) - (15_2)$ and $(16_1) - (16_2)$, the formulas $(18_1) - (18_2)$ lead to the equalities

$$\{\kappa_n, \omega_m^{(1)}\} = \{\kappa_n, \omega_m^{(2)}\} = \delta_{n,m} \qquad (1 \le n,m < +\infty). \qquad (19_1)$$

So, the desired assertions (17_1) and (17_2) are proved in the case when $1 \le n,m < +\infty$.

As the next step we shall extend the biorthogonality (19_1) on the cases $-(s-1) \le n \le 0$ and $-s \le n \le 0$, assuming that $1 \le m < +\infty$. To this end we again use identities (4_1) and (4_2) of Lemma 7.3-1 in which we replace the functions $\Omega_{m,h}^{(1)}(\tau;s)$ and $\Omega_{m,h}^{(2)}(\tau;s)$ respectively by $\overline{\omega_{m,h}^{(1)}(\tau)}$ and $\overline{\omega_{m,h}^{(2)}(\tau)}$ in accordance with $(13_1) - (13_2)$. The right-hand sides of the identities (4_1) and (4_2) have zeros at the point $z = 0$, and the orders are s and $s + 1$ respectively. Hence their derivatives, of orders $s - 1$ and s, correspondingly, vanish at $z = 0$ and, on the other hand,

$$\left.\frac{d^{-n}}{dz^{-n}}\left\{E_{s+1/2}\left(\alpha_s^{h+1/2} z\tau^{\frac{2}{2s+1}}; \mu\right)\tau^{\mu-1}\right\}\right|_{z=0} = \kappa_{n,h}(\tau)$$
$$= \alpha_s^{-(h+1/2)n}\frac{\Gamma(1-n)}{\Gamma\left(\mu - \frac{2n}{2s+1}\right)}\tau^{-\frac{2n}{2s+1}+\mu-1}, \qquad -\infty < n \le 0,\ -s \le h \le s. \qquad (20)$$

Therefore the differentiation of the mentioned identities leads to the equalities

$$\sum_{h=-s}^{s} \int_0^\sigma \kappa_{n,h}(\tau)\overline{\omega_{m,h}^{(1)}(\tau)}d\tau = \{\kappa_n, \omega_m^{(1)}\} = 0, \qquad -(s-1) \le n \le 0, 1 \le m < +\infty,$$

$$\sum_{h=-s}^{s} \int_0^\sigma \kappa_{n,h}(\tau)\overline{\omega_{m,h}^{(2)}(\tau)}d\tau = \{\kappa_n, \omega_m^{(2)}\} = 0, \qquad -s \le n \le 0,\ 1 \le m < +\infty.$$

Thus, together with (19_1), we have

$$\{\kappa_n, \omega_m^{(1)}\} = \delta_{n,m}, \qquad -(s-1) \le n < +\infty,\ 1 \le m < +\infty,$$
$$\{\kappa_n, \omega_m^{(2)}\} = \delta_{n,m}, \qquad -s \le n < +\infty,\ 1 \le m < +\infty. \qquad (19_2)$$

To obtain one more extension of the biorthogonality $(19_1) - (19_2)$, we use identities (7_1) and (7_2) of Lemma 7.3-2, replacing the functions $\Omega^{(1)}_{m,h}(\tau; s)$ and $\Omega^{(2)}_{m,h}(\tau; s)$ correspondingly by $\overline{\omega^{(1)}_{m,h}(\tau)}$ and $\overline{\omega^{(2)}_{m,h}(\tau)}$. We take $z = \mu_n (1 \le n < +\infty)$ in these identities and, using definition (14_2) of $\kappa_{n,h}(\tau)$, arrive at the equalities

$$\sum_{h=-s}^{s} \int_0^{\sigma} \kappa_{n,h}(\tau)\overline{\omega^{(1)}_{m,h}(\tau)}d\tau = \{\kappa_n, \omega^{(1)}_m\} = 0,$$

$$1 \le n < +\infty, \ -(s-1) \le m \le 0,$$

$$\sum_{h=-s}^{s} \int_0^{\sigma} \kappa_{n,h}(\tau)\overline{\omega^{(2)}_{m,h}(\tau)}d\tau = \{\kappa_n, \omega^{(2)}_m\} = 0,$$ $\qquad (19_3)$

$$1 \le n < +\infty, \ -s \le m \le 0.$$

Finally, concluding the extension of the biorthogonality, observe that the right-hand sides of identities (7_1) and (7_2) have at $z = 0$ zeros of order $-m$, where $-(s-1) \le m \le 0$ or $-s \le m \le 0$ respectively. Besides, it is easy to verify that

$$\frac{d^{-n}}{dz^{-n}}\left\{\frac{\Gamma(1+\nu)}{\Gamma(1-m)}z^{-m}E_{1/2}\left(-\sigma^2 z^{2s+1}; 1+\nu\right)\right\}\bigg|_{z=0} = \delta_{n,m} \qquad (21)$$

in both considered cases

$$-(s-1) \le m, n \le 0 \qquad \text{and} \qquad -s \le m, n \le 0. \qquad (21')$$

Hence the differentiation of the left-hand sides of (7_1) and (7_2) leads, in view of (20) and (21), to the relations

$$\{\kappa_n, \omega^{(1)}_m\} = \delta_{n,m}, \quad -(s-1) \le n, m \le 0,$$

$$\{\kappa_n, \omega^{(2)}_m\} = \delta_{n,m}, \quad -s \le n, m \le 0. \qquad (19_4)$$

The obtained equalities $(19_1), (19_2), (19_3)$ and (19_4) are equivalent to the desired assertions (17_1) and (17_2).

7.4 Theorems on completeness and basis property

In the previous section two pairs of systems of $2s+1$-dimensional vector functions biorthogonal in the sense of the inner product 7.3(11) were defined by formulas (1_1) and $(6_1), (1_2)$ and $(6_2), (14_1)$ and $(14_2), (13_1)$ and (13_2) and (15_1) and (15_2). Namely, if $\nu \in \Delta_s(1^\circ)$, we considered the systems

$$\kappa_n(\tau) = \{\kappa_{n,h}(\tau)\}^s_{-s}, \quad \omega^{(1)}_n(\tau) = \{\omega^{(1)}_{n,h}(\tau)\}^s_{-s}, \qquad -(s-1) \le n < +\infty, \quad (1)$$

and, if $\nu \in \Delta_s(2^\circ)$, we considered the systems

$$\kappa_n(\tau) = \{\kappa_{n,h}(\tau)\}^s_{-s}, \quad \omega^{(2)}_n(\tau) = \{\omega^{(2)}_{n,h}(\tau)\}^s_{-s}, \qquad -s \le n < +\infty. \qquad (2)$$

(a) First we shall prove that all these systems are complete in the space $L_2^{2s+1}(0, \sigma)$ of $2s+1$-dimensional vector functions

$$\varphi(\tau) = \{\varphi_h(\tau)\}^s_{-s}. \qquad (3)$$

Theorem 7.4-1. 1°. *If $\nu \in \Delta_s(1^\circ)$, then the systems of vector functions* $\{\kappa_n(\tau)\}^\infty_{-(s-1)}$ *and* $\{\omega_n^{(1)}(\tau)\}^\infty_{-(s-1)}$ *are complete in* $L_2^{2s+1}(0,\sigma)$, *if* $\omega \in (-1,1)$ *or* $\omega \in (0,1)$ *respectively.*

2°. *If $\nu \in \Delta_s(2^\circ)$, then the systems of vector functions* $\{\kappa_n(\tau)\}^\infty_{-s}$ *and* $\{\omega_n^{(2)}(\tau)\}^\infty_{-s}$ *are complete in* $L_2^{2s+1}(0,\sigma)$, *if* $\omega \in (-1,1)$ *or* $\omega \in (-1,0)$ *respectively.*

Proof. Suppose $\varphi(\tau) = \{\varphi_h(\tau)\}^s_{-s} \in L_2^{2s+1}(0,\sigma)$ is any vector function, and introduce the function

$$\Phi(z;\varphi) = \sum_{h=-s}^{s} \int_0^\sigma E_{s+1/2}\left(\alpha_s^{h+1/2} z\tau^{\frac{2}{2s+1}};\mu\right) \tau^{\mu-1}\overline{\varphi_h(\tau)}d\tau, \tag{4}$$

assuming, as everywhere, that

$$\alpha_s = \exp\left\{i\frac{2\pi}{2s+1}\right\} \quad \text{and} \quad \mu = \frac{3/2+s+\omega}{2s+1} \quad (-1 < \omega < 1).$$

According to the Wiener-Paley type Theorem $2.4 - 1(1^\circ)$, $\Phi(z;\varphi)$ is an entire function of class $W^{2,\omega}_{s+1/2,\sigma}$. What is more, by formulas $7.3(20)$ and $7.3(14_1)-(14_2)$,

$$\frac{d^{-n}}{dz^{-n}}\Phi(z;\varphi)\bigg|_{z=0} = \sum_{h=-s}^{s} \int_0^\sigma \left\{\alpha_s^{-(h+1/2)n}\frac{\Gamma(1-n)}{\Gamma\left(\mu - \frac{2n}{2s+1}\right)}\tau^{-\frac{2n}{2s+1}+\mu-1}\right\}\overline{\varphi_h(\tau)}d\tau$$

$$= \sum_{h=-s}^{s} \int_0^\sigma \kappa_{n,h}(\tau)\overline{\varphi_h(\tau)}d\tau = \{\kappa_n,\varphi\} = \overline{\{\varphi,\kappa_n\}} \quad (-\infty < n \leq 0) \tag{5}$$

and

$$\Phi(\mu_n;\varphi) = \sum_{h=-s}^{s} \int_0^\sigma E_{s+1/2}\left(\alpha_s^{h+1/2}\mu_n\tau^{\frac{2}{2s+1}};\mu\right) \tau^{\mu-1}\overline{\varphi_h(\tau)}d\tau \tag{6}$$

$$=\{\kappa_n,\varphi\} = \overline{\{\varphi,\kappa_n\}} \quad (1 \leq n < +\infty).$$

Now suppose that the vector function $\varphi(\tau)$ satisfies the conditions

$$\{\varphi,\kappa_n\} = 0 \quad (-(s-1) \leq n < +\infty), \tag{7_1}$$

if $\nu \in \Delta_s(1^\circ)$, or the conditions

$$\{\varphi,\kappa_n\} = 0 \quad (-s \leq n < +\infty), \tag{7_2}$$

if $\nu \in \Delta_s(2^\circ)$. Then (5) and (6) imply correspondingly

$$\Phi^{(k)}(0;\varphi) = 0 \quad (0 \leq k \leq s-1), \quad \Phi(\mu_n;\varphi) = 0 \quad (1 \leq n < +\infty), \tag{8_1}$$

and
$$\Phi^{(k)}(0;\varphi) = 0 \quad (0 \leq k \leq s), \quad \Phi(\mu_n;\varphi) = 0 \quad (1 \leq n < +\infty). \tag{8_2}$$

Hence, by the uniqueness Theorem 6.2-3 proved for the functions of $W^{2,\omega}_{s+1/2,\sigma}$, we conclude that $\Phi(z;\varphi) \equiv 0$, but assertion 2° of Theorem 2.4-1 gives $\varphi_h(\tau) = 0 (-s \leq h \leq s)$ almost everywhere in $(0,\sigma)$, i.e., $\varphi(\tau) = 0$. This completes the proof of both desired assertions relating to the systems $\{\kappa_{n,h}(\tau)\}^s_{-s}$ $(-(s-1) \leq n < +\infty)$ and $\{\kappa_{n,h}(\tau)\}^s_{-s}$ $(-s \leq n < +\infty)$. To prove that the right-hand side systems of (1) and (2) are also complete in $L_2^{2s+1}(0,\sigma)$, we return to the formulas of Section 7.3 defining these systems. Then we conclude that:

(i) When $\nu \in_s$ (1°), the functions of $\overline{\{w^{(1)}_{n,h}(\tau)\}^s}_{-s}$ differ from the functions of $\{\kappa_{n,h}(\sigma - \tau)\}^s_{-s}$ only by constant multipliers, if in the latter we substitute $\mu = (3/2 + s + \omega)/(2s + 1)$ $(-1 < \omega < 1)$ by

$$\eta_{2,0} = \frac{3}{2s+1} + \nu - \mu = \frac{3/2 + s + \omega_0}{2s+1},$$

where, as is easy to observe, $\omega_0 \in (0,1)$. Besides, it is easy to verify that

$$\nu \in \left(\frac{2s + \omega_0}{2s+1}, \frac{2s + \omega_0 + 1}{2s+1} \right)$$

when $\omega \in (0,1)$, i.e., ν belongs to the interval $\Delta_s(1°)$ constructed by use of ω_0.

(ii) When $\nu \in \Delta_s(2°)$, the functions of $\overline{\{w^{(2)}_{n,h}(\tau)\}^s}_{-s}$ differ from the functions of $\{\kappa_{n,h}(\sigma - \tau)\}^s_{-s}$ only by constant multipliers, if in the latter we substitute μ by

$$\eta_{2,1} = \frac{1}{2s+1} + \nu - \mu = \frac{3/2 + s + \omega_1}{2s+1},$$

where now $\omega_1 \in (-1,0)$. Similarly, it is easy to verify that

$$\nu \in \left(\frac{2s + \omega_1 + 1}{2s+1}, \frac{2s + \omega_1 + 2}{2s+1} \right)$$

when $\omega \in (-1,0)$, i.e., ν belongs to the interval $\Delta_s(2°)$ constructed by use of ω_1. From the conclusions we made it immediately follows that the right-hand side systems of (1) and (2) are also complete in $L_2^{2s+1}(0,\sigma)$.

Remark. We have already proved the completeness of the second systems of (1) and (2) in $L_2^{2s+1}(0,\sigma)$ assuming that $\omega \in (0,1)$ and $\omega \in (-1,0)$ respectively. This differs from our usual assumption that $\omega \in (-1,1)$. Nevertheless, we shall see in Theorem $7.4 - 3(1°)$ that the mentioned particular restrictions on ω turn out not to be essential.

(b) According to Theorem 2.4-1, the space $W^{2,\omega}_{s+1/2,\sigma}$ coincides with the set of entire functions $\Phi(z)$ representable in the form

$$\Phi(z) = \sum_{h=-s}^{s} \int_0^{\sigma} E_{s+1/2}\left(\alpha_s^{h+1/2} z \tau^{\frac{2}{2s+1}}; \mu\right) \tau^{\mu-1}\overline{\varphi_h(\tau)}d\tau, \tag{9}$$

where $\mu = (3/2 + s + \omega)/(2s + 1)(-1 < \omega < 1)$ and $\varphi(\tau) = \{\varphi_h(\tau)\}_{-s}^{s}$ is an arbitrary vector function of $L_2^{2s+1}(0,\sigma)$. It is necessary to prove one more assertion on functions of this class.

Lemma 7.4-1. *If $\Phi(z) \in W^{2,\omega}_{s+1/2,\sigma}$ is any function and $\varphi(\tau)$ is the vector function of its representation (9), then the following two-sided estimates are true:*

$$\|\Phi; \Gamma_{2s+1}\|_{2,\omega} \asymp \|\varphi\| = \left\{\sum_{h=-s}^{s} \int_0^{\sigma} |\varphi_h(\tau)|^2 d\tau\right\}^{1/2}, \tag{10}$$

where the suitable constants are independent of Φ and φ.

Proof. First we use formulas 2.4(6)-(7) of Theorem 2.4-1 which give an explicit form of the components of the vector function $\overline{\varphi(\tau)}$ of the representation (9). By these formulas,

$$\frac{1}{\sqrt{2\pi}(s+1/2)}\left\{e^{i\frac{\pi}{2}(1-\mu)}\Psi_{h+1}(-\tau) + e^{-i\frac{\pi}{2}(1-\mu)}\Psi_h(\tau)\right\}$$

$$= \begin{cases} \overline{\varphi_h(\tau)}, & \tau \in (0,\sigma) \\ 0, & \tau \in (\sigma, +\infty) \end{cases} \quad (-s \leq h \leq s),$$

where

$$\Psi_h(\tau) = \frac{1}{\sqrt{2\pi}}\frac{d}{d\tau}\int_0^{+\infty} \frac{e^{-i\tau t}-1}{-it}\Phi\left(\alpha_s^{-h}t^{\frac{2}{2s+1}}\right)t^{\mu-1}dt \quad (-s \leq h \leq s)$$

and $\Psi_{s+1}(\tau) = \Psi_{-s}(\tau)$. Hence, by Parseval's equalities,

$$\int_0^{\sigma} |\varphi_h(\tau)|^2 d\tau \leq \frac{4}{\pi(2s+1)^2}\int_{-\infty}^{+\infty}\left\{|\Psi_{h+1}(-\tau)|^2 + |\Psi_h(\tau)|^2\right\}d\tau$$

$$= \frac{4}{\pi(2s+1)^2}\int_{-\infty}^{+\infty}\left\{\left|\Phi\left(\alpha_s^{-(h+1)}t^{\frac{2}{2s+1}}\right)\right|^2 + \left|\Phi\left(\alpha_s^{-h}t^{\frac{2}{2s+1}}\right)\right|^2\right\}t^{2(\mu-1)}dt$$

$$= \frac{2}{\pi(2s+1)}\int_0^{+\infty}\left\{\left|\Phi\left(\alpha_s^{-(h+1)}r\right)\right|^2 + \left|\Phi\left(\alpha_s^{-h}r\right)\right|^2\right\}r^{\omega}dr \quad (-s \leq h \leq s).$$

Summing these equalities over h, we get

$$\|\varphi\|^2 = \sum_{h=-s}^{s}\int_0^{\sigma}|\varphi_h(\tau)|^2 d\tau$$

$$\leq \frac{4}{\pi(2s+1)}\sum_{h=-s}^{s}\int_0^{+\infty}\left|\Phi\left(\alpha_s^{-h}r\right)\right|^2 r^{\omega}dr = \frac{4}{\pi(2s+1)}\|\Phi;\Gamma_{2s+1}\|^2_{2,\omega}. \tag{10_1}$$

To prove the converse inequality, we introduce the entire functions

$$\Phi_h(z) = \int_0^\sigma E_{s+1/2}\left(\alpha_s^{h+1/2} z \tau^{\frac{2}{2s+1}}; \mu\right) \tau^{\mu-1} \overline{\varphi_h(\tau)} d\tau \qquad (-s \le h \le s) \qquad (11)$$

and observe that

$$\Phi(z) = \sum_{h=-s}^{s} \Phi_h(z). \qquad (12)$$

Further, introducing the set

$$\delta_h = \left\{\zeta \in \mathbb{C} : |\zeta| = 1, \ \left|\zeta - \alpha_s^{-(h+1/2)}\right| \ge 2\sin\frac{\pi}{2(2s+1)}\right\} \qquad (-s \le h \le s)$$

of closed arcs of the unit circle, we see that, by Theorem 2.5-7,

$$\sup_{\zeta \in \delta_h}\left\{\int_0^{+\infty} |\Phi_h(r\zeta)|^2 r^\omega dr\right\} \le M_\mu \int_0^\sigma |\varphi_h(\tau)|^2 d\tau \qquad (-s \le h \le s).$$

But, since $\cap_{-s}^{s}\delta_h = \{\alpha_s^j\}_{-s}^{s}$, in the same way we can estimate simultaneously the integrals of all functions Φ_h $(-s \le h \le s)$ when $\zeta = \alpha_s^j$ $(-s \le j \le s)$. Thus

$$\int_0^{+\infty}\left|\Phi_h(\alpha_s^j r)\right|^2 r^\omega dr \le M_\mu \int_0^\sigma |\varphi_h(\tau)|^2 d\tau \qquad (-s \le h, j \le s).$$

Hence, by (12),

$$\|\Phi; \Gamma_{2s+1}\|_{2,\omega} = \left\{\sum_{j=-s}^{s}\int_0^{+\infty}\left|\Phi(\alpha_s^j r)\right|^2 r^\omega dr\right\}^{1/2}$$

$$\le (2s+1)\left\{\sum_{j=-s}^{s}\sum_{h=-s}^{s}\int_0^{+\infty}\left|\Phi_h(\alpha_s^j r)\right|^2 r^\omega dr\right\}^{1/2} \qquad (10_2)$$

$$\le (2s+1)^{3/2}M_\mu^{1/2}\left\{\sum_{h=-s}^{s}\int_0^\sigma |\varphi_h(\tau)|^2 d\tau\right\}^{1/2} = M_{\mu,s}\|\varphi\|.$$

The two-sided inequality (10) follows from (10_1) and (10_2), and the proof is complete.

(c) Finally, we shall establish that each of the considered biorthogonal systems is a basis of $L_2^{2s+1}(0,\sigma)$, but first we note that, similar to Section 6.2, the sequence of numbers $\{\varphi_n\}_{-r}^\infty$ $(0 \le r \le s)$ is of Hilbert space $L_{2,\kappa_{-s}}^{(r)}$, if

$$\|\{\varphi_n\}_{-r}^\infty\|_{2,\kappa_{-s}} \equiv \left\{\sum_{n=-r}^{0}\left|\frac{\varphi_n}{|n|!}\right|^2 + \sum_{n=1}^{\infty}|\varphi_n|^2(1+n)^{\kappa_{-s}}\right\}^{1/2} < +\infty, \qquad (13)$$

where

$$\kappa_{-s} = \frac{1+2(\omega-s)}{2s+1} = 2(\mu-1). \qquad (14)$$

Theorem 7.4-2. *1°. If $\nu \in \Delta_s(1°)$, then the vector series*

$$\varphi(\tau) = \sum_{n=-(s-1)}^{\infty} \varphi_n \omega_n^{(1)}(\tau) \tag{15}$$

are convergent in the norm of $L_2^{2s+1}(0,\sigma)$ and represent continuous one-to-one mapping of the space $L_{2,\kappa_{-s}}^{(s-1)}$ of sequences $\{\varphi_n\}_{-(s-1)}^{\infty}$ onto the space $L_2^{2s+1}(0,\sigma)$ of $2s+1$ — dimensional vector functions $\varphi(\tau)$. Besides, for every vector series of form (15),

$$\varphi_n = \{\varphi, \kappa_n\}, \qquad -(s-1) \le n < +\infty \tag{16}$$

and

$$\|\varphi\| = \{\varphi, \varphi\}^{1/2} \asymp \|\{\varphi_n\}_{-(s-1)}^{\infty}\|_{2,\kappa_{-s}}. \tag{17}$$

2°. If $\nu \in \Delta_s(2°)$, then the vector series

$$\varphi(\tau) = \sum_{n=-s}^{+\infty} \varphi_n \omega_n^{(2)}(\tau), \tag{18}$$

which are convergent in the norm of $L_2^{2s+1}(0,\sigma)$, represent continuous one-to-one mapping of the space $L_{2,\kappa_{-s}}^{(s)}$ of sequences $\{\varphi_n\}_{-s}^{\infty}$ onto the space $L_2^{2s+1}(0,\sigma)$ of $2s+1$ — dimensional vector functions $\varphi(\tau)$. Besides, for every vector series of the form (18),

$$\varphi_n = \{\varphi, \kappa_n\}, \qquad -s \le n < +\infty \tag{19}$$

and

$$\|\varphi\| = \{\varphi, \varphi\}^{1/2} \asymp \|\{\varphi_n\}_{-s}^{\infty}\|_{2,\kappa_{-s}}. \tag{20}$$

Proof. We shall consider the series of the form

$$\begin{aligned}
\Phi(z) =& \Gamma(1+\nu) \left(\sum_{n=-r}^{0} \frac{\overline{\varphi_n}}{|n|!} z^{-n} \right) \mathcal{E}_{s+1/2,\sigma}(z;\nu) \\
& + \sum_{n=1}^{\infty} \overline{\varphi_n} \frac{z^{r+1}\mathcal{E}_{s+1/2,\sigma}(z;\nu)}{\mu_n^{r+1}\mathcal{E}'_{s+1/2,\sigma}(\mu_n;\nu)(z-\mu_n)},
\end{aligned} \tag{21}$$

assuming that $r = s - 1$, if $\nu \in \Delta_s(1°)$, and $r = s$, if $\nu \in \Delta_s(2°)$. Using Theorems 6.2-1 and 6.2-2 (where we take $p = 2$) we can conclude that a series of the form (21) represents in both considered cases a continuous one-to-one mapping of the space $L_{2,\kappa_{-s}}^{(r)}$ of sequences $\{\varphi_n\}_{-r}^{\infty}$ onto the space $W_{s+1/2,\sigma}^{2,\omega}$ of entire functions $\Phi(z)$. In addition, we can conclude that in both cases $r = s - 1$ and $r = s$

$$\|\Phi; \Gamma_{2s+1}\|_{2,\omega} \asymp \|\{\varphi_n\}_{-r}^{\infty}\|_{2,\kappa_{-s}} \tag{22}$$

and

$$\Phi^{(k)}(0) = \overline{\varphi_{-k}} \quad (0 \le k \le r), \qquad \Phi(\mu_n) = \overline{\varphi_n} \quad (1 \le n < +\infty). \qquad (23)$$

On the other hand, according to Wiener-Paley type Theorem 2.4-1, the formula

$$\Phi(z) = \sum_{h=-s}^{s} \int_0^{\sigma} E_{s+1/2}\left(\alpha_s^{h+1/2} z \tau^{\frac{2}{2s+1}}; \mu\right) \tau^{\mu-1} \overline{\varphi_h(\tau)} d\tau$$

represents a one-to-one mapping of $L_2^{2s+1}(0,\sigma)$ onto $W_{s+1/2,\sigma}^{2,\omega}$. And the two-sided estimates

$$\|\Phi; \Gamma_{2s+1}\|_{2,\omega} \asymp \|\varphi\| = \left\{ \sum_{h=-s}^{s} |\varphi_h(\tau)|^2 d\tau \right\}^{1/2}$$

are true according to Lemma 7.4-1. Thus in both cases $r = s - 1$ and $r = s$ there exists a canonical homeomorphism $\{\varphi_n\}_{-r}^{\infty} \to \varphi$ between the spaces $L_{2,\kappa_{-s}}^{(r)}$ and $L_2^{2s+1}(0,\sigma)$, and the two-sided inequalities (17) and (20) are true. To be convinced that formulas (16) and (19) are true, we note first that

$$\Phi(\mu_n) = \sum_{h=-s}^{s} \int_0^{\sigma} \kappa_{n,h}(\tau)\overline{\varphi_h(\tau)} d\tau = \{\kappa_n, \varphi\} = \overline{\{\varphi, \kappa_n\}} \quad (1 \le n < +\infty), \qquad (24)$$

according to (9) and to definitions 7.3(14_2) of functions $\kappa_{n,h}(\tau)$ $(1 \le n < +\infty,$ $-s \le h \le s)$. Hence the interpolation data (23) leads to formulas (16) and (19), but only for $1 \le n < +\infty$. Next, we use formula 7.3(20), where $-\infty < n \le 0$ and $-s \le h \le s$. Then the representations (9) and (21) of $\Phi(z)$ and the interpolation data (23) for the cases $r = s - 1$ and $r = s$ lead correspondingly to formulas (16), where $-(s-1) \le n \le 0$, and (19), where $-s \le n \le 0$. Thus, to complete the proofs of both assertions $1°$ and $2°$, we have to show that the mentioned homeomorphism between the spaces $L_{2,\kappa_{-s}}^{(r)}$ and $L_2^{2s+1}(0,\sigma)$ $(r = s - 1$ or $r = s)$ can be given also by means of the vector series (15) and (18) converging in the norm of $L_2^{2s+1}(0,\sigma)$. To this end, observe that, if

$$\rho_m(\tau) = \varphi(\tau) - \sum_{n=-r}^{m} \{\varphi, \kappa_n\} \omega_n^{(i_r)}(\tau), m \ge 0,$$

where $i_r = 1$ when $r = s - 1$ and $i_r = 2$ when $r = s$, then using estimate (17) (when $r = s - 1$) and estimate (20) (when $r = s$), we obtain

$$\|\rho_m\| \asymp \left(\sum_{n=m+1}^{\infty} |\{\varphi, \kappa_n\}|^2 (1+n)^{\kappa-s} \right)^{1/2}.$$

Obviously $\|\rho_m\| \to 0$ as $m \to +\infty$. Hence the vector series (15) and (18) converge to φ in norm of $L_2^{2s+1}(0,\sigma)$, and the proof is complete.

(d) Now we shall prove that Theorem 7.4-2 can be formulated in the following way.

Theorem 7.4-3. *1°. If $\nu \in \Delta_s(1°)$, or $\nu \in \Delta_s(2°)$, then the systems*

$$\{\omega_n^{(1)}(\tau)\}^{\infty}_{-(s-1)} \ \text{or} \ \{\omega_n^{(2)}(\tau)\}^{\infty}_{-s} \tag{25}$$

are correspondingly complete in the space $L_2^{2s+1}(0,\sigma)$ of vector functions.
2°. The systems (25) become Riesz bases of $L_2^{2s+1}(0,\sigma)$ after the suitable normalizations.

Proof. Assertion 1° follows immediately from Theorem 7.4-2.
2°. It is appropriate to move from the biorthogonal vector systems (1) to the systems

$$K_n(\tau) = \{K_{n,h}(\tau)\}^s_{-s}, \quad \tilde{\Omega}_n^{(1)}(\tau) = \{\tilde{\Omega}_{n,h}^{(1)}(\tau)\}^s_{-s} \quad (-(s-1) \le n < +\infty), \tag{26}$$

where

$$K_{n,h}(\tau) = (1 + |n|)^{\kappa - s/2} \kappa_{n,h}(\tau), \quad \tilde{\Omega}_{n,h}^{(1)}(\tau) = (1 + |n|)^{-\kappa - s/2} \omega_{n,h}^{(1)}(\tau). \tag{26'}$$

These systems of vector functions are also biorthogonal, since obviously

$$\{K_n, \tilde{\Omega}_m^{(1)}\} = \{\kappa_n, \omega_m^{(1)}\} = \delta_{n,m}, \quad (-(s-1) \le n, m < +\infty).$$

Further, the series similar to (15) takes the form

$$\varphi(\tau) = \sum_{n=-(s-1)}^{\infty} \{\varphi, K_n\} \tilde{\Omega}_n^{(1)}(\tau), \tag{27}$$

and the estimates (17) turn into

$$\|\varphi\| \asymp \left\{ \sum_{n=-(s-1)}^{\infty} |\{\varphi, K_n\}|^2 \right\}^{1/2} \tag{28}$$

since, by the notations (26'), $\{\varphi, K_n\}\tilde{\Omega}_n^{(1)}(\tau) = \{\varphi, \kappa_n\}\omega_n^{(1)}(\tau)$ and $|\{\varphi, K_n\}|^2 = |\{\varphi, \kappa_n\}|^2 (1+|n|)^{\kappa - s}$ when $-(s-1) \le n < +\infty$. But, if the biorthogonal expansion (27) and the two-sided estimates (28) are true for any function $\varphi(\tau) \in L_2^{2s+1}(0,\sigma)$, then by the well-known definition, $\{\tilde{\Omega}_n^{(1)}(\tau)\}^{\infty}_{-(s-1)}$ is a Riesz basis of $L_2^{2s+1}(0,\sigma)$.

When $\nu \in \Delta_s(2°)$, the proof is similar. Namely, we introduce the systems

$$K_n(\tau) = \{K_{n,h}(\tau)\}^s_{-s}, \quad \tilde{\Omega}_n^{(2)}(\tau) = \{\tilde{\Omega}_{n,h}^{(2)}(\tau)\}^s_{-s} \quad (-s \le n < +\infty), \tag{29}$$

where

$$K_{n,h}(\tau) = (1 + |n|)^{\kappa - s/2} \kappa_{n,h}(\tau), \quad \tilde{\Omega}_{n,h}^{(2)}(\tau) = (1 + |n|)^{-\kappa - s/2} \omega_{n,h}^{(2)}(\tau). \tag{29'}$$

Then, evidently,

$$\{K_n, \tilde{\Omega}_m^{(2)}\} = \{\kappa_n, \omega_m^{(2)}\} = \delta_{n,m}, \quad (-s \le n, m < +\infty).$$

The series similar to (18) now takes the form

$$\varphi(\tau) = \sum_{n=-s}^{\infty} \{\varphi, K_n\} \tilde{\Omega}_n^{(2)}(\tau), \tag{30}$$

and the corresponding two-sided estimates are

$$\|\varphi\| \asymp \left\{ \sum_{n=-s}^{\infty} |\{\varphi, K_n\}|^2 \right\}^{1/2}. \tag{31}$$

Hence we conclude that the system $\{\tilde{\Omega}_n^{(2)}(\tau)\}_{-s}^{\infty}$ is a Riesz basis of $L_2^{2s+1}(0,\sigma)$, and this completes the proof.

(e) Concluding this chapter, we prove a general theorem on the basis property of the constructed systems of vector functions.

Theorem 7.4-4. 1°. *If* $\nu \in \Delta_s(1^\circ)$, *then the systems*

$$\left\{ \{K_{n,h}(\tau)\}_{h=-s}^{s} \right\}_{-(s-1)}^{\infty} = \left\{ \left\{ (1+|n|)^{\kappa-s/2} K_{n,h}(\tau) \right\}_{h=-s}^{s} \right\}_{-(s-1)}^{\infty} \tag{32_1}$$

and

$$\left\{ \left\{ \tilde{\Omega}_{n,h}^{(1)}(\tau) \right\}_{h=-s}^{s} \right\}_{-(s-1)}^{\infty} = \left\{ \left\{ (1+|n|)^{-\kappa-s/2} \omega_{n,h}^{(1)}(\tau) \right\}_{h=-s}^{s} \right\}_{-(s-1)}^{\infty} \tag{32_2}$$

are biothogonal and both are Riesz bases of $L_2^{2s+1}(0,\sigma)$.
2°. *If* $\nu \in \Delta_s(2^\circ)$, *then the systems*

$$\left\{ \{K_{n,h}(\tau)\}_{h=-s}^{s} \right\}_{-s}^{\infty} = \left\{ \left\{ (1+|n|)^{\kappa-s/2} K_{n,h}(\tau) \right\}_{h=-s}^{s} \right\}_{-s}^{\infty} \tag{33_1}$$

and

$$\left\{ \left\{ \tilde{\Omega}_{n,h}^{(2)}(\tau) \right\}_{h=-s}^{s} \right\}_{-s}^{\infty} = \left\{ \left\{ (1+|n|)^{-\kappa-s/2} \omega_{n,h}^{(2)}(\tau) \right\}_{h=-s}^{s} \right\}_{-s}^{\infty} \tag{33_2}$$

are biorthogonal and both are Riesz bases of $L_2^{2s+1}(0,\sigma)$.

Proof. 1°. As it was proved above, the systems (32_1) and (32_2) are biorthogonal. Besides, it was proved that the system (32_2) is a Riesz basis of $L_2^{2s+1}(0,\sigma)$. But the system (32_1) is complete in this space according to Theorem $7.4 - 1(1^\circ)$. Hence, by the well-known theorem on basis property in the Riesz sense of biorthogonal systems, which was mentioned in the proof of Theorem 5.3-1, the system (32_1) is also a Riesz basis of $L_2^{2s+1}(0,\sigma)$.
2°. The basis property of the systems (33_1) and (33_2) follows similarly from the corresponding assertions of Theorems 7.4-3 and 7.4-1.

7.5 Notes
The assertions of the main theorems of this chapter are obtained by M.M. Djrbashian-S.G. Raphaelian [4, §4–5] in the simplest case $s = 1$.

8 Interpolation series expansions in spaces $W_{s,\sigma}^{p;\omega}$ of entire functions

8.1 Introduction

In this chapter we establish interpolation series expansions in the Banach spaces $W_{s,\sigma}^{p;\omega}$ of entire functions $\Phi(z)$ of arbitrary natural order $s \geq 1$ and of type $\leq \sigma$, satisfying the condition

$$\left\{ \sum_{j=0}^{2s-1} \int_0^{+\infty} \left| \Phi\left(e^{i\frac{\pi j}{s}} r \right) \right|^p r^\omega \, dr \right\}^{1/p} < +\infty, \tag{1}$$

where it is assumed, as always, that

$$1 < p < +\infty \quad \text{and} \quad -1 < \omega < p - 1. \tag{2}$$

As in Chapter 6, the main result is established here by use of the concluding Theorems 4.4-1 and 4.4-2 of Chapter 4 establishing in two different ways expansions of entire functions of $W_{1/2,\sigma}^{p;\omega}$ in interpolation series. Thus, the auxiliary apparatus used here is similar to that of Chapter 6. Nevertheless, in contrast to the pair of *essentially different* Theorems 6.2-1 and 6.2-2 of Chapter 6, proved for the classes $W_{s+1/2,\sigma}^{p;\omega}$, in this chapter we arrive at a *single* theorem on expansion in interpolation series in the class $W_{s,\sigma}^{p;\omega}$. Finally note, that in this chapter we use mainly the notations introduced in Chapter 6.

8.2 The formulation of the main interpolation theorem

It is necessary first to introduce some definitions and notations which we use all over this chapter.

(a) We denote by

$$\Gamma_{2s} = \bigcup_{j=0}^{2s-1} \Gamma_{1,j} \qquad (s \geq 1) \tag{1}$$

the sum of rays

$$\Gamma_{1,j} = \left\{ z = r \, \exp\left\{ i\frac{\pi}{s}j \right\} : \ 0 \leq r < +\infty \right\} \tag{2}$$

and observe that the set Γ_2 ($s = 1$) coincides with the real axis. Further we introduce the Banach space $W_{s,\sigma}^{p;\omega}$ of entire functions $\Phi(z)$ of natural order $s \geq 1$ and of type $\leq \sigma$ as the space of those functions whose norm

$$\|\Phi; \Gamma_{2s}\|_{p,\omega} \equiv \left\{ \int_{\Gamma_{2s}} |\Phi(z)|^p |z|^\omega |dz| \right\}^{1/p}$$

$$= \left\{ \sum_{j=0}^{2s-1} \int_0^{+\infty} \left| \Phi\left(e^{i\frac{\pi j}{s}} r \right) \right|^p r^\omega \, dr \right\}^{1/p} \tag{3}$$

is finite.

(b) Using the function $\mathcal{E}_\sigma(z;\nu)$ defined earlier we introduce the entire function

$$\mathcal{E}_{s,\sigma}(z;\nu) \equiv \mathcal{E}_\sigma\left(z^{2s};\nu\right) = E_{1/2}\left(-\sigma^2 z^{2s};1+\nu\right) \tag{4}$$

which is obviously of order s and of type σ. According to Theorem 1.4-3, the zeros $\{\lambda_k\}_1^\infty$ ($0 < \lambda_k < \lambda_{k+1}, k \geq 1$) of $\mathcal{E}_\sigma(z;\nu)$ are simple and positive when $\nu \in [0,2)$. Hence, if $\nu \in [0,2)$, all the zeros of $\mathcal{E}_{s,\sigma}(z;\nu)$ are also simple and are situated on Γ_{2s}. Now denoting

$$\beta_s = \exp\left\{i\frac{\pi}{s}\right\}, \tag{5}$$

we can state that the set of zeros of $\mathcal{E}_{s,\sigma}(z;\nu)$ situated on each ray $\Gamma_{1,j} = [0,\beta_s^j\infty)$ is the sequence

$$\{\mu_{j,k}\}_1^\infty = \left\{\beta_s^j \lambda_k^{1/2s}\right\}_1^\infty \subset \Gamma_{1,j} \qquad (0 \leq j \leq 2s - 1), \tag{6}$$

and, obviously,

$$\mu_{j,k}^{2s} = \lambda_k \qquad (0 \leq j \leq 2s - 1, 1 \leq k < +\infty). \tag{7}$$

To introduce a universal numeration for all zeros of $\mathcal{E}_{s,\sigma}(z;\nu)$ ($\nu \in [0,2)$), we define the sequence

$$\mu_{2s(k-1)+j+1} = \mu_{j,k} \qquad (0 \leq j \leq 2s - 1, 1 \leq k < +\infty). \tag{8}$$

Then (6) and (7) give the two-sided inequalities

$$|\mu_{j,k}| \asymp (1 + k)^{1/s}, \quad |\mu_n| \asymp (1 + n)^{1/s} \quad (0 \leq j \leq 2s - 1, 1 \leq k, n < +\infty), \tag{9}$$

since $\lambda_k \asymp (1 + k)^2$ ($1 \leq k < +\infty$) according to Theorem 1.4-3.

(c) Let $\{\Phi_n\}_{-(s-1)}^\infty$ be an arbitrary sequence of complex numbers. We shall call the quantity

$$\|\{\Phi_n\}_{-(s-1)}^\infty\|_{p,\kappa_0} \equiv \left\{\sum_{n=0}^{s-1}\left|\frac{\Phi_{-n}}{n!}\right|^p + \sum_{n=1}^\infty |\Phi_n|^p (1+n)^{\kappa_0}\right\}^{1/p}, \tag{10}$$

where $\kappa_0 = (1+\omega)/s - 1$, *the norm* of $\{\Phi_n\}_{-(s-1)}^\infty$ and say that this sequence is of class $L^{(s-1)}_{p,\kappa_0}$, if the mentioned quantity is finite. Further, we denote

$$\gamma = \frac{1+\omega}{sp} \quad \left(0 < \gamma < \frac{1}{s} \leq 1\right), \tag{11}$$

then introduce the interval

$$\Delta_s = \left(\gamma + 1 - \frac{1}{s}, \gamma + 1\right) \subset (0,2) \tag{12}$$

and observe that, if $\nu \in \Delta_s$, the function $\mathcal{E}_{s,\sigma}(z;\nu)$ has a countable set of simple zeros $\{\mu_n\}_1^\infty \subset \Gamma_{2s}$.

(d) Now we can formulate the main interpolation theorem of this chapter, which we prove later, in the concluding Section 8.5.

Theorem 8.2-1. *Let $\{\mu_n\}_1^\infty \subset \Gamma_{2s}$ be the zeros of the function*

$$\mathcal{E}_{s,\sigma}(z;\nu), \quad \nu \in \Delta_s. \tag{13}$$

Then the series of the form

$$\Phi(z) = \Gamma(1+\nu)\left\{\sum_{n=0}^{s-1} \frac{\Phi_{-n}}{n!} z^n\right\} \mathcal{E}_{s,\sigma}(z;\nu) + \sum_{n=1}^{\infty} \Phi_n \frac{z^s \mathcal{E}_{s,\sigma}(z;\nu)}{\mu_n^s \mathcal{E}'_{s,\sigma}(\mu_n;\nu)(z-\mu_n)} \tag{14}$$

represents a continuous one-to-one mapping of the space $L_{p,\kappa_0}^{(s-1)}$ ($\kappa_0 = (1+\omega)/s-1$) of sequences $\{\Phi_n\}_{-(s-1)}^{\infty}$ onto the space $W_{s,\sigma}^{p,\omega}$ of entire functions $\Phi(z)$. In addition,
1°. A series (14) converges to its sum $\Phi(z)$ in the norm $\|.; \Gamma_{2s}\|_{p,\omega}$ of the space $W_{s,\sigma}^{p,\omega}$ and it also converges uniformly to the same limit in any disk $|z| \le R < +\infty$.
2°. The following two-sided inequalities are true:

$$\|\Phi; \Gamma_{2s}\|_{p,\omega} \asymp \|\{\Phi_n\}_{-(s-1)}^{\infty}\|_{p,\kappa_0}. \tag{15}$$

Here, as always, the suitable constants are independent of both estimated elements of Banach spaces.
3°. The function $\Phi(z)$ has the interpolation data

$$\Phi^{(n)}(0) = \Phi_{-n} \ (0 \le n \le s-1), \qquad \Phi(\mu_n) = \Phi_n \ (1 \le n < +\infty). \tag{16}$$

The following uniqueness theorem is an immediate consequence of both expansion (14) and the two-sided estimates (15) of the previous theorem.

Theorem 8.2-2. *If $\Phi(z) \in W_{s,\sigma}^{p,\omega}$ and $\{\mu_n\}_1^\infty \subset \Gamma_{2s}$ is the sequence of zeros of the function $\mathcal{E}_{s,\sigma}(z;\nu)$ ($\nu \in \Delta_s$), then equalities*

$$\Phi^{(n)}(0) = 0 \ (0 \le n \le s-1), \qquad \Phi(\mu_n) = 0 \ (1 \le n < +\infty) \tag{17}$$

imply the identity

$$\Phi(z) \equiv 0. \tag{18}$$

(e) Now we shall mention some particular cases and consequences of the main interpolation Theorem 8.2-1, which are of independent interest.
Note that for any natural $s \ge 1$

$$\nu = 1 \in \Delta_s = \left(\frac{1+\omega}{sp} + 1 - \frac{1}{s}, \frac{1+\omega}{sp} + 1\right) \tag{19}$$

and

$$\mathcal{E}_{s,\sigma}(z;1) = E_{1/2}\left(-\sigma^2 z^{2s}; 2\right) = \frac{\sin \sigma z^s}{\sigma z^s}. \tag{20}$$

Therefore, the sequence of zeros $\{\mu_n\}_1^\infty \subset \Gamma_{2s}$ of the last function can be expressed in the following explicit form:

$$\{\mu_n\}_1^\infty = \bigcup_{j=0}^{2s-1} \{\mu_{j,k}\}_1^\infty, \tag{21}$$

where

$$\mu_{j,k} = e^{i\frac{\pi j}{s}} \left(\frac{\pi k}{\sigma}\right)^{1/s} \qquad (0 \le j \le 2s-1, 1 \le k < +\infty). \tag{22}$$

As follows from formulas (20) and (22), the following equalities are true for any j and $k(0 \le j \le 2s-1, 1 \le k < +\infty)$:

$$\mu_{j,k}^s \mathcal{E}_{s,\sigma}'(\mu_{j,k};1) = (-1)^k s e^{i\pi(1-1/s)j} \left(\frac{\pi k}{\sigma}\right)^{1-1/s}, \tag{23}$$

$$\frac{z^s \mathcal{E}_{s,\sigma}(z;1)}{\mu_{j,k}^s \mathcal{E}_{s,\sigma}'(\mu_{j,k};1)(z - \mu_{j,k})}$$
$$= (-1)^k e^{-i\pi(1-1/s)j} \frac{\sin \sigma z^s}{\sigma s \left(\frac{\pi}{\sigma}k\right)^{1-1/s} \left[z - e^{i\pi j/s} \left(\frac{\pi}{\sigma}k\right)^{1/s}\right]}. \tag{24}$$

The first particular case we mention is when $s = 1$.

Theorem 8.2-3. 1°. *Let $\Phi(z) \in W_{1,\sigma}^{p,\omega}$ and consequently*

$$\|\Phi; \Gamma_2\|_{p,\omega} = \left\{\int_{-\infty}^{+\infty} |\Phi(x)|^p |x|^\omega dx\right\}^{1/p} < +\infty. \tag{25}$$

Then this function can be expanded in the series

$$\Phi(z) = \Phi(0)\frac{\sin \sigma z}{\sigma z} + \sum_{n \ne 0} (-1)^n \Phi\left(\frac{\pi n}{\sigma}\right) \frac{\sin \sigma z}{\sigma z - \pi n}$$
$$= \sum_{n=-\infty}^{+\infty} (-1)^n \Phi\left(\frac{\pi n}{\sigma}\right) \frac{\sin \sigma z}{\sigma z - \pi n} \tag{26}$$

converging in the norm (25) and converging uniformly to the same limit in any disk $|z| \le R < +\infty$. And

$$\|\Phi; \Gamma_2\|_{p,\omega}^p \asymp \sum_{n=-\infty}^{+\infty} \left|\Phi\left(\frac{\pi n}{\sigma}\right)\right|^p (1 + |n|)^\omega. \tag{27}$$

$2°$. *Let $\Phi(z)$ be an entire function of exponential type $\leq \sigma$ and also let*

$$\int_{|x|\geq 1}\left|\frac{\Phi(x)}{x}\right|^p |x|^\omega dx < +\infty. \tag{28}$$

Then this function can be expanded in the series

$$\Phi(z) = \Phi(0)\frac{\sin\sigma z}{\sigma z} + \Phi'(0)\frac{\sin\sigma z}{\sigma} + \sigma z \sum_{n\neq 0}(-1)^n\Phi\left(\frac{\pi n}{\sigma}\right)\frac{\sin\sigma z}{\pi n(\sigma z - \pi n)} \tag{29}$$

which converges uniformly in any disk $|z| \leq R < +\infty$. And

$$\int_{-\infty}^{+\infty}\left|\frac{\Phi(x) - \Phi(0)\frac{\sin\sigma x}{\sigma x}}{x}\right|^p |x|^\omega dx \asymp |\Phi'(0)|^p + \sum_{n\neq 0}\left|\Phi\left(\frac{\pi n}{\sigma}\right)\right|^p (1+|n|)^{\omega-p}. \tag{30}$$

Proof. $1°$. If $s = 1$, then we can numerate the set of numbers (22) as follows:

$$\mu_{0,k} = \mu_{2k-1} = \frac{\pi k}{\sigma}, \quad \mu_{1,k} = \mu_{2k} = -\frac{\pi k}{\sigma} \quad (1 \leq k < +\infty). \tag{31}$$

Then (24) becomes

$$\frac{z\mathcal{E}_{1,\sigma}(z;1)}{\mu_{j,k}\mathcal{E}'_{1,\sigma}(\mu_{j,k};1)(z-\mu_{j,k})} = \frac{(-1)^k\sin\sigma z}{\sigma\left(z-(-1)^j\frac{\pi k}{\sigma}\right)} \quad (j = 0,1; 1 \leq k < +\infty). \tag{32}$$

Therefore, expansion (14) of Theorem 8.2-1 can be written down in the form (26), and the two-sided inequalities (15) in the form (27).

$2°$. Consider the function

$$\Psi(z) = \frac{1}{z}\left\{\Phi(z) - \Phi(0)\frac{\sin\sigma z}{\sigma z}\right\} \tag{33}$$

satisfying the conditions

$$\Psi(0) = \Phi'(0), \quad \Psi\left(\frac{\pi n}{\sigma}\right) = \Phi\left(\frac{\pi n}{\sigma}\right)\Big/\frac{\pi n}{\sigma} \quad (n \neq 0). \tag{34}$$

Relations (33) and (28) imply $\Psi(z) \in W^{p,\omega}_{1,\sigma}$. Hence, according to the assertion $1°$ which is already proved,

$$\frac{1}{z}\left\{\Phi(z) - \Phi(0)\frac{\sin\sigma z}{\sigma z}\right\} = \sum_{n=-\infty}^{+\infty}(-1)^n\Psi\left(\frac{\pi n}{\sigma}\right)\frac{\sin\sigma z}{\sigma z - \pi n}. \tag{35}$$

Expansion (29) and the two-sided inequalities (30) now follow from formulas (34) and (35).

Corollary. *Let $\Phi(z)$ be an entire function of exponential type $\leq \sigma$ and let*

$$\Phi(x) = O(|x|^{\alpha}), \qquad |x| \to +\infty, \tag{36}$$

where $\alpha < 1$ is a given number. Then $\Phi(z)$ can be expanded in the interpolation series (29). Indeed, it is enough to observe that integral (28) is finite, if (36) is true and $\omega \in (-1, p(1 - \alpha) - 1) \cap (-1, p - 1)$.

(f) As the second particular case of the main interpolation Theorem 8.2-1 we consider the case $s = 2$.

Theorem 8.4-2. *Let $\Phi(z) \in W_{2,\sigma}^{p,\omega}$ and consequently*

$$\|\Phi; \Gamma_4\|_{p,\omega} = \left\{ \int_{-\infty}^{+\infty} |\Phi(x)|^p |x|^{\omega} dx + \int_{-\infty}^{+\infty} |\Phi(iy)|^p |y|^{\omega} dy \right\}^{1/p} < +\infty. \tag{37}$$

Then this function can be expanded in the series

$$\Phi(z) = \Phi(0) \frac{\sin \sigma z^2}{\sigma z^2} + \Phi'(0) \frac{\sin \sigma z^2}{\sigma z}$$

$$+ \sin \sigma z^2 \sum_{n \neq 0} (-1)^n \frac{\operatorname{sign} n}{2\sqrt{\pi |n|}} \left\{ \frac{\Phi\left(\sqrt{\frac{\pi |n|}{\sigma}} \operatorname{sign} n\right)}{\sqrt{\sigma} z - \sqrt{\pi |n|} \operatorname{sign} n} \right. \tag{38}$$

$$\left. - \frac{i\Phi\left(i\sqrt{\frac{\pi |n|}{\sigma}} \operatorname{sign} n\right)}{\sqrt{\sigma} z - i\sqrt{\pi |n|} \operatorname{sign} n} \right\}$$

converging to $\Phi(z)$ in the norm (37) and also uniformly in any disk $|z| \leq R < +\infty$. In addition, the following two-sided inequalities are true:

$$\|\Phi; \Gamma_4\|_{p,\omega}^p \asymp |\Phi(0)|^p + |\Phi'(0)|^p$$

$$+ \sum_{n \neq 0} \left\{ \left| \Phi\left(\sqrt{\frac{\pi |n|}{\sigma}} \operatorname{sign} n\right) \right|^p + \left| \Phi\left(i\sqrt{\frac{\pi |n|}{\sigma}} \operatorname{sign} n\right) \right| \right\} (1 + |n|)^{\frac{\omega - 1}{2}}. \tag{39}$$

Proof. If $s = 2$, then (24) implies

$$\frac{z^2 \mathcal{E}_{2,\sigma}(z; 1)}{\mu_{j,k}^2 \mathcal{E}'_{2,\sigma}(\mu_{j,k}; 1)(z - \mu_{j,k})} = (-1)^k \frac{e^{-i\pi j/2} \sin \sigma z^2}{2\sqrt{\pi k}\left(\sqrt{\sigma} z - e^{i\pi j/2}\sqrt{\pi k}\right)} \tag{40}$$

for any $j, k (j = 0, 1, 2, 3; k \geq 1)$. But the sequence $\{\mu_n\}_1^{\infty} \subset \Gamma_4$ of zeros of the function

$$E_{1/2}\left(-\sigma^2 z^4; 2\right) = \frac{\sin \sigma z^2}{\sigma z^2}$$

can be expressed in the form $\{\mu_n\}_1^\infty = \bigcup_{j=0}^3 \{\mu_{j,k}\}_1^\infty$. Thus expansion (14) turns into

$$\Phi(z) = \Phi(0)\frac{\sin\sigma z^2}{\sigma z^2} + \Phi'(0)\frac{\sin\sigma z^2}{\sigma z}$$
$$+ \sum_{k=1}^\infty \sum_{j=0}^3 \frac{z^2 \mathcal{E}_{2,\sigma}(z;1)\Phi(\mu_{j,k})}{\mu_{j,k}^2 \mathcal{E}'_{2,\sigma}(\mu_{j,k};1)(z-\mu_{j,k})}. \tag{41}$$

If we denote by $S_k(z)$ the sum over $j(0 \le j \le 3)$ and take into account formulas (40), then we arrive at the equality

$$S_k(z) = (-1)^k \Bigg\{ \frac{\Phi\left(\sqrt{\frac{\pi k}{\sigma}}\right)}{2\sqrt{\pi k}(\sqrt{\sigma}z - \sqrt{\pi k})} - \frac{\Phi\left(-\sqrt{\frac{\pi k}{\sigma}}\right)}{2\sqrt{\pi k}(\sqrt{\sigma}z + \sqrt{\pi k})}$$
$$- \frac{i\Phi\left(i\sqrt{\frac{\pi k}{\sigma}}\right)}{2\sqrt{\pi k}(\sqrt{\sigma}z - i\sqrt{\pi k})} + \frac{i\Phi\left(-i\sqrt{\frac{\pi k}{\sigma}}\right)}{2\sqrt{\pi k}(\sqrt{\sigma}z + i\sqrt{\pi k})} \Bigg\} \sin\sigma z^2 \quad (1 \le k < +\infty).$$

Now, if we insert this into (41) and do the suitable simplifications, then we obtain expansion (38). The two-sided inequalities (39) follow from (15).

8.3 Auxiliary relations and lemmas

(a) Let

$$\Phi(z) = \sum_{n=0}^\infty c_n z^n \tag{1}$$

be an entire function of natural order $s \ge 1$ and of type σ. Then the relations 6.2(2) become

$$\limsup_{n\to\infty} \frac{n\ln n}{\ln 1/|c_n|} = s, \quad \limsup_{n\to\infty} n|c_n|^{s/n} = e\sigma s. \tag{2}$$

Now introduce the set $\{\varphi_j(w)\}_0^{2s-1}$ of $2s$ entire functions

$$\varphi_j(w) = \sum_{n=0}^\infty c_{n_j} w^n, \quad n_j = 2sn + j \ (0 \le j \le 2s-1, \ 0 \le n < +\infty) \tag{3}$$

and observe that, if ρ_j and σ_j correspondingly are the order and the type of $\varphi_j(w)$, then (2) and (3) imply $0 \le \rho_j \le 1/2$ and $0 \le \sigma_j \le \sigma$ when $\rho_j = 1/2$. The following lemma, which is based on the elementary equalities

$$\sum_{h=0}^{2s-1} \beta_s^{kh} = \begin{cases} 2s, & \text{if } k \equiv 0(\mod 2s) \\ 0, & \text{if } k \not\equiv 0(\mod 2s) \end{cases} \quad \beta_s = \exp\left\{i\frac{\pi}{s}\right\} \tag{4}$$

(see 1.2(13)), relates to the connection between the functions $\Phi(z)$ and $\varphi_j(w)(0 \le j \le 2s-1)$.

Lemma 8.3-1. $1°$. *The representations*

$$z^j \varphi_j(z^{2s}) = \frac{1}{2s} \sum_{h=0}^{2s-1} \beta_s^{-jh} \Phi\left(\beta_s^h z\right) \quad (0 \le j \le 2s - 1) \tag{5}$$

and their inversion

$$\Phi(z) = \sum_{j=0}^{2s-1} z^j \varphi_j\left(z^{2s}\right) \tag{6}$$

are true.

$2°$. *There exists in* $\{\varphi_j(w)\}_0^{2s-1}$ *at least one function of order* $1/2$ *and type* σ.

Proof. $1°$. Representations (5) and (6) follow from (1) and (3) and from equalities (4).

$2°$. If $\rho_j < 1/2$, or $\rho_j = 1/2$ and $\sigma_j < \sigma$ for all $j(0 \le j \le 2s - 1)$, then identity (6) leads to a contradiction $-\Phi(z)$ has either an order $\rho < s$, or else its order is $\rho = s$ in which case its type is less than σ. This completes the proof.

Note that, if we take $\Phi(z) = E_s(z; \mu)$, then formulas (5) and (6) pass to the identities 1.2(15) written for $\rho = 1/2$. Hence the following lemma is true.

Lemma 8.3-2. *If* $s \ge 1$ *is any natural number, then*

$$z^j E_{1/2}\left(z^{2s}; \mu + \frac{j}{s}\right) = \frac{1}{2s} \sum_{h=0}^{2s-1} \beta_s^{-jh} E_s\left(\beta_s^h z; \mu\right) \quad (0 \le j \le 2s - 1) \tag{7}$$

and

$$E_s(z; \mu) = \sum_{j=0}^{2s-1} z^j E_{1/2}\left(z^{2s}; \mu + \frac{j}{s}\right). \tag{8}$$

(b) Remember that in 4.1(1) we had accepted the quantity

$$\|\varphi; \Gamma_1\|_{p,\omega} = \|\varphi\|_{p,\omega}^+ = \left\{\int_0^{+\infty} |\varphi(r)|^p r^\omega dr\right\}^{1/p} \tag{9}$$

to be the norm of a function $\varphi(w) \in W^{p,\omega}_{1/2,\sigma}$ $(1 < p < +\infty, -1 < \omega < p - 1)$. Further, we introduce the parameters

$$\omega_j = \frac{\omega - 2s + pj + 1}{2s} \quad (0 \le j \le 2s - 1) \tag{10}$$

and prove the following analog of Lemma 6.3-3.

Lemma 8.3-3. $1°$. *The class $W_{s,\sigma}^{p,\omega}$ ($s \geq 1$) coincides with the set of functions representable in the form*

$$\Phi(z) = \sum_{j=0}^{2s-1} z^j \varphi_j\left(z^{2s}\right), \tag{11}$$

where

$$\varphi_j(w) \in W_{1/2,\sigma}^{p,\omega_j} \quad (0 \leq j \leq 2s-1) \text{ and } \varphi_j\left(z^{2s}\right) = \frac{1}{2s} \sum_{h=0}^{2s-1} \beta_s^{-jh} z^{-j} \Phi\left(\beta_s^h z\right). \tag{12}$$

$2°$. *If (11) and (12) are true, then*

$$\|\Phi;\Gamma_{2s}\|_{p,\omega}^{\delta} \asymp \sum_{j=0}^{2s-1} \|\varphi_j;\Gamma_1\|_{p,\omega_j}^{\delta} = \sum_{j=0}^{2s-1} \left\{\|\varphi_j\|_{p,\omega_j}^{+}\right\}^{\delta} \quad (\delta = 1,p). \tag{13}$$

Proof. $1°$. It is obvious that

$$\int_0^{+\infty} \left|r^j \varphi_j\left(r^{2s}\right)\right|^p r^\omega dr = (2s)^{-1}\|\varphi_j;\Gamma_1\|_{p,\omega_j}^p \quad (0 \leq j \leq 2s-1).$$

but $z^{2s} = r^{2s}$ when $z \in \Gamma_{2s}$, i.e., when $z = \beta_s^j r$ ($0 \leq r < +\infty, 0 \leq j \leq 2s-1$). Hence the last equalities can be written in the form

$$\int_{\Gamma_{2s}} \left|z^j \varphi_j\left(z^{2s}\right)\right|^p |z|^\omega |dz| = \|\varphi_j;\Gamma_1\|_{p,\omega_j}^p \quad (0 \leq j \leq 2s-1). \tag{14}$$

If now we suppose that the inclusions of (12) and representation (11) are true, then using Minkowski's inequality and equalities (14) we obtain

$$\|\Phi;\Gamma_{2s}\|_{p,\omega} \leq \sum_{j=0}^{2s-1} \|\varphi_j;\Gamma_1\|_{p,\omega_j}. \tag{15}$$

Thus $\Phi(z) \in W_{s,\sigma}^{p,\omega}$. Further, suppose $\Phi(z) \in W_{s,\sigma}^{p,\omega}$ is any function. Then, using formula (3), we introduce the functions $\varphi_j(w)$ ($0 \leq j \leq 2s-1$) which can be represented by $\Phi(z)$ as in (5), according to Lemma 8.3-1. Using once again (14) and Minkowski's inequality we obtain

$$\|\varphi_j;\Gamma_1\|_{p,\omega_j} = (2s)^{-1} \left\{\int_{\Gamma_{2s}} \left|\sum_{h=0}^{2s-1} \beta_s^{-jh} \Phi\left(\beta_s^h z\right)\right|^p |z|^\omega |dz|\right\}^{1/p}$$

$$\leq (2s)^{-1} \sum_{h=0}^{2s-1} \left\{\int_{\Gamma_{2s}} \left|\Phi\left(\beta_s^h z\right)\right|^p |z|^\omega |dz|\right\}^{1/p} \quad (0 \leq j \leq 2s-1).$$

And, since $w = \beta_s^h z$ maps the sum of rays Γ_{2s} into itself for any $h(0 \leq h \leq 2s-1)$, we arrive at the inequalities

$$\|\varphi_j; \Gamma_1\|_{p,\omega_j}^{\delta} \leq \|\Phi; \Gamma_{2s}\|_{p,\omega}^{\delta} \qquad (0 \leq j \leq 2s-1, \delta = 1, p).$$

Consequently, $\varphi_j(w) \in W_{1/2,\sigma}^{p,\omega_j}$ $(0 \leq j \leq 2s-1)$ and, besides,

$$\|\Phi; \Gamma_{2s}\|_{p,\omega}^{\delta} \geq (2s)^{-1} \sum_{j=0}^{2s-1} \|\varphi_j; \Gamma_1\|_{p,\omega_j}^{\delta} \, (\delta = 1, p). \tag{16}$$

$2°$. If we observe that (15) gives

$$\|\Phi; \Gamma_{2s}\|_{p,\omega}^{p} \leq (2s)^p \sum_{j=0}^{2s-1} \|\varphi_j; \Gamma_1\|_{p,\omega_j}^{p}, \tag{15'}$$

then we conclude that the two-sided inequalities (13) are true.

Remark 1. If any of the functions $\varphi_j(w)$ $(0 \leq j \leq 2s-1)$ of representation (11) of $\Phi(z) \in W_{s,\sigma}^{p,\omega}$ is of order $\rho_j < 1/2$, then it turns out that $\varphi_j(w) \equiv 0$.

Remark 2. According to definitions (10),

$$\frac{pj}{2s} - 1 < \omega_j < \frac{p(j+1)}{2s} - 1 \qquad (0 \leq j \leq 2s-1). \tag{17}$$

Hence the intervals in which parameters ω_j vary have no common points, and the sum of these intervals and their common endpoints $pj/2s - 1$ $(1 \leq j \leq 2s-1)$ coincides with the whole interval $(-1, p-1)$ in which the parameter ω varies.

(c) Now we put $z = \lambda_k^{1/2s}$ and $z = \mu_{h,k} = \beta_s^h \lambda_k^{1/2s}$ correspondingly in formulas (5) and (6) of Lemma 8.3-1 and obtain the following pair of relations

$$\varphi_j(\lambda_k) = \frac{1}{2s} \sum_{h=0}^{2s-1} \mu_{h,k}^{-j} \Phi(\mu_{h,k}) \qquad (0 \leq j \leq 2s-1, 1 \leq k < +\infty), \tag{18}$$

$$\Phi(\mu_{h,k}) = \sum_{j=0}^{2s-1} \mu_{h,k}^{j} \varphi_j(\lambda_k) \qquad (0 \leq h \leq 2s-1, 1 \leq k < +\infty) \tag{19}$$

connecting the sequences $\{\Phi(\mu_{h,k})\}_1^{\infty}$ $(0 \leq h \leq 2s-1)$ and $\{\varphi_j(\lambda_k)\}_1^{\infty}$ $(0 \leq j \leq 2s-1)$. If

$$\{\Phi_{h,k}\}_1^{\infty} \quad (0 \leq h \leq 2s-1) \quad \text{and} \quad \{\varphi_{j,k}\}_1^{\infty} \quad (0 \leq j \leq 2s-1) \tag{20}$$

are any sequences of complex numbers, then the following analog of Lemma 6.3-4 is true.

Lemma 8.3-4. *If one of the equalities*

$$\varphi_{j,k} = \frac{1}{2s} \sum_{h=0}^{2s-1} \mu_{h,k}^{-j} \Phi_{h,k}, \tag{21}$$

$$\Phi_{h,k} = \sum_{j=0}^{2s-1} \mu_{h,k}^{j} \varphi_{j,k} \tag{22}$$

is true, the other one is also true.

Proof. Indeed, (22) follows from (21), since using the equalities

$$\sum_{j=0}^{2s-1} \left(\frac{\mu_{h,k}}{\mu_{i,k}} \right)^{j} = \sum_{j=0}^{2s-1} \beta_s^{(h-i)j} = \begin{cases} 2s, & \text{when } i = h \\ 0, & \text{when } i \neq h \end{cases}$$

we obtain

$$\sum_{j=0}^{2s-1} \mu_{h,k}^{j} \varphi_{j,k} = \frac{1}{2s} \sum_{j=0}^{2s-1} \mu_{h,k}^{j} \sum_{i=0}^{2s-1} \mu_{i,k}^{-j} \Phi_{i,k}$$

$$= \frac{1}{2s} \sum_{i=0}^{2s-1} \Phi_{i,k} \sum_{j=0}^{2s-1} \left(\frac{\mu_{h,k}}{\mu_{i,k}} \right)^{j} = \Phi_{h,k}.$$

Conversely, if (22) is true, then using the equalities

$$\sum_{h=0}^{2s-1} \mu_{h,k}^{i-j} = \lambda_k^{\frac{i-j}{2s}} \sum_{h=0}^{2s-1} \beta_s^{(i-j)h} = \begin{cases} 2s, & \text{when } i = j \\ 0, & \text{when } i \neq j \end{cases}$$

we obtain (21) in a similar way.

(d) The relations

$$\Phi_n = \Phi_{h,k}, \quad n = 2s(k-1) + h + 1 \ (0 \leq h \leq 2s - 1, 1 \leq k < +\infty) \tag{23}$$

enumerate the set of sequences of any complex numbers $\{\Phi_{h,k}\}_1^\infty$ $(0 \leq h \leq 2s-1)$. And, obviously,

$$\{\Phi_n\}_1^\infty = \bigcup_{h=0}^{2s-1} \{\Phi_{h,k}\}_1^\infty. \tag{24}$$

Remember now that, as was defined earlier, a sequence of complex numbers $\{c_n\}_k^\infty$ $(k = 0, 1)$ is of class $l^{p,\kappa}$, if

$$\|\{c_n\}_k^\infty\|_{p,\kappa} = \left\{ \sum_{n=k}^{\infty} |c_n|^p (1 + n)^\kappa \right\}^{1/p} < +\infty \qquad (k = 0, 1). \tag{25}$$

Further, we introduce the parameters

$$\kappa_j = 1 + 2\omega_j = \frac{1 + \omega + pj}{s} - 1 \ (0 \leq j \leq 2s - 1), \quad \kappa_0 = \frac{1 + \omega}{s} - 1 \tag{26}$$

and prove the following assertion.

Lemma 8.3-5. *If the sequences $\{\Phi_{h,k}\}_1^\infty$ $(0 \le h \le 2s-1)$ and $\{\varphi_{j,k}\}_1^\infty$ $(0 \le j \le 2s-1)$ are connected by formulas (21)-(22), and $\{\Phi_n\}_1^\infty$ is defined by (23), then:*
1°. The following two-sided inequalities are true:

$$\|\{\Phi_n\}_1^\infty\|_{p,\kappa_0}^\delta \asymp \sum_{j=0}^{2s-1} \|\{\varphi_{j,k}\}_1^\infty\|_{p,\kappa_j}^\delta \qquad (\delta = 1,p). \tag{27}$$

2°. The inclusions

$$\{\Phi_n\}_1^\infty \in l^{p,\kappa_0} \quad \text{and} \quad \{\varphi_{j,k}\}_1^\infty \in l^{p,\kappa_j} \quad (0 \le j \le 2s-1) \tag{28}$$

are equivalent.

Proof. 1°. Using the relations 8.2(9) we obtain

$$|\varphi_{j,k}|^p \le C_1 \sum_{h=0}^{2s-1} |\Phi_{h,k}|^p (1+k)^{-pj/s} \quad (0 \le j \le 2s-1),$$

$$|\Phi_{h,k}|^p \le C_2 \sum_{j=0}^{2s-1} |\varphi_{j,k}|^p (1+k)^{pj/s} \quad (0 \le h \le 2s-1), \tag{29}$$

where the constants $C_{1,2} > 0$ depend only on p, s, σ and ν (such constants, even if they depend also on ω, will be denoted below by C_m $(m = 3,4,\ldots)$). But $\kappa_j = \kappa_0 + pj/s$ $(0 \le j \le 2s-1)$ by (26). So (29) implies

$$|\varphi_{j,k}|^p(1+k)^{\kappa_j} \le C_1 \sum_{h=0}^{2s-1} |\Phi_{h,k}|^p (1+k)^{\kappa_0} \quad (0 \le j \le 2s-1, 1 \le k < +\infty), \tag{30_1}$$

$$|\Phi_{h,k}|^p(1+k)^{\kappa_0} \le C_2 \sum_{j=0}^{2s-1} |\varphi_{j,k}|^p (1+k)^{\kappa_j} \quad (0 \le h \le 2s-1, 1 \le k < +\infty). \tag{30_2}$$

Further, we sum up the inequalities (30_1) over $k(1 \le k < +\infty)$ and, taking into account (24) and (23), we obtain that for any $j(0 \le j \le 2s-1)$

$$\sum_{k=1}^\infty |\varphi_{j,k}|^p(1+k)^{\kappa_j} = \|\{\varphi_{j,k}\}_1^\infty\|_{p,\kappa_j}^p \le C_1 \sum_{k=1}^\infty \sum_{h=0}^{2s-1} |\Phi_{h,k}|^p(1+k)^{\kappa_0}$$

$$\le C_3 \sum_{k=1}^\infty \sum_{h=0}^{2s-1} |\Phi_{h,k}|^p \left(1+2s(k-1)+h+1\right)^{\kappa_0} \le C_3 \sum_{n=1}^\infty |\Phi_n|^p(1+n)^{\kappa_0}. \tag{31}$$

Now we sum up these inequalities over $j(0 \le j \le 2s-1)$ and arrive at the estimates

$$\sum_{j=0}^{2s-1} \|\{\varphi_{j,k}\}_1^\infty\|_{p,\kappa_j}^\delta \le C_4 \|\{\Phi_n\}_1^\infty\|_{p,\kappa_0}^\delta \qquad (\delta = 1,p).$$

To be convinced that an estimate converse to the last one is also true (whence follows the two-sided estimate (27)), we sum up the inequalities (30_2) over $k(1 \le k < +\infty)$ and over $h(0 \le h \le 2s-1)$ and use relations (23) and (24).

2°. The equivalence of the inclusions (28) follows from the estimates (27).

(e) The following lemma concludes this section.

Lemma 8.3-6. *If $\Phi(z) \in W_{s,\sigma}^{p,\omega}$, and the set of functions $\{\varphi_j(w)\}_0^{2s-1}$ is defined by formula (5) of Lemma 8.3-1, then*

$$\varphi_{j,0} \equiv \varphi_j(0) = \frac{\Phi^{(j)}(0)}{j!} \qquad (0 \le j \le 2s-1). \tag{32}$$

Proof. Obviously

$$\Omega_{s,j}(z) \equiv \frac{d^j}{dz^j}\left\{ z^j \varphi_j\left(z^{2s}\right)\right\}$$

$$= \sum_{k=0}^{j} C_j^k j(j-1)\dots(j-k+1) z^{j-k} \frac{d^{j-k}}{dz^{j-k}}\left\{\varphi_j\left(z^{2s}\right)\right\}.$$

Hence $\Omega_{s,j}(0) = j!\varphi_j(0)(0 \le j \le 2s-1)$. But by representation (5),

$$\Omega_{s,j}(z) = \frac{1}{2s} \sum_{h=0}^{2s-1} \Phi^{(j)}\left(\beta_s^h z\right).$$

This gives $\Omega_{s,j}(0) = \Phi^{(j)}(0)$ $(0 \le j \le 2s-1)$, whence follows formula (32).

Later we shall associate with a given function $\Phi(z) \in W_{s,\sigma}^{p,\omega}$ not only the sequence $\{\varphi_{j,0}\}_0^{2s-1}$, but also the numbers

$$\Phi_{-k} \equiv \Phi^{(k)}(0) \qquad (k = 0, 1, 2, \dots). \tag{33}$$

Then formulas (32) and (33) give

$$P_{s-1}(z; \Phi) \equiv \sum_{j=0}^{s-1} \varphi_{j,0} z^j = \sum_{j=0}^{s-1} \Phi_{-j} \frac{z^j}{j!}. \tag{34}$$

8.4 Further auxiliary results

Here we shall frequently use the notations introduced in the previous sections of this chapter. Suitable references will be given whenever necessary.

(a) We introduce the following sets of intervals:

$$\begin{aligned}
\Delta_j &= \left(\frac{2(1+\omega_j)}{p} - 1, \frac{2(1+\omega_j)}{p}\right) \\
\Delta_j^* &= \left(\frac{2(1+\omega_j)}{p}, 1 + \frac{2(1+\omega_j)}{p}\right)
\end{aligned} \qquad (0 \le j \le 2s-1), \tag{1}$$

where $\omega_j = (\omega + 1 + pj)/2s - 1$ $(0 \le j \le 2s-1)$. But we had also introduced the parameters $\kappa_j = 1 + 2\omega_j = (\omega + 1 + pj)/s - 1$. Using them we can write

$2(1+\omega_j)/p = (1+\kappa_j)/p$ $(0 \le j \le 2s-1)$. Thus the intervals (1) can also be written in the form

$$
\begin{aligned}
\Delta_j &= \left(\frac{1+\kappa_j}{p} - 1, \frac{1+\kappa_j}{p}\right) = \left(\gamma + \frac{j}{s} - 1, \gamma + \frac{j}{s}\right) \\
\Delta_j^* &= \left(\frac{1+\kappa_j}{p}, 1 + \frac{1+\kappa_j}{p}\right) = \left(\gamma + \frac{j}{s}, 1 + \gamma + \frac{j}{s}\right)
\end{aligned} \quad (0 \le j \le 2s-1), \quad (2)
$$

where

$$
\gamma = \frac{1+\omega}{ps} \quad \text{and} \quad 0 < \gamma < \frac{1}{s} \le 1 \quad (s = 1, 2, \dots), \tag{3}
$$

since $-1 < \omega < p-1$. By the notations of Chapter 4,

$$
\Delta_j \cap [0,2) = \Delta(\kappa_j, p), \quad \Delta_j^* \cap [0,2) = \Delta^*(\kappa_j, p) \quad (0 \le j \le 2s-1). \tag{1'}
$$

Further, we suppose

$$
J \subset \{j\}_0^{2s-1} \tag{4}
$$

is an arbitrary set of indices and denote

$$
\Delta_J = \bigcap_{j \in J} \Delta_j, \quad \Delta_J^* = \bigcap_{j \in J} \Delta_j^*. \tag{5}
$$

We also denote by

$$
J^* = \{j\}_0^{2s-1} \backslash J \tag{6}
$$

the complementary set of indices and prove an assertion which will play a role similar to that of Lemma 6.4-1.

Lemma 8.4-1. *The intersection of the sums of intervals $\Delta_{J^*}^*$ and Δ_J is not empty, i.e.,*

$$
\Delta_{J^*}^* \bigcap \Delta_J \ne \emptyset \tag{7}
$$

only in the case when

$$
J^* = \{j\}_0^{s-1} \quad \text{and} \quad J = \{j\}_s^{2s-1}, \tag{8}
$$

and in this case

$$
\Delta_s \equiv \left\{\bigcap_{j=0}^{s-1} \Delta_j^*\right\} \bigcap \left\{\bigcap_{j=s}^{2s-1} \Delta_j\right\} = \left(\gamma + 1 - \frac{1}{s}, \gamma + 1\right) \subset (0,2). \tag{9}
$$

Proof. If $J = (j_1, j_2, \ldots, j_r)$, where $0 \le j_1 < j_2 < \ldots < j_r \le 2s - 1$, then, by (2), $\Delta_J^* = \Delta_J = \emptyset$ in the case when $j_r - j_1 > s - 1$. Particularly, it is so when $r > s$. Therefore, if (7) is true, then $J = (j_1, j_2, \ldots j_s)$, $J^* = (j_1^*, j_2^*, \ldots, j_s^*)$ and $j_{k+1} = j_k + 1, j_{k+1}^* = j_k^* + 1$ $(1 \le k < s)$. Consequently,

$$J = \{j\}_0^{s-1} \quad \text{and} \quad J^* = \{j\}_s^{2s-1} \tag{10}$$

or we shall have case (8). But (7) is not true when (10) is valid, since $\Delta_i \cap \Delta_j^* = \emptyset$ when $0 \le i \le s - 1$ and $s \le j \le 2s - 1$. So only (8) remains to be true. Formula (9) can be easily verified.

(b) The following theorem is an immediate consequence of the main results of Chapter 4 and of the preceding Lemma 8.4-1. Besides, it is similar to Theorems 6.4-1 and 6.4-2.

Theorem 8.4-1. *Let*

$$\nu \in \Delta_s = \left(\gamma + 1 - \frac{1}{s}, \gamma + 1 \right), \quad \gamma = \frac{1+\omega}{ps} \quad (s \ge 1). \tag{12}$$

Then the series

$$\varphi_j(w) = \varphi_{j,0}\Gamma(1+\nu)\mathcal{E}_\sigma(w;\nu) + \sum_{k=1}^{\infty} \varphi_{j,k} \frac{w\mathcal{E}_\sigma(w;\nu)}{\lambda_k \mathcal{E}_\sigma'(\lambda_k;\nu)(w - \lambda_k)} \quad (0 \le j \le s-1), \tag{13}$$

and

$$\varphi_j(w) = \sum_{k=1}^{\infty} \varphi_{j,k} \frac{\mathcal{E}_\sigma(w;\nu)}{\mathcal{E}_\sigma'(\lambda_k;\nu)(w - \lambda_k)} \quad (s \le j \le 2s - 1) \tag{14}$$

represent continuous one-to-one mappings of the spaces of sequences

$$\{\varphi_{j,k}\}_0^\infty \in l^{p,\kappa_j} \quad (0 \le j \le s - 1) \quad \text{and} \quad \{\varphi_{j,k}\}_1^\infty \in l^{p,\kappa_j} \quad (s \le j \le 2s - 1) \tag{15}$$

onto the spaces $W_{1/2,\sigma}^{p,\omega_j}$ $(0 \le j \le 2s - 1)$ *of entire functions. Each of these series converges to its sum* $\varphi_j(w)$ *in the norm of the corresponding space* $W_{1/2,\sigma}^{p,\omega_j}$, *and it also converges uniformly to the same limit in any disk* $|z| \le R < +\infty$.

Besides, the sums of series (13) and (14) have the following interpolation data:

$$\begin{aligned} \varphi_j(\lambda_k) &= \varphi_{j,k} \quad (0 \le j \le s - 1, 0 \le k < +\infty), \\ \varphi_j(\lambda_k) &= \varphi_{j,k} \quad (s \le j \le 2s - 1, 1 \le k < +\infty), \end{aligned} \tag{16}$$

and the following two-sided inequalities are true:

$$\begin{aligned} \|\varphi_j\|_{p,\omega_j}^+ &\asymp \|\{\varphi_{j,k}\}_0^\infty\|_{p,\kappa_j} \quad (0 \le j \le s - 1), \\ \|\varphi_j\|_{p,\omega_j}^+ &\asymp \|\{\varphi_{j,k}\}_1^\infty\|_{p,\kappa_j} \quad (s \le j \le 2s - 1). \end{aligned} \tag{17}$$

Using this theorem, the assertions of Lemma 8.3-3 can be complemented in the following way.

Lemma 8.4-2. *If the function* $\Phi(z) \in W_{s,\sigma}^{p,\omega}$ *is representable in the form*

$$\Phi(z) = \sum_{j=0}^{2s-1} z^j \varphi_j\left(z^{2s}\right), \quad \varphi_j(w) \in W_{1/2,\sigma}^{p,\omega_j} \quad (0 \le j \le 2s - 1),$$

and $\nu \in \Delta_s$, *then the following two-sided inequalities are true:*

$$\|\Phi; \Gamma_{2s}\|_{p,\omega}^{\delta} \asymp \sum_{j=0}^{s-1} \|\{\varphi_j(\lambda_k)\}_0^{\infty}\|_{p,\kappa_j}^{\delta} + \sum_{j=s}^{2s-1} \|\{\varphi_j(\lambda_k)\}_1^{\infty}\|_{p,\kappa_j}^{\delta} \quad (\delta = 1, p). \quad (18)$$

Proof. The two-sided inequalities (18) follow immediately from Lemma $8.3-3(2^\circ)$ by virtue of assertions (16) and (17) of Theorem 8.4-1.

(c) Let $\{\varphi_{j,k}\}_1^{\infty}$ $(0 \le j \le 2s - 1)$ be a set of arbitrary sequences of complex numbers. We associate with it the set of sequences $\{\Phi_{h,k}\}_1^{\infty}$ $(0 \le h \le 2s - 1)$ defined by the formula

$$\Phi_{h,k} = \sum_{j=0}^{2s-1} \mu_{h,k}^j \varphi_{j,k} \quad (0 \le h \le 2s - 1, 1 \le k < +\infty), \quad (19)$$

where $\mu_{h,k} = \beta_s^h \lambda_k^{1/2s}$ as always. Conversely, if we have a set of sequences $\{\Phi_{h,k}\}_1^{\infty}$ $(0 \le h \le 2s - 1)$, then, according to Lemma 8.3-4 the corresponding set of sequences $\{\varphi_{j,k}\}_1^{\infty}$ $(0 \le j \le 2s-1)$ connected with $\{\Phi_{h,k}\}_1^{\infty}$ by (19), may be deduced as follows:

$$\varphi_{j,k} = \frac{1}{2s} \sum_{h=0}^{2s-1} \mu_{h,k}^{-j} \Phi_{h,k} \quad (0 \le j \le 2s - 1, 1 \le k < +\infty). \quad (20)$$

Further, we introduce the sums

$$R_{s,k}(z) = \sum_{j=0}^{s-1} z^{2s+j} \varphi_{j,k} + \lambda_k \sum_{j=s}^{2s-1} z^j \varphi_{j,k} \quad (1 \le k < +\infty), \quad (21)$$

and prove the following analog of Lemma 6.4-3.

Lemma 8.4-3. *The following identity is true for any* $k(1 \le k < +\infty)$:

$$R_{s,k}(z) = \frac{z^s(z^{2s} - \lambda_k)}{2s} \sum_{h=0}^{2s-1} \frac{\Phi_{h,k}}{\mu_{h,k}^{s-1}(z - \mu_{h,k})}, \quad z \in \mathbb{C}. \quad (21')$$

Proof. If we insert representations (20) of $\varphi_{j,k}$ into formula (21) and change the order of summations over j and h, then we obtain

$$R_{s,k}(z) = \frac{1}{2s}\left\{\sum_{h=0}^{2s-1}\Phi_{h,k}z^{2s}\sum_{j=0}^{s-1}\left(\frac{z}{\mu_{h,k}}\right)^j + \lambda_k\sum_{h=0}^{2s-1}\Phi_{h,k}\sum_{j=s}^{2s-1}\left(\frac{z}{\mu_{h,k}}\right)^j\right\}.$$

Hence, calculating the sums over j and using the equalities $\mu_{h,k}^{2s} = \lambda_k$ $(0 \leq h \leq 2s - 1)$, we arrive at the identity (21').

(d) A one-to-one correspondence between the sequences of numbers

$$\{\Phi_n\}_1^\infty, \quad \{\Phi_{h,k}\}_1^\infty \quad (0 \leq h \leq 2s - 1), \quad \{\varphi_{j,k}\}_1^\infty \quad (0 \leq j \leq 2s - 1) \tag{22}$$

was mentioned in Section 8.3(d). Besides, the two-sided estimates

$$\begin{aligned}
\sum_{n=1}^\infty |\Phi_n|^p(1+n)^{\kappa_0} &\asymp \sum_{h=0}^{2s-1}\sum_{k=1}^\infty |\Phi_{h,k}|^p(1+k)^{\kappa_0} \\
&\asymp \sum_{j=0}^{2s-1}\sum_{k=1}^\infty |\varphi_{j,k}|^p(1+k)^{\kappa_j}
\end{aligned} \tag{23}$$

were particularly obtained in the proof of Lemma 8.3-5. Hence the estimated quantities may be finite only simultaneously.

Now we add to a given set of sequences $\{\varphi_{j,k}\}_1^\infty$ $(0 \leq j \leq 2s - 1)$ any set of numbers $\{\varphi_{j,0}\}$ $(0 \leq j \leq 2s - 1)$. Then, obviously, the norms

$$\|\{\varphi_{j,k}\}_1^\infty\|_{p,\kappa_j} \quad \text{and} \quad \|\{\varphi_{j,k}\}_0^\infty\|_{p,\kappa_j} \quad (0 \leq j \leq 2s - 1) \tag{24}$$

also may be finite only simultaneously. Further, we add to the first two sequences of (22), i.e., to $\{\Phi_n\}_1^\infty = \bigcup_{h=0}^{2s-1}\{\Phi_{h,k}\}_1^\infty$, an arbitrary set of numbers $\{\Phi_n\}_{-(s-1)}^\circ$ and obtain the sequence

$$\{\Phi_n\}_{-(s-1)}^\infty \equiv \{\Phi_n\}_{-(s-1)}^\circ \bigcup\left\{\bigcup_{h=0}^{2s-1}\{\Phi_{h,k}\}_1^\infty\right\}. \tag{25}$$

As is obvious, the norms

$$\|\{\Phi_n\}_1^\infty\|_{p,\kappa_0} = \left\{\sum_{n=1}^\infty |\Phi_n|^p(1+n)^{\kappa_0}\right\}^{1/p},$$

$$\|\{\Phi_n\}_{-(s-1)}^\infty\|_{p,\kappa_0} = \left\{\sum_{n=0}^{s-1}\left|\frac{\Phi_{-n}}{n!}\right|^p + \sum_{n=1}^\infty |\Phi_n|^p(1+n)^{\kappa_0}\right\}^{1/p} \tag{26}$$

may be finite only simultaneously. Further, assuming that

$$\varphi_{j,0} = \Phi_{-j}/j! \quad (0 \le j \le s-1), \tag{27}$$

we establish a correspondence between the sets of numbers $\{\varphi_{j,0}\}$ $(0 \le j \le s-1)$ and $\{\Phi_n\}_{-(s-1)}^{\circ}$. So, we have established a correspondence between the sequences

$$\{\Phi_n\}_{-(s-1)}^{\infty} \quad \text{and} \quad \bigcup_{j=0}^{s-1}\{\varphi_{j,0}\}\bigcup\left\{\bigcup_{j=0}^{2s-1}\{\varphi_{j,k}\}_1^{\infty}\right\}, \tag{28}$$

and now we are ready to prove the last theorem of this section.

Theorem 8.4-2. *Let $\Phi(z) \in W_{s,\sigma}^{p,\omega}$, $\nu \in \Delta_s$ and let also*

$$\Phi_{-n} = \Phi^{(n)}(0)(n=0,1,2,\dots), \quad \Phi_n = \Phi(\mu_n) \quad (n=1,2,\dots). \tag{29}$$

Then the following two-sided inequalities are true:

$$\|\Phi;\Gamma_{2s}\|_{p,\omega} \asymp \|\{\Phi_n\}_{-(s-1)}^{\infty}\|_{p,\kappa_0}, \quad \kappa_0 = \frac{1+\omega}{s} - 1. \tag{30}$$

Proof. The case $\delta = p$ of inequalities (18) gives

$$\int_{\Gamma_{2s}} |\Phi(z)|^p |z|^{\omega}|dz| \asymp \sum_{j=0}^{s-1}|\varphi_j(\lambda_0)|^p + \sum_{j=0}^{2s-1}\sum_{k=1}^{\infty}|\varphi_j(\lambda_k)|^p(1+k)^{\kappa_j}.$$

But, according to 8.3(32), $\varphi_j(\lambda_0) = \varphi_j(0) = \varphi_{j,0} = \Phi^{(j)}(0)/j!$ $(0 \le j \le 2s-1)$. Hence, by notation (29),

$$\sum_{j=0}^{s-1}|\varphi_j(0)|^p = \sum_{n=0}^{s-1}\left|\frac{\Phi_{-n}}{n!}\right|^p. \tag{31}$$

Further, by formulas 8.3(18)-(19) and by Lemma 8.3-5, the two-sided inequalities (23) are true particularly for the sequences $\{\Phi(\mu_n)\}_1^{\infty}$, $\{\Phi(\mu_{h,k})\}_1^{\infty}$ $(0 \le h \le 2s-1)$ and $\{\varphi_j(\lambda_k)\}_1^{\infty}$ $(0 \le j \le 2s-1)$. Therefore the inequalities

$$\sum_{j=0}^{2s-1}\sum_{k=1}^{\infty}|\varphi_j(\lambda_k)|^p(1+k)^{\kappa_j} \asymp \sum_{n=1}^{\infty}|\Phi(\mu_n)|^p(1+k)^{\kappa_0}$$

are true along with (23). Hence, by (31), the inequalities (30) follow.

8.5 The proof of the main interpolation theorem

Finally we pass to the proof of the main Theorem 8.2-1 of this chapter which relates to the representability of the class $W^{p,\omega}_{s,\sigma}$ $(s \geq 1)$ of entire functions by interpolation series with points of interpolation at zeros $\{\mu_n\}^\infty_1 \subset \Gamma_{2s}$ of the entire function

$$\mathcal{E}_{s,\sigma}(z;\nu) = E_{1/2}\left(-\sigma^2 z^{2s}; 1+\nu\right), \qquad \nu \in [0,2). \tag{1}$$

(a) As the first step we shall prove the following expansion theorem.

Theorem 8.5-1. *Let*

$$\nu \in \Delta_s = \left(\frac{1+\omega}{ps} + 1 - \frac{1}{s}, \frac{1+\omega}{ps} + 1\right). \tag{2}$$

Then any function $\Phi(z) \in W^{p,\omega}_{s,\sigma}$ *can be expanded in the series*

$$\Phi(z) = \Gamma(1+\nu)P_{s-1}(z;\Phi)\mathcal{E}_{s,\sigma}(z;\nu) + \sum_{n=1}^\infty \Phi(\mu_n)\frac{z^s \mathcal{E}_{s,\sigma}(z;\nu)}{\mu_n^s \mathcal{E}'_{s,\sigma}(\mu_n;\nu)(z-\mu_n)}, \tag{3}$$

where

$$P_{s-1}(z;\Phi) = \sum_{n=0}^{s-1} \frac{\Phi^{(n)}(0)}{n!} z^n. \tag{4}$$

This series converges to $\Phi(z)$ *in the norm* $\|.;\Gamma_{2s}\|_{p,\omega}$ *of the space* $W^{p,\omega}_{s,\sigma}$, *and it converges uniformly to the same limit in any disk* $|z| \leq R < +\infty$. *In addition, the following two-sided inequalities are true:*

$$\|\Phi;\Gamma_{2s}\|_{p,\omega} \asymp \|\{\Phi_n\}^\infty_{-(s-1)}\|_{p,\kappa_0}, \tag{5}$$

where

$$\Phi_{-n} = \Phi^{(n)}(0) \quad (0 \leq n \leq s-1), \qquad \Phi_n = \Phi(\mu_n) \quad (1 \leq n < +\infty) \tag{6}$$

and $\kappa_0 = (1+\omega)/s - 1$.

Proof. According to Lemma 8.3-3,

$$\Phi(z) = \sum_{j=0}^{2s-1} z^j \varphi_j(z^{2s}), \tag{7}$$

where

$$\varphi_j(w) \in W^{p,\omega_j}_{1/2,\sigma}, \qquad \omega_j = \frac{\omega - 2s + 1 + pj}{2s} \quad (0 \leq j \leq 2s-1). \tag{8}$$

And the equality

$$\int_{\Gamma_{2s}} \left| z^j \varphi_j \left(z^{2s} \right) \right|^p |z|^\omega |dz| = \int_0^{+\infty} |\varphi_j(r)|^p r^{\omega_j}\, dr < +\infty \qquad (0 \le j \le 2s - 1) \quad (9)$$

obviously implies

$$z^j \varphi_j \left(z^{2s} \right) \in W^{p,\omega}_{s,\sigma} \qquad (0 \le j \le 2s - 1). \tag{10}$$

Now note that, according to Theorem 8.4-1, the following two expansions are true when $\nu \in \Delta_s$:

$$z^j \varphi_j \left(z^{2s} \right) = \Gamma(1 + \nu)\varphi_j(0) z^j \mathcal{E}_\sigma \left(z^{2s}; \nu \right)$$
$$+ \sum_{k=1}^{\infty} \varphi_j(\lambda_k) \frac{z^{2s+j}\mathcal{E}_\sigma \left(z^{2s}; \nu \right)}{\lambda_k \mathcal{E}'_\sigma(\lambda_k; \nu)(z^{2s} - \lambda_k)} \qquad (0 \le j \le s - 1), \tag{11}$$

$$z^j \varphi_j \left(z^{2s} \right) = \sum_{k=1}^{\infty} \varphi_j(\lambda_k) \frac{z^j \mathcal{E}_\sigma \left(z^{2s}; \nu \right)}{\mathcal{E}'_\sigma(\lambda_k; \nu)(z^{2s} - \lambda_k)} \qquad (s \le j \le 2s - 1), \tag{12}$$

where the series converge to their sums in the norm of the space $W^{p,\omega}_{s,\sigma}$ and, also, they converge uniformly to their sums in any disk $|z| \le R < +\infty$. We insert these expansions into the right-hand side of representation (7) of $\Phi(z)$ and change the order of summations over j and k. Then we arrive at the expansion

$$\Phi(z) = \Gamma(1 + \nu) \left\{ \sum_{j=0}^{s-1} \varphi_j(0) z^j \right\} \mathcal{E}_\sigma(z^{2s}; \nu) + \sum_{k=1}^{\infty} \frac{\mathcal{E}_\sigma(z^{2s}; \nu)\tilde{R}_{s,k}(z)}{\lambda_k \mathcal{E}'_\sigma(\lambda_k; \nu)(z^{2s} - \lambda_k)}, \tag{13}$$

where

$$\tilde{R}_{s,k}(z) = \sum_{j=0}^{s-1} z^{2s+j}\varphi_j(\lambda_k) + \lambda_k \sum_{j=s}^{2s-1} z^j \varphi_j(\lambda_k), \tag{14}$$

and series (13) converges in the required sense. To be convinced that expansion (13) coincides with expansion (3), we use formulas 8.4(21) − (21′) and obtain

$$\tilde{R}_{s,k}(z) = \frac{z^s (z^{2s} - \lambda_k)}{2s} \sum_{h=0}^{2s-1} \frac{\Phi(\mu_{h,k})}{\mu_{h,k}^{s-1}(z - \mu_{h,k})} \tag{14'}$$

since, according to relations 8.3(18)-(19), in the considered case

$$\varphi_{j,k} = \varphi_j(\lambda_k) = \frac{1}{2s} \sum_{h=0}^{2s-1} \mu_{h,k}^{-j}\Phi(\mu_{h,k}).$$

Now (13) and (14$'$) give

$$
\Phi(z) = \Gamma(1+\nu)\left\{\sum_{j=0}^{s-1}\varphi_j(0)z^j\right\}\mathcal{E}_\sigma\left(z^{2s};\nu\right)
$$
$$
+\sum_{k=1}^{\infty}\frac{z^s\mathcal{E}_\sigma\left(z^{2s};\nu\right)}{2s\lambda_k\mathcal{E}'_\sigma(\lambda_k;\nu)}\sum_{h=0}^{2s-1}\frac{\Phi(\mu_{h,k})}{\mu_{h,k}^{s-1}(z-\mu_{h,k})}. \tag{13$'$}
$$

But $\mathcal{E}_{s,\sigma}(z;\nu) \equiv \mathcal{E}_\sigma(z^{2s};\nu)$ and $\mathcal{E}'_{s,\sigma}(z;\nu) \equiv 2sz^{2s-1}\mathcal{E}'_\sigma(z^{2s};\nu)$ according to definition 8.2(4) of the function $\mathcal{E}_{s,\sigma}(z;\nu)$. We put $z = \mu_{h,k} = \beta_s^h\lambda_k^{1/2s}$ in the second of these identities and obtain

$$
2s\lambda_k\mathcal{E}'_\sigma(\lambda_k;\nu) = \mu_{h,k}\mathcal{E}'_{s,\sigma}(\mu_{h,k};\nu) \qquad (0 \le h \le 2s-1, 1 \le k < +\infty). \tag{15}
$$

Also we note that, according to 8.3(32),

$$
\sum_{j=0}^{s-1}\varphi_j(0)z^j = \sum_{j=0}^{s-1}\frac{\Phi^{(j)}(0)}{j!}z^j = P_{s-1}(z;\Phi). \tag{16}
$$

If we now replace the factor $2s\lambda_k\mathcal{E}'_\sigma(\lambda_k;\nu)$ of the denominator of the first sum of the expansion (13$'$) to the denominator of its second sum, then, using (15) and (16), we arrive at the expansion of the form

$$
\Phi(z) = \Gamma(1+\nu)P_{s-1}(z;\Phi)\mathcal{E}_{s,\sigma}(z;\nu)
$$
$$
+\sum_{k=1}^{\infty}z^s\mathcal{E}_{s,\sigma}(z;\nu)\sum_{h=0}^{2s-1}\frac{\Phi(\mu_{h,k})}{\mu_{h,k}^s\mathcal{E}'_{s,\sigma}(\mu_{h,k};\nu)(z-\mu_{h,k})}. \tag{3$'$}
$$

Expansion (3) follows by the definition of the sequence $\{\mu_n\}_1^\infty = \bigcup_{h=0}^{2s-1}\{\mu_{h,k}\}_1^\infty \subset \Gamma_{2s}$. Finally, it remains to observe that the two-sided inequalities (5) were established earlier in Theorem 8.4-2.

Remark. Of course we need an additional argument to prove that the double sum (3$'$) can be represented also by the series (3) which converges in the same sense. We omit this, as a similar argument was fully illustrated earlier in the proof of Theorem 6.5-1.

(b) Proof of Theorem 8.2-1 According to Theorem 8.5-1, any function $\Phi(z) \in W_{s,\sigma}^{p,\omega}$ is representable by a series 8.2(14), and assertions $1^\circ, 2^\circ, 3^\circ$ of Theorem 8.2-1 are true. Thus, to complete the proof, it remains only to show that, if $\{\Phi_n\}_{-(s-1)}^\infty$ is any sequence of $L_{p,\kappa_0}^{(s-1)}$, then there exists an entire function $\Phi(z) \in W_{s,\sigma}^{p,\omega}$, such that

$$
\Phi^{(n)}(0) = \Phi_{-n} \quad (0 \le n \le s-1), \qquad \Phi(\mu_n) = \Phi_n \quad (1 \le n < +\infty). \tag{17}
$$

So, let $\{\Phi_n\}_{-(s-1)}^{\infty} \equiv \{\Phi_n\}_{-(s-1)}^{\circ} \bigcup \{\Phi_n\}_1^{\infty} \in L_{p,\kappa_0}^{(s-1)}$ be an arbitrary sequence. Then $\{\Phi_n\}_1^{\infty} \equiv \bigcup_{h=0}^{2s-1} \{\Phi_{h,k}\}_1^{\infty}$ according to 8.3(24). Now, as always, using formulas 8.3(21)-(22), we put $\{\Phi_{h,k}\}_1^{\infty}$ $(0 \le h \le 2s-1)$ into correspondence with a new set of sequences $\{\varphi_{j,k}\}_1^{\infty}$ $(0 \le j \le 2s-1)$. Then, by Lemma 8.3-5, the two-sided inequalities

$$\sum_{j=0}^{2s-1} \sum_{k=1}^{\infty} |\varphi_{j,k}|^p (1+k)^{\kappa_j} \asymp \sum_{n=1}^{\infty} |\Phi_n|^p (1+n)^{\kappa_0} \tag{18}$$

are true. Further, we add to $\{\varphi_{j,k}\}_1^{\infty}$ $(0 \le j \le 2s-1)$ the set of numbers $\varphi_{j,0} = \Phi_{-j}/j! (0 \le j \le s-1)$. We have put $\{\Phi_n\}_{-(s-1)}^{\infty}$ into correspondence with the sequence

$$\{\varphi_{j,0}\}_0^{s-1} \bigcup \left\{ \bigcup_{j=0}^{2s-1} \{\varphi_{j,k}\}_1^{\infty} \right\} \tag{19}$$

as was done in 8.4(28). But $\{\Phi_n\}_{-(s-1)}^{\infty} \in L_{p,\kappa_0}^{(s-1)}$; therefore (18) gives

$$\sum_{j=0}^{s-1} \|\{\varphi_{j,k}\}_0^{\infty}\|_{p,\kappa_j} < +\infty, \qquad \sum_{j=s}^{2s-1} \|\{\varphi_{j,k}\}_1^{\infty}\|_{p,\kappa_j} < +\infty.$$

Now, using the series 8.4(13)-(14) of Theorem 8.4-1, we can construct for sequence (19) the entire functions $\varphi_j(w) \in W_{1/2,\sigma}^{p,\omega_j}$ $(0 \le j \le 2s-1)$. Then, obviously, $z^j \varphi_j(z^{2s})$ $(0 \le j \le 2s-1)$ are entire functions of the class $W_{s,\sigma}^{p,\omega}$. Consequently,

$$\Phi(z) \equiv \sum_{j=0}^{2s-1} z^j \varphi_j \left(z^{2s}\right) \in W_{s,\sigma}^{p,\omega}. \tag{20}$$

Thus we have put each sequence $\{\Phi_n\}_{-(s-1)}^{\infty} \in L_{p,\kappa_0}^{(s-1)}$ into correspondence with a definite function $\Phi(z) \in W_{s,\sigma}^{p,\omega}$. The further steps of the proof are similar to those used in the proof of Theorem 8.5-1. Namely, first we obtain the representation of $\Phi(z)$ by an expansion of the form (13)-(14) in which $\{\varphi_j(\lambda_k)\}_1^{\infty}$ and $\varphi_j(0)$ are replaced respectively by $\{\varphi_{j,k}\}_1^{\infty}$ and $\varphi_{j,0}$. Next we use formulas 8.4(21)-(21′) and, after suitable simplifications, we obtain for $\Phi(z)$ an expansion (3′) in which $\{\Phi(\mu_{h,k})\}_1^{\infty}$ and $\Phi^{(j)}(0)$ are replaced respectively by $\{\Phi_{h,k}\}_1^{\infty}$ and Φ_{-j}. In other words, we arrive at the expansion

$$\Phi(z) = \Gamma(1+\nu) \left\{ \sum_{j=0}^{s-1} \frac{\Phi_{-j}}{j!} z^j \right\} \mathcal{E}_{s,\sigma}(z;\nu)$$
$$+ \sum_{k=1}^{\infty} z^s \mathcal{E}_{s,\sigma}(z;\nu) \sum_{h=0}^{2s-1} \frac{\Phi_{h,k}}{\mu_{h,k}^s \mathcal{E}_{s,\sigma}'(\mu_{h,k};\nu)(z-\mu_{h,k})}, \tag{21}$$

where $\{\mu_n\}_1^\infty = \bigcup_{h=0}^{2s-1}\{\mu_{h,k}\}_1^\infty$ is the set of zeros of the entire function $\mathcal{E}_{s,\sigma}(z;\nu)$. Hence relations (17) immediately follow. Indeed, if we take $z = \mu_n$ $(1 \le n < +\infty)$ in expansion (21), then we obtain $\Phi(\mu_n) = \Phi_n$ $(1 \le n < +\infty)$. And, since

$$\mathcal{E}_{s,\sigma}(0;\nu) = \frac{1}{\Gamma(1+\nu)}, \qquad \frac{d^h}{dz^h}\left\{\sum_{j=0}^{s-1}\frac{\Phi_{-j}}{j!}z^j\right\}\Bigg|_{z=0} = \Phi_{-h} \qquad (0 \le h \le s-1)$$

and

$$\frac{d^h}{dz^h}\left\{\frac{z^s\mathcal{E}_{s,\sigma}(z;\nu)}{z-\mu_n}\right\}\Bigg|_{z=0} = 0 \qquad (0 \le h \le s-1, 1 \le n < +\infty),$$

we obtain $\Phi^{(h)}(0) = \Phi_{-h}$ $(0 \le h \le s-1)$.

8.6 Notes

The results of this chapter were obtained by use of a different method in the papers of S.G. Raphaelian [1–4] considering the simple case $s = 1$. In these papers interpolation interpolation expansions in several classes of entire functions of order ρ $(1 \le \rho < 2)$ were established under a more general hypothesis relating to the points of interpolation. In another simple case $s = 2$ the results of this chapter were obtained by M.M. Djrbashian [7,8].

9 Basic Fourier type systems in L_2 spaces of even-dimensional vector functions

9.1 Introduction

This chapter is similar to Chapter 7. Here we pass from the interpolation Theorem 8.2-1 (the case $p = 2$), relating to the classes $W_{s,\sigma}^{2;\omega}$ ($s \geq 1$) of entire functions, to theorems on the basis property of some systems of even-dimensional vector functions which are biorthogonal in a Hilbert space $L_2^{2s}(0,\sigma)$ of $2s$-dimensional vector functions defined on $(0,\sigma)$. First we construct the mentioned biorthogonal systems. Then, using Theorem 8.2-1 and Theorem 2.4-2 on parametric representations of the classes $W_{s,\sigma}^{2;\omega}$ ($s \geq 1$), we establish the completeness and the basis property in the Riesz sense of these systems in the space $L_2^{2s}(0,\sigma)$. Note that a reformulation of the results of this chapter leads later in Chapter 12 to an explicit and complete apparatus of Fourier type systems of entire functions. These systems prove to be bases of the weighted space L_2 considered over the sum of $2s$ ($s \geq 1$) segments of the same length having a common endpoint at the origin and forming equal angles of the opening π/s in the complex plane.

9.2 Some identities

(a) The first of the fundamental identities 1.2(10) takes the following form when $\rho = s$ and $s \geq 1$ is assumed, as everywhere, to be an arbitrary natural number:

$$
\begin{aligned}
J_s(z; \lambda; \eta_1; \eta_2) &\equiv \int_0^\sigma E_s\left(z\tau^{1/s}; \eta_1\right) \tau^{\eta_1 - 1} E_s\left(\lambda(\sigma - \tau)^{1/s}; \eta_2\right) (\sigma - \tau)^{\eta_2 - 1} d\tau \\
&= \frac{E_s\left(\sigma^{1/s} z; \eta_3\right) - E_s\left(\sigma^{1/s}\lambda; \eta_3\right)}{z - \lambda} \sigma^{\eta_4} \qquad (\eta_1, \eta_2 \geq 0),
\end{aligned}
\tag{1}
$$

where

$$
\eta_3 = \eta_1 + \eta_2 - \frac{1}{s}, \qquad \eta_4 = \eta_3 - 1.
\tag{2}
$$

Further, according to formula 8.3(7),

$$
z^j E_{1/2}\left(z^{2s}; \mu + \frac{j}{s}\right) = \frac{1}{2s} \sum_{h=0}^{2s-1} \beta_s^{-jh} E_s\left(\beta_s^h z; \mu\right) \qquad (0 \leq j \leq 2s - 1),
\tag{3}
$$

where $\beta_s = \exp\{i\pi/s\}$. Besides, if in identity (1) we assume $0 \leq j \leq 2s - 1$ and

$$
\eta_1 = \eta_{1,j} = \mu, \qquad \eta_2 = \eta_{2,j} = \frac{s + 1 - j}{s} + \nu - \mu,
\tag{4_1}
$$

then, by (2),

$$
\eta_3 = \eta_{3,j} = 1 + \nu - \frac{j}{s}, \qquad \eta_4 = \eta_{4,j} = \eta_{3,j} - 1 = \nu - \frac{j}{s}.
\tag{4_2}
$$

In addition, replacing z and λ in (1) by $\beta_s^{h+1/2}z$ and $\beta_s^{h+1/2}\lambda$ respectively we arrive at the following identities which are true for any $j(0 \leq j \leq 2s-1)$:

$$
\begin{aligned}
V_{s,h}^{(j)}(z;\lambda) &\equiv J_s\left(\beta_s^{h+1/2}z;\beta_s^{h+1/2}\lambda;\mu;\eta_{2,j}\right) \\
&= \int_0^\sigma E_s\left(\beta_s^{h+1/2}z\tau^{1/s};\mu\right)\tau^{\mu-1}E_s\left(\beta_s^{h+1/2}\lambda(\sigma-\tau)^{1/s};\eta_{2,j}\right)(\sigma-\tau)^{\eta_{2,j}-1}d\tau \\
&= \beta_s^{-(h+1/2)}\frac{E_s\left(\sigma^{1/s}\beta_s^{h+1/2}z;\eta_{3,j}\right)-E_s\left(\sigma^{1/s}\beta_s^{h+1/2}\lambda;\eta_{3,j}\right)}{z-\lambda}\sigma^{\eta_{4,j}}.
\end{aligned}
\tag{5}
$$

Now in identity (3) we replace z by $\beta_s^{1/2}\sigma^{1/s}z$ and μ by $\eta_{3,j}$. Then we obtain the second group of identities:

$$
z^j E_{1/2}\left(-\sigma^2 z^{2s};1+\nu\right) = \frac{\sigma^{-j/s}}{2s}\sum_{h=0}^{2s-1}\beta_s^{-(h+1/2)j}E_s\left(\sigma^{1/s}\beta_s^{h+1/2}z;1+\nu-\frac{j}{s}\right)
$$
$$
(0 \leq j \leq 2s-1).
\tag{6}
$$

Further we shall use identities (5) and (6) together with notations (4_1) and (4_2) only in the case when $j = s$. In this case

$$
\begin{aligned}
V_{s,h}(z;\lambda) &\equiv V_{s,h}^{(s)}(z;\lambda) \\
&= \int_0^\sigma E_s\left(\beta_s^{h+1/2}z\tau^{1/s};\mu\right)\tau^{\mu-1}E_s\left(\beta_s^{h+1/2}\lambda(\sigma-\tau)^{1/s};\eta_{2,s}\right)(\sigma-\tau)^{\eta_{2,s}-1}d\tau \\
&= \beta_s^{-(h+1/2)}\frac{E_s\left(\sigma^{1/s}\beta_s^{h+1/2}z;\nu\right)-E_s\left(\sigma^{1/s}\beta_s^{h+1/2}\lambda;\nu\right)}{z-\lambda}\sigma^{\nu-1},
\end{aligned}
\tag{5_1}
$$

where $\eta_{2,s} = 1/s + \nu - \mu$, and

$$
z^s E_{1/2}\left(-\sigma^2 z^{2s};1+\nu\right) = \frac{\sigma^{-1}}{2s}\sum_{h=0}^{2s-1}\beta_s^{-(h+1/2)s}E_s\left(\sigma^{1/s}\beta_s^{h+1/2}z;\nu\right).
\tag{6_1}
$$

Obviously, identity (5_1) is true, if only $\mu \geq 0$ and $\eta_{2,s} = 1/s + \nu - \mu \geq 0$, and identity (6_1) is true for any ν. But, as in Chapter 7, we need to indicate more precisely what conditions the parameters should satisfy to ensure the applicability of the mentioned identities. As was deduced in Chapter 8, the condition

$$
\nu \in \Delta_s = \left(\frac{1+\omega}{sp}+1-\frac{1}{s},\frac{1+\omega}{sp}+1\right) \subset (0,2),
$$

where $1 < p < +\infty$ and $-1 < \omega < p-1$, is sufficient for the validity of Theorem 8.2-1 relating to expansions of the entire functions of class $W_{s,\sigma}^{p,\omega}$ in interpolation series whose points of interpolation are $\{\mu_n\}_1^\infty$, i.e., the zeros of the function

$$
\mathcal{E}_{s,\sigma}(z;\nu) = E_{1/2}\left(-\sigma^2 z^{2s};1+\nu\right).
\tag{7}
$$

We shall assume everywhere that $p = 2$ and $-1 < \omega < 1$. Thus,

$$\nu \in \Delta_s = \left(\frac{\omega - 1}{2s} + 1, \frac{\omega + 1}{2s} + 1 \right) \subset (0, 2). \tag{8}$$

In addition, we assume everywhere

$$\mu = \frac{s + \omega + 1}{2s} \quad (s \geq 1, -1 < \omega < 1). \tag{9}$$

This implies $1/2 < \mu < 1/2 + 1/s$ in all cases. So, if $\nu \in \Delta_s$, then always

$$\mu > \frac{1}{2} \text{ and } \eta_{2,s} = \frac{1}{s} + \nu - \mu = \frac{1 - s - \omega}{2s} + \nu > \frac{1}{2}.$$

Therefore, for any $\sigma \in (0, +\infty)$

$$\tau^{\mu-1} \in L_2(0, \sigma) \text{ and } (\sigma - \tau)^{\eta_{2,s}-1} \in L_2(0, \sigma). \tag{10}$$

From this it follows that the function integrated in (5_1) is of $L_1(0, \sigma)$, if $\nu \in \Delta_s$.

(b) Now we introduce the function

$$V_s(z; \lambda) \equiv \sum_{h=0}^{2s-1} \beta_s^{-(h+1/2)(s-1)} V_{s,h}(z; \lambda)$$

$$= \sum_{h=0}^{2s-1} \beta_s^{-(h+1/2)(s-1)} \int_0^\sigma E_s \left(\beta_s^{h+1/2} z \tau^{1/s}; \mu \right) \tau^{\mu-1} \tag{11}$$

$$\times E_s \left(\beta_s^{h+1/2} \lambda (\sigma - \tau)^{1/s}; \eta_{2,s} \right) (\sigma - \tau)^{\eta_{2,s}-1} d\tau,$$

where $\eta_{2,s} = 1/s + \nu - \mu$.

Lemma 9.2-1. *If $\nu \in \Delta_s$, then the identity*

$$V_s(z; \lambda) \equiv \frac{2s\sigma^\nu}{z - \lambda} \left\{ z^s E_{1/2} \left(-\sigma^2 z^{2s}; 1 + \nu \right) - \lambda^s E_{1/2} \left(-\sigma^2 \lambda^{2s}; 1 + \nu \right) \right\} \tag{12}$$

is true for any $z, \lambda \in \mathbb{C}$.

Proof. We multiply identity (5_1) by $\beta_s^{-(h+1/2)(s-1)}$ and sum both its sides over $h(0 \leq h \leq 2s - 1)$. Then the use of (6_1) gives identity (12).

Obviously notation (7) may be used in identity (12). Further, it is useful to remember that all the zeros of the function $\mathcal{E}_{s,\sigma}(z; \nu), \nu \in [0, 2)$ are simple and are situated on the sum of rays

$$\Gamma_{2s} = \bigcup_{j=0}^{2s-1} \Gamma_{1,j}, \quad \Gamma_{1,j} = [0, \beta_s^j \infty) \quad (0 \leq j \leq 2s - 1). \tag{13}$$

Remember also that the universal numeration

$$\mu_{2s(k-1)+j+1} = \mu_{j,k}, \quad \mu_{j,k} = \beta_s^j \lambda_k^{1/2s} \quad (0 \leq j \leq 2s - 1, \quad 1 \leq k < +\infty) \tag{14}$$

was introduced for the zeros $\{\mu_n\}_1^\infty \subset \Gamma_{2s}$ of the mentioned function. Here $\{\lambda_k\}_1^\infty$ $(0 < \lambda_k < \lambda_{k+1}, k \geq 1)$, as always, are the zeros of the function

$$\mathcal{E}_\sigma(z; \nu) = E_{1/2} \left(-\sigma^2 z; 1 + \nu \right), \nu \in [0, 2), \tag{15}$$

which are also simple.

9.3 The construction of biorthogonal systems of even-dimensional vector functions.

(a) We assume

$$\nu \in \Delta_s = \left(\frac{2s + \omega - 1}{2s}, \frac{2s + \omega + 1}{2s} \right) \tag{1}$$

and introduce on $(0, \sigma)$ $(0 < \sigma < +\infty)$ the sequence of vector functions

$$\{\Omega_{m,h}(\tau)\}_0^{2s-1}, \quad (-(s-1) \le m < +\infty) \tag{2}$$

in the following way:

$$\Omega_{m,h}(\tau) \equiv \beta_s^{(h+1/2)m} \frac{\sigma^{-\nu} \Gamma(1+\nu)}{2s\Gamma(1-m)} \frac{(\sigma - \tau)^{(s+m-1)/s + \eta_{2,s} - 1}}{\Gamma\left(\eta_{2,s} + \frac{s+m-1}{s}\right)} \tag{2_1}$$

when $-(s-1) \le m \le 0$ and $0 \le h \le 2s - 1$, and

$$\Omega_{m,h}(\tau) \equiv \beta_s^{(h+1/2)(1-s)} \frac{\sigma^{-\nu}}{2s\mu_m^s \mathcal{E}_{s,\sigma}'(\mu_m; \nu)}$$
$$\times E_s\left(\beta_s^{h+1/2} \mu_m (\sigma - \tau)^{1/s}; \eta_{2,s}\right) (\sigma - \tau)^{\eta_{2,s} - 1} \tag{2_2}$$

when $1 \le m < +\infty$ and $0 \le h \le 2s - 1$. Here, as before, we assume

$$\eta_{2,s} = \frac{1}{s} + \nu - \mu. \tag{2_3}$$

Noting also that

$$\mu = \frac{s + \omega + 1}{2s} \quad (-1 < \omega < 1), \tag{3}$$

we prove two lemmas.

Lemma 9.3-1. *The following identities are true:*

$$\sum_{h=0}^{2s-1} \int_0^\sigma E_s\left(\beta_s^{h+1/2} z\tau^{1/s}; \mu\right) \tau^{\mu-1} \Omega_{m,h}(\tau) d\tau = \frac{z^s \mathcal{E}_{s,\sigma}(z; \nu)}{\mu_m^s \mathcal{E}_{s,\sigma}'(\mu_m; \nu)(z - \mu_m)}, \tag{4}$$

$$(z \in \mathbb{C}, \ 1 \le m < +\infty).$$

Proof. Using representations 9.2(11) and 9.2(12) of the function $V_s(z; \lambda)$, where we take $\lambda = \mu_m (1 \le m < +\infty)$, we obtain the following identities which are true for any $z \in \mathbb{C}$:

$$\sum_{h=0}^{2s-1} \beta_s^{-(h+1/2)(s-1)} \int_0^\sigma E_s\left(\beta_s^{h+1/2} z\tau^{1/s}; \mu\right) \tau^{\mu-1}$$
$$\times E_s\left(\beta_s^{h+1/2} \mu_m (\sigma - \tau)^{1/s}; \eta_{2,s}\right) (\sigma - \tau)^{\eta_{2,s} - 1} d\tau \tag{5}$$
$$= 2s\sigma^\nu \frac{z^s E_{1/2}\left(-\sigma^2 z^{2s}; 1 + \nu\right)}{z - \mu_m} \quad (1 \le m < +\infty).$$

The use of (2_2) now gives identities (4).

Lemma 9.3-2. *The following identities are true:*

$$\sum_{h=0}^{2s-1} \int_0^\sigma E_s\left(\beta_s^{h+1/2} z\tau^{1/s}; \mu\right) \tau^{\mu-1}\Omega_{m,h}(\tau)d\tau$$

$$= \frac{\Gamma(1+\nu)}{\Gamma(1-m)} z^{-m} E_{1/2}\left(-\sigma^2 z^{2s}; 1+\nu\right) \quad (z \in \mathbb{C}, \quad -(s-1) \le m \le 0). \tag{6}$$

Proof. Using representation 9.2(11) of the function $V_s(z;\lambda)$, we obtain

$$\left.\frac{d^{s+m-1}}{d\lambda^{s+m-1}} V_s(z;\lambda)\right|_{\lambda=0} = \sum_{h=0}^{2s-1} \int_0^\sigma E_s\left(\beta_s^{h+1/2} z\tau^{1/s}; \mu\right) \tau^{\mu-1}$$

$$\times \left\{ \frac{\beta_s^{(h+1/2)m}\Gamma(s+m)(\sigma-\tau)^{(s+m-1)/s+\eta_{2,s}-1}}{\Gamma\left(\eta_{2,s} + \frac{s+m-1}{s}\right)} \right\} d\tau \tag{7_1}$$

$$(-(s-1) \le m \le 0).$$

Further, the second term of the right-hand side of identity 9.2(12), considered as a function of λ, obviously has a zero of order s in the point $\lambda = 0$ when $z \ne 0$. Therefore, by representation 9.2(12),

$$\left.\frac{d^{s+m-1}}{d\lambda^{s+m-1}} V_s(z;\lambda)\right|_{\lambda=0} = 2s\sigma^\nu \Gamma(s+m) z^{-m} E_{1/2}\left(-\sigma^2 z^{2s}; 1+\nu\right) \tag{7_2}$$

for any $m(-(s-1) \le m \le 0)$. Equating the right-hand sides of (7_1) and (7_2) and using (2_1), we obtain identity (6).

(b) If

$$y(\tau) = \{y_h(\tau)\}_0^{2s-1} \text{ and } z(\tau) = \{z_h(\tau)\}_0^{2s-1} \quad (s \ge 1) \tag{8}$$

is a pair of arbitrary vector functions whose components are all functions of $L_2(0,\sigma)$, then we shall say that these vector functions are of class $L_2^{2s}(0,\sigma)$. We define the inner product of these vector functions in the following way:

$$\{y, z\} = \sum_{h=0}^{2s-1} \int_0^\sigma y_h(\tau)\overline{z_h(\tau)}d\tau. \tag{9}$$

Then, evidently, the norm of a vector-function $y(\tau) \in L_2^{2s}(0,\sigma)$ is the quantity

$$\|y\| = \{y, y\}^{1/2} = \left\{ \sum_{h=0}^{2s-1} \int_0^\sigma |y_h(\tau)|^2 d\tau \right\}^{1/2} < +\infty, \tag{10}$$

and $\|y\| = 0$ only in the case when $y_h(\tau) = 0$ $(0 \le h \le 2s-1)$ almost everywhere in $(0,\sigma)$.

(**c**) Now we introduce

$$\{\kappa_{n,h}(\tau)\}_0^{2s-1} \quad (-(s-1) \leq n < +\infty) \tag{11}$$

as the second system of vector functions of $L_2^{2s}(0,\sigma)$ setting

$$\kappa_{n,h}(\tau) \equiv \beta_s^{-(h+1/2)n} \frac{\Gamma(1-n)}{\Gamma\left(\mu - \frac{n}{s}\right)} \tau^{-\frac{n}{s}+\mu-1} \tag{11_1}$$
$$(-(s-1) \leq n \leq 0, 0 \leq h \leq 2s-1),$$

$$\kappa_{n,h}(\tau) \equiv E_s\left(\beta_s^{h+1/2}\mu_n\tau^{1/s};\mu\right)\tau^{\mu-1} \quad (1 \leq n < +\infty,\ 0 \leq h \leq 2s-1). \tag{11_2}$$

We also introduce the system

$$\{\omega_{n,h}(\tau)\}_0^{2s-1} \quad (-(s-1) \leq n < +\infty), \tag{12}$$

where

$$\omega_{n,h}(\tau) = \overline{\Omega_{n,h}(\tau)} \quad (-(s-1) \leq n < +\infty,\ 0 \leq h \leq 2s-1) \tag{12_1}$$

and prove the main theorem on biorthogonality of the introduced systems.

Theorem 9.3-1. *The systems of vector functions (11) and (12) are biorthogonal in $L_2^{2s}(0,\sigma)$. In other words, if we denote*

$$\kappa_n(\tau) = \{\kappa_{n,h}(\tau)\}_0^{2s-1}, \omega_n(\tau) = \{\omega_{n,h}(\tau)\}_0^{2s-1} \quad (-(s-1) \leq n < +\infty), \tag{13}$$

then

$$\{\kappa_n, \omega_m\} = \delta_{n,m} \quad (-(s-1) \leq n, m < +\infty). \tag{14}$$

Proof. As we have mentioned, $\{\mu_n\}_1^\infty \subset \Gamma_{2s}$ is the sequence of zeros of the function $\mathcal{E}_{s,\sigma}(z;\nu)$, and all the zeros of this function are simple. Hence

$$\frac{z^s \mathcal{E}_{s,\sigma}(z;\nu)}{\mu_m^s \mathcal{E}'_{s,\sigma}(\mu_m;\nu)(z-\mu_m)}\bigg|_{z=\mu_n} = \delta_{n,m} \quad (1 \leq n,m < +\infty). \tag{15}$$

Thus, if we put $z = \mu_n (1 \leq n < +\infty)$ in identity (4) of Lemma 9.3-1, then, using notations (12_1) and (11_1) and also (15), we obtain equalities (14), but only in the case when $1 \leq n, m < +\infty$. Assuming $1 \leq m < +\infty$, we expand the obtained equalities to the case when $-(s-1) \leq n \leq 0$. To this end, according to (12_1), we replace $\Omega_{m,h}(\tau)$ in identities (4) by $\overline{\omega_{m,h}(\tau)}$. But the right-hand sides of the obtained identities have at $z = 0$ zeros of order s. Therefore their derivatives, up to the order $s - 1$, vanish at $z = 0$. At the same time

$$\frac{d^{-n}}{dz^{-n}}\left\{E_s\left(\beta_s^{h+1/2}z\tau^{1/s};\mu\right)\tau^{\mu-1}\right\}\bigg|_{z=0} \equiv \kappa_{n,h}(\tau)$$
$$= \beta_s^{-(h+1/2)n}\frac{\Gamma(1-n)}{\Gamma\left(\mu-\frac{n}{s}\right)}\tau^{-\frac{n}{s}+\mu-1} \quad (-(s-1) \leq n \leq 0,\ 0 \leq h \leq 2s-1). \tag{16}$$

Thus the conclusion is that

$$\sum_{h=0}^{2s-1} \int_0^\sigma \kappa_{n,h}(\tau)\overline{\omega_{m,h}(\tau)}d\tau = \{\kappa_n,\omega_m\} = 0$$

when $-(s-1) \le n \le 0$ and $1 \le m < +\infty$. So equalities (14) are true also in this case. To prove (14) in the case when $-(s-1) \le m \le 0$ and $1 \le n < +\infty$, we put $z = \mu_n$ in identities (6). Then we obtain

$$\sum_{h=0}^{2s-1} \int_0^\sigma E_s\left(\beta_s^{h+1/2}\mu_n\tau^{1/s};\mu\right)\tau^{\mu-1}\Omega_{m,h}(\tau)d\tau = 0$$

when $-(s-1) \le m \le 0$ and $1 \le n < +\infty$. And, according to notations (11_2) and (12_1), equalities (14) are true in the same case. Now it remains to prove equalities (14) only in the case when $-(s-1) \le m,n \le 0$. In this case the right-hand side of identity (6) has a zero of order $-m$ at the point $z = 0$. And, obviously,

$$\frac{d^{-n}}{dz^{-n}}\left\{\frac{\Gamma(1+\nu)}{\Gamma(1-m)}z^{-m}E_{1/2}\left(-\sigma^2 z^{2s};1+\nu\right)\right\}\bigg|_{z=0} = \delta_{n,m}. \tag{17}$$

If we calculate the $(-n)$-th derivative of the left-hand side of (6) and use (16) and (17), the proof will be completed.

9.4 Theorems on completeness and basis property

(a) First we shall prove a theorem on completeness of the biorthogonal systems 9.3(13) of vector functions in $L_2^{2s}(0,\sigma)$.

Theorem 9.4-1. *The systems 9.3(13) are complete in $L_2^{2s}(0,\sigma)$.*

Proof. Suppose

$$\varphi(\tau) = \{\varphi_h(\tau)\}_0^{2s-1} \in L_2^{2s}(0,\sigma) \tag{1}$$

is an arbitrary vector function and introduce the entire function

$$\Phi(z;\varphi) \equiv \sum_{h=0}^{2s-1} \int_0^\sigma E_s\left(\beta_s^{h+1/2}z\tau^{1/s};\mu\right)\tau^{\mu-1}\overline{\varphi_h(\tau)}d\tau, \tag{2}$$

where, as always,

$$\beta_s = \exp\left\{i\frac{\pi}{s}\right\}, \quad \mu = \frac{s+w+1}{2s} \quad (-1 < w < 1). \tag{3}$$

Then, by Theorem $2.4-2(1^\circ)$, $\Phi(z;\varphi) \in W_{s,\sigma}^{2,w}$. In addition, using formulas 9.3(16) and $9.3(11_1)-(11_2)$, we obtain from (2)

$$\frac{d^{-n}}{dz^{-n}}\Phi(z;\varphi)\bigg|_{z=0} = \sum_{h=0}^{2s-1} \int_0^\sigma \kappa_{n,h}(\tau)\overline{\varphi_h(\tau)}d\tau$$

$$= \{\kappa_n,\varphi\} = \overline{\{\varphi,\kappa_n\}}, \quad (-(s-1) \le n \le 0) \tag{4_1}$$

and also

$$\Phi(\mu_n;\varphi) = \sum_{h=0}^{2s-1} \int_0^\sigma \kappa_{n,h}(\tau)\overline{\varphi_h(\tau)}d\tau \tag{4_2}$$

$$= \{\kappa_n, \varphi\} = \overline{\{\varphi, \kappa_n\}} \quad (1 \le n < +\infty).$$

So, if we suppose

$$\{\varphi, \kappa_n\} = 0 \qquad (-(s-1) \le n < +\infty), \tag{5}$$

then (4_1) and (4_2) yield

$$\Phi^{(k)}(0) = 0 \quad (0 \le k \le s-1) \text{ and } \Phi(\mu_n) = 0 \quad (1 \le n < +\infty). \tag{6}$$

Hence $\Phi(z;\varphi) \equiv 0$ by the uniqueness Theorem 8.2-2. And, therefore, $\varphi_h(\tau) = 0$ $(0 \le h \le 2s-1)$ almost everywhere in $(0,\sigma)$, according to Theorem 2.4 – 2(2°). Thus the first system of 9.3(13) is complete in $L_2^{2s}(0,\sigma)$. To prove that the second system of 9.3(13) is also complete in $L_2^{2s}(0,\sigma)$, observe that, according to definitions of Section 9.3, the functions of the system $\overline{\{\omega_{n,h}(\tau)\}_0^{2s-1}}$ $(-(s-1) \le n < +\infty)$ differ from the functions of the system $\{\kappa_{n,h}(\sigma-\tau)\}_0^{2s-1}$ $(-(s-1) \le n < +\infty)$ only by constant multipliers, if only the parameter $\mu = (s+\omega+1)/2s$ $(-1 < \omega < 1)$ is replaced in the last ones by $\eta_{2,s} = 1/s + \nu - \mu$. But $\eta_{2,s} = (s+\omega_0+1)/2s$, where $\omega_0 = 2s\nu - 2s - \omega \in (-1,1)$ since $\nu \in \Delta_s$. And, besides, one can easily verify that the condition

$$\nu \in \left(\frac{2s+\omega_0-1}{2s}, \frac{2s+\omega_0+1}{2s}\right)$$

is also satisfied, i.e., ν is in the interval Δ_s constructed by the use of ω_0. Hence it follows that the second system of 9.3(13) is also complete in $L_2^{2s}(0,\sigma)$.

(b) To prove the second main theorem of this section, first recall the definition of the Hilbert space $L_{2,\kappa_0}^{(s-1)}$ of sequences $\{\varphi_n\}_{-(s-1)}^\infty$ of complex numbers given in Section 8.2. Such a sequence was said to be of $L_{2,\kappa_0}^{(s-1)}$, if its norm

$$\|\{\varphi_n\}_{-(s-1)}^\infty\|_{2,\kappa_0} \equiv \left\{ \sum_{n=-(s-1)}^0 \left|\frac{\varphi_n}{|n|!}\right|^2 + \sum_{n=1}^\infty |\varphi_n|^2(1+n)^{\kappa_0} \right\}^{1/2}, \tag{7}$$

where $\kappa_0 = (1+\omega)/s - 1$ and $-1 < \omega < 1$, is finite. Next recall that, according to Theorem 2.4-2, the class $W_{s,\sigma}^{2,\omega}$ coincides with the set of those functions $\Phi(z)$ which are representable in the form

$$\Phi(z) = \sum_{h=0}^{2s-1} \int_0^\sigma E_s\left(\beta_s^{h+1/2} z\tau^{1/s}; \mu\right) \tau^{\mu-1}\overline{\varphi_h(\tau)}d\tau, \tag{8}$$

where $\mu = (s+\omega+1)/2s$ $(-1 < \omega < 1)$ and $\varphi(\tau) = \{\varphi_h(\tau)\}_0^{2s-1}$ is an arbitrary vector function of $L_2^{2s}(0,\sigma)$. Finally, note that the following assertion is true for the classes $W_{s,\sigma}^{2,\omega}$.

Lemma 9.4-1. *If $\Phi(z) \in W_{s,\sigma}^{2,\omega}$ is an arbitrary function, then*

$$\|\Phi;\Gamma_{2s}\|_{2,\omega} \asymp \|\varphi\| = \left\{ \sum_{h=0}^{2s-1} \int_0^\sigma |\varphi_h(\tau)|^2 d\tau \right\}^{1/2}, \tag{9}$$

where the suitable constants are independent of $\Phi(z)$ and $\varphi(\tau)$.

Proof. The proof is similar to that of Lemma 7.4-1. It is based essentially on Theorem 2.5-7, the only difference being the use of formulas 2.4(12)-(13) of Theorem 2.4-2.

(c) Now we are ready to prove the second main theorem of this section.

Theorem 9.4-2. *If $\nu \in \Delta_s$, then the vector series of the form*

$$\varphi(\tau) = \sum_{n=-(s-1)}^{\infty} \varphi_n \omega_n(\tau) \tag{10}$$

are convergent in the norm of $L_2^{2s}(0,\sigma)$ and represent a continuous one-to-one mapping of the space $L_{2,\kappa_0}^{(s-1)}$ of sequences $\{\varphi_n\}_{-(s-1)}^{\infty}$ onto the space $L_2^{2s}(0,\sigma)$ of vector functions $\varphi(\tau)$. In addition, the equalities

$$\varphi_n = \{\varphi, \kappa_n\} \quad (-(s-1) \le n < +\infty) \tag{11}$$

and the two-sided inequalities

$$\|\varphi\| = \{\varphi, \varphi\}^{1/2} \asymp \|\{\varphi_n\}_{-(s-1)}^{\infty}\|_{2,\kappa_0} \tag{12}$$

are true.

Proof. By Theorem 8.2-1 (where we take $p = 2$), the series

$$\Phi(z) = \Gamma(1+\nu) \left\{ \sum_{n=-(s-1)}^{0} \frac{\overline{\varphi_n}}{|n|!} z^{-n} \right\} \mathcal{E}_{s,\sigma}(z;\nu)$$
$$+ \sum_{n=1}^{\infty} \overline{\varphi_n} \frac{z^s \mathcal{E}_{s,\sigma}(z;\nu)}{\mu_n^s \mathcal{E}_{s,\sigma}'(\mu_n;\nu)(z-\mu_n)} \tag{13}$$

represents a continuous one-to-one mapping of the space $L_{2,\kappa_0}^{(s-1)}$ of sequences $\{\varphi_n\}_{-(s-1)}^{\infty}$ onto the space $W_{s,\sigma}^{2,\omega}$ of entire functions $\Phi(z)$. At the same time the following two-sided estimates are true:

$$\|\Phi;\Gamma_{2s}\|_{2,\omega} \asymp \|\{\varphi_n\}_{-(s-1)}^{\infty}\|_{2,\kappa_0}. \tag{14}$$

Besides, by the interpolation data of the sum of the series (13) established in Theorem 8.2-1,

$$\Phi^{(k)}(0) = \overline{\varphi_{-k}} \quad (0 \le k \le s-1) \text{ and } \Phi(\mu_n) = \overline{\varphi_n} \quad (1 \le n < +\infty). \tag{15}$$

On the other hand, Wiener-Paley type Theorem 2.4-2 states that (8) represents a one-to-one mapping of the space $L_2^{2s}(0,\sigma)$ onto the space $W_{s,\sigma}^{2,\omega}$. Besides, the two-sided inequalities (9) were already proved in Lemma 9.4-1. Consequently, this mapping and its inverse are both continuous. We can conclude that there exists a canonical homeomorphism $\{\varphi_n\}_{-(s-1)}^{\infty} \to \varphi$ between the spaces $L_{2,\kappa_0}^{(s-1)}$ and $L_2^{2s}(0,\sigma)$, and the two-sided inequalities (12) are true. To be convinced that equalities (11) are also true, we note first that

$$\Phi(\mu_n) = \sum_{h=0}^{2s-1} \int_0^{\sigma} \kappa_{n,h}(\tau)\overline{\varphi_h(\tau)}d\tau = \{\kappa_n, \varphi\} = \overline{\{\varphi, \kappa_n\}} \quad (1 \le n < +\infty),$$

as follows from representation (8) of $\Phi(z)$ and from the definition $9.3(11_2)$ of the functions $\kappa_{n,h}(\tau)$ $(1 \le n < +\infty, 0 \le h \le 2s-1)$. Hence the interpolation data (15) implies formulas (11), but only in the case when $1 \le n < +\infty$. As to the case $-(s-1) \le n \le 0$, the desired formulas (11) easily follow from 9.3(16), from representation (8) and from (15). Now, to complete the proof, we have to show only that the mentioned homeomorphism between $L_{2,\kappa_0}^{(s-1)}$ and $L_2^{2s}(0,\sigma)$ can also be given by means of the vector series (10) *which are convergent in the norm of* $L_2^{2s}(0,\sigma)$. Indeed, if $\varphi(\tau) \in L_2^{2s}(0,\sigma)$ and

$$r_m(\tau) = \varphi(\tau) - \sum_{n=-(s-1)}^{m} \{\varphi, \kappa_n\}\omega_n(\tau) \quad (m \ge 0),$$

then inequalities (12) obviously give

$$\|r_m\| \asymp \left\{ \sum_{n=m+1}^{\infty} |\{\varphi, \kappa_n\}|^2 (1+n)^{\kappa_0} \right\}^{1/2}.$$

Hence $\|r_m\| \to 0$ as $m \to +\infty$, and consequently the vector series (10) converges to φ in the norm of $L_2^{2s}(0,\sigma)$. Thus, the proof is complete.

(d) Now we show that Theorem 9.4-2 can be formulated also in the following way, if we assume again that $\nu \in \Delta_s$.

Theorem 9.4-3. *If* $\nu \in \Delta_s$, *then:*
1°. *The system*

$$\{\omega_n(\tau)\}_{-(s-1)}^{\infty} \tag{16}$$

is complete in $L_2^{2s}(0,\sigma)$.
2°. *This system becomes a Riesz basis of* $L_2^{2s}(0,\sigma)$ *after suitable normalization.*

Proof. Assertion $1°$ is already established in Theorem 9.4-1. Besides, it follows immediately from Theorem 9.4-2. To prove assertion $2°$, we pass to the systems

$$K_n(\tau) = \{K_{n,h}(\tau)\}_0^{2s-1} \text{ and } \Omega_n^*(\tau) = \{\Omega_{n,h}^*(\tau)\}_0^{2s-1}, \tag{17}$$

where

$$K_{n,h}(\tau) = (1+|n|)^{\kappa_0/2}\kappa_{n,h}(\tau), \quad \Omega_{n,h}^*(\tau) = (1+|n|)^{-\kappa_0/2}\omega_{n,h}(\tau) \tag{18}$$

for any n and $h(-(s-1) \le n < +\infty, \quad 0 \le h \le 2s-1)$. As it is easy to verify, these systems are also biorthogonal, i.e.,

$$\{K_n, \Omega_m^*\} = \{\kappa_n, \omega_m\} = \delta_{n,m} \quad (-(s-1) \le n, m < +\infty). \tag{19}$$

The passage to systems (17) leads to the expansion

$$\varphi(\tau) = \sum_{n=-(s-1)}^{\infty} \{\varphi, K_n\}\Omega_n^*(\tau) \tag{20}$$

which is similar to (10), and the estimates (12) pass to

$$\|\varphi\| \asymp \left\{ \sum_{n=-(s-1)}^{\infty} |\{\varphi, K_n\}|^2 \right\}^{1/2}. \tag{21}$$

Indeed, by notations (18), $\{\varphi, K_n\}\Omega_n^*(\tau) = \{\varphi, \kappa_n\}\omega_n(\tau)$ and $|\{\varphi, K_n\}|^2 = |\{\varphi, \kappa_n\}|^2(1+|n|)^{\kappa_0} \ (-(s-1) \le n < +\infty)$. Hence relations (20) and (21) follow. These relations show that, according to the well-known definition, the system $\{\Omega_n^*(\tau)\}_{-(s-1)}^{\infty}$ is a Riesz basis of $L_2^{2s}(0,\sigma)$.

(e) We conclude this chapter with a general theorem relating to the basis properties of two last biorthogonal systems.

Theorem 9.4-4. *The systems of vector functions*

$$\left\{\{K_{n,h}(\tau)\}_{h=0}^{2s-1}\right\}_{-(s-1)}^{\infty} = \left\{\left\{(1+|n|)^{\kappa_0/2}\kappa_{n,h}(\tau)\right\}_{h=0}^{2s-1}\right\}_{-(s-1)}^{\infty},$$
$$\left\{\{\Omega_{n,h}^*(\tau)\}_{h=0}^{2s-1}\right\}_{-(s-1)}^{\infty} = \left\{\left\{(1+|n|)^{-\kappa_0/2}\omega_{n,h}(\tau)\right\}_{h=0}^{2s-1}\right\}_{-(s-1)}^{\infty} \tag{22}$$

are biorthogonal, and both are Riesz bases of $L_2^{2s}(0,\sigma)$.

Proof. Systems (22) are biorthogonal in $L_2^{2s}(0,\sigma)$ according to (19). The second system of (22) is a Riesz basis of $L_2^{2s}(0,\sigma)$, while the first system of (22) is complete in $L_2^{2s}(0,\sigma)$ by Theorem 9.4-1. Hence, by the well-known theorem on the Riesz basis property of biorthogonal systems, the first system of (22) is also a Riesz basis of $L_2^{2s}(0,\sigma)$.

9.5 Notes

As for the results of this chapter for each of the cases $s \geq 3$ and $s = 1, 2$ we can say nearly the same as for the results of Chapter 8, and we can refer to the same papers as in Notes 8.6.

10 The simplest Cauchy type problems and the boundary value problems connected with them

10.1 Introduction

As is well known, the trigonometric systems

$$\left\{ \sqrt{\frac{2}{\pi}} \sin \frac{\pi k}{\sigma} x \right\}_1^\infty \quad \text{and} \quad \left\{ \frac{1}{\sqrt{\sigma}}, \left\{ \sqrt{\frac{2}{\pi}} \cos \frac{\pi k}{\sigma} x \right\}_1^\infty \right\} \tag{1}$$

are orthogonal bases of $L_2(0, \sigma)$ and at the same time they are systems of eigenfunctions of definite boundary value problems. These boundary value problems are connected with the simplest differential equation

$$\frac{d^2}{dx^2} y(x) + \lambda y(x) = 0, \qquad x \in (0, \sigma) \tag{2}$$

with the suitable boundary conditions at the ends of the interval $(0, \sigma)$. As was mentioned in the introduction to Chapter 5, the basic biorthogonal systems of Mittag-Leffler type functions, constructed there for $L_2(0, \sigma)$, also turn out to be systems of eigenfunctions. But in this case they are systems of eigenfunctions of some non-ordinary boundary value problems formulated in terms of integro-differential operators of fractional orders.

This chapter deals with the interpretation of the biorthogonal expansions established in Theorems 5.3-1 and 5.3-2 as expansions in terms of eigenfunctions of the mentioned non-ordinary boundary value problems. Two associated non-ordinary boundary value problems on $(0, \sigma)$ are considered in Sections 10.3 and 10.4. The solutions of these problems bring us to the same sequence of eigenvalues $\{\lambda_k\}_1^\infty (0 < \lambda_1 < \lambda_2 < \dots)$ and to two corresponding systems of eigenfunctions which are biorthogonal in $L_2(0, \sigma)$:

$$\left\{ E_{1/2} \left(-\lambda_k x^2; \mu \right) x^{\mu-1} \right\}_1^\infty \quad \text{and} \quad \left\{ E_{1/2} \left(-\lambda_k (\sigma - x)^2; \mu^* \right) (\sigma - x)^{\mu^*-1} \right\}_1^\infty . \tag{3}$$

Also, we realize that the eigenfunctions

$$\frac{x^{\mu-1}}{\Gamma(\mu)} \quad \text{and} \quad \frac{(\sigma - x)^{\mu^*-1}}{\Gamma(\mu^*)} \tag{4}$$

must be added correspondingly to systems (3) when $\lambda_0 = 0$ is also an eigenvalue. In the concluding Section 10.5 by use of the main results of Chapter 5 we prove that, if the parameters μ and μ^* belong to the suitable intervals, then the considered boundary value problems are completely solvable. This enables us to establish suitable eigenfunction expansion theorems.

In this and in the following chapters we shall frequently use the main definitions and propositions of the theory of Riemann-Liouville fractional integro-differentiation. Therefore, in the next section we give a brief survey of some fundamental propositions of this theory together with their proofs.

10.2 Riemann-Liouville fractional integrals and derivatives

(a) Let $f(x)$ be an arbitrary function of class $L_1(0,\sigma)(0 < \sigma < +\infty)$. Then *the integral of order $\alpha(0 < \alpha < +\infty)$ of $f(x)$ with the origin at the point $x = 0$ is taken to be the function*

$$D^{-\alpha}f(x) \equiv \frac{1}{\Gamma(\alpha)} \int_0^x (x-t)^{\alpha-1} f(t)dt, \quad x \in (0,\sigma). \tag{1}$$

Similarly, *the integral of order $\alpha(0 < \alpha < +\infty)$ of $f(x)$ with the end at the point $x = \sigma$ is taken to be the function*

$$D_\sigma^{-\alpha}f(x) \equiv \frac{1}{\Gamma(\alpha)} \int_x^\sigma (t-x)^{\alpha-1} f(t)dt, \quad x \in (0,\sigma). \tag{2}$$

Note that
$$D_\sigma^{-\alpha}f(x) = D^{-\alpha} \left[f(\sigma - t)\right](y)\big|_{y=\sigma-x}, \quad x \in (0,\sigma). \tag{3}$$

Note also that, if $\alpha = n \geq 1$ is an integer, then $D^{-\alpha}f(x)$ is the n-th primitive of $f(x)$, i.e.,

$$D^{-n}f(x) = \int_0^x dt_1 \int_0^{t_1} dt_2 \ldots \int_0^{t_{n-1}} f(t_n)dt_n. \tag{4}$$

Hence for almost all $x \in (0,\sigma)$

$$\frac{d^n}{dx^n} D^{-n}f(x) = f(x). \tag{5}$$

Using Fubini's well-known theorem, we arrive at the inequalities

$$\int_0^\sigma |D^{-\alpha}f(x)|dx \leq \frac{1}{\Gamma(1+\alpha)} \int_0^\sigma (\sigma - t)^\alpha |f(t)|dt, \tag{6_1}$$

$$\int_0^\sigma |D_\sigma^{-\alpha}f(x)|dx \leq \frac{1}{\Gamma(1+\alpha)} \int_0^\sigma t^\alpha |f(t)|dt \tag{6_2}$$

which imply the following proposition.

1°. *If $f(x) \in L_1(0,\sigma)$, then for any $\alpha \in (0,+\infty)$ the functions $D^{-\alpha}f(x)$ and $D_\sigma^{-\alpha}f(x)$ are defined almost everywhere in $(0,\sigma)$ and are of $L_1(0,\sigma)$.*

In connection with this proposition note that, if $\gamma \in (-1,+\infty)$, then the functions $x^\gamma/\Gamma(1+\gamma)$ and $(\sigma - x)^\gamma/\Gamma(1+\gamma)$ are both of class $L_1(0,\sigma)$ and, in addition, it appears that

$$D^{-\alpha}\left\{\frac{x^\gamma}{\Gamma(1+\gamma)}\right\} = \frac{x^{\gamma+\alpha}}{\Gamma(1+\gamma+\alpha)}, \quad D_\sigma^{-\alpha}\left\{\frac{(\sigma - x)^\gamma}{\Gamma(1+\gamma)}\right\} = \frac{(\sigma - x)^{\gamma+\alpha}}{\Gamma(1+\gamma+\alpha)} \tag{7}$$

for any $\alpha \in (0,+\infty)$, and these functions are also of $L_1(0,\sigma)$.

Let $f(x) \in L_1(0, \sigma)$ be a given function. We shall denote by E_f the set of those points $x \in (0, \sigma)$, where the functions $f(x)$ and $|f(x)|$ are both the derivatives of their primitives. Then the measure of $(0, \sigma)/E_f$ is zero according to the well-known Lebesgue theorem. In addition, note that any Lebesgue point of $f(x)$ automatically belongs to E_f. Lastly, it is obvious that, if $\tilde{f}(x) = f(\sigma - x)$, then $x \in E_f$, if and only if $\sigma - x \in E_{\tilde{f}}$.

2^0. *If* $f(x) \in L_1(0, \sigma)$, *then for any* $\alpha \in (0, +\infty)$ *the integrals* $D^{-\alpha} f(x)$ *and* $D_\sigma^{-\alpha} f(x)$ *exist in every point* $x \in E_f$ *and*

$$\lim_{\alpha \to +0} D^{-\alpha} f(x) = f(x), \quad \lim_{\alpha \to +0} D_\sigma^{-\alpha} f(x) = f(x), \quad x \in E_f. \tag{8}$$

We shall prove only the first of these relations, as the proof of the second one is similar. To this end, we suppose $x_0 \in E_f$ is any point and denote

$$F_0(t) = \int_0^t f(x_0 - \tau)d\tau, \quad \Phi_0(t) = \int_0^t |f(x_0 - \tau)|d\tau, \quad t \in [0, x_0]. \tag{9}$$

Then, obviously, $F_0(t)/t \to f(x_0)$ and $\Phi_0(t)/t \to |f(x_0)|$ as $t \to +0$. Therefore

$$F_0(t) = t\left[f(x_0) + \omega_0(t)\right], \tag{9_1}$$

where

$$\omega_0(t) \to 0 \quad \text{and} \quad \Phi_0(t) = O(t) \tag{9_2}$$

as $t \to +0$. To be convinced that the integral

$$\int_0^{x_0} (x_0 - t)^{\alpha - 1}|f(t)|dt = \int_0^{x_0} |f(x_0 - \tau)|\tau^{\alpha - 1}d\tau \tag{10}$$

is convergent for any $\alpha \in (0, +\infty)$, we observe that for any $\delta \in (0, x_0)$

$$\int_\delta^{x_0} |f(x_0 - \tau)|\tau^{\alpha - 1}d\tau = \int_\delta^{x_0} \Phi_0'(\tau)\tau^{\alpha - 1}d\tau$$
$$= \Phi_0(\tau)\tau^{\alpha - 1}\Big|_\delta^{x_0} - (\alpha - 1)\int_\delta^{x_0} \Phi_0(\tau)\tau^{\alpha - 2}d\tau, \tag{11}$$

where, according to (9_2), the right-hand side tends to a definite limit as $\delta \to +0$. Hence integral (10) converges. Now we take any $\epsilon > 0$ and choose $\delta \equiv \delta(\epsilon) \in (0, x_0)$ for which $|\omega_0(t)| < \epsilon$ when $0 < t < \delta$. Then, using (9_1) and (9_2), we can write the following equalities for any $\alpha \in (0, +\infty)$:

$$D^{-\alpha} f(x_0) = \frac{1}{\Gamma(\alpha)} \int_0^{x_0} (x_0 - t)^{\alpha - 1}f(t)dt = \frac{1}{\Gamma(\alpha)} \int_0^{x_0} f(x_0 - \tau)\tau^{\alpha - 1}d\tau$$

$$= \frac{1}{\Gamma(\alpha)} \int_0^{x_0} F_0'(\tau)\tau^{\alpha - 1}d\tau = \frac{F_0(\tau)\tau^{\alpha - 1}}{\Gamma(\alpha)}\Big|_0^{x_0} - \frac{\alpha - 1}{\Gamma(\alpha)} \int_0^{x_0} F_0(\tau)\tau^{\alpha - 2}d\tau$$

$$= \frac{F_0(x_0)}{\Gamma(\alpha)}x_0^{\alpha - 1} - \frac{\alpha - 1}{\Gamma(1 + \alpha)}f(x_0)x_0^\alpha - \frac{\alpha - 1}{\Gamma(\alpha)} \int_0^\delta \omega_0(\tau)\tau^{\alpha - 1}d\tau$$

$$- \frac{\alpha - 1}{\Gamma(\alpha)} \int_\delta^{x_0} \omega_0(\tau)\tau^{\alpha - 1}d\tau.$$

Hence we obtain

$$\limsup_{\alpha \to +0} |D^{-\alpha} f(x_0) - f(x_0)| \le \epsilon,$$

and the passage $\epsilon \to +0$ completes the proof.

In view of proposition 2°, which is already established, it is reasonable to extend the definitions (1) and (2) of operators $D^{-\alpha}$ and $D_\sigma^{-\alpha}$ to the value $\alpha = 0$, setting

$$D^{-0} f(x) = D_\sigma^{-0} f(x) = f(x), \quad x \in (0, \sigma). \tag{12}$$

3°. *Let $f(x) \in L_1(0, \sigma)$ and let $\alpha_1, \alpha_2 \in [0, +\infty)$ be arbitrary numbers. Then the equalities*

$$D^{-\alpha_2} \left(D^{-\alpha_1} f(x) \right) = D^{-\alpha_1} \left(D^{-\alpha_2} f(x) \right) = D^{-(\alpha_1 + \alpha_2)} f(x), \tag{13_1}$$

$$D_\sigma^{-\alpha_2} \left(D_\sigma^{-\alpha_1} f(x) \right) = D_\sigma^{-\alpha_1} \left(D_\sigma^{-\alpha_2} f(x) \right) = D_\sigma^{-(\alpha_1 + \alpha_2)} f(x) \tag{13_2}$$

are true for almost all $x \in (0, \sigma)$.

Indeed,

$$D^{-\alpha_2} \left(D^{-\alpha_1} f(x) \right) = \frac{1}{\Gamma(\alpha_2)} \int_0^x (x - t_2)^{\alpha_2 - 1} \left\{ \frac{1}{\Gamma(\alpha_1)} \int_0^{t_2} (t_2 - t_1)^{\alpha_1 - 1} f(t_1) dt_1 \right\} dt_2$$

$$= \frac{1}{\Gamma(\alpha_1)\Gamma(\alpha_2)} \int_0^x f(t_1) \left\{ \int_{t_1}^x (x - t_2)^{\alpha_2 - 1}(t_2 - t_1)^{\alpha_1 - 1} dt_2 \right\} dt_1$$

$$= \frac{1}{\Gamma(\alpha_1 + \alpha_2)} \int_0^x (x - t_1)^{\alpha_1 + \alpha_2 - 1} f(t_1) dt_1 = D^{-(\alpha_1 + \alpha_2)} f(x),$$

where all the operations are permitted for almost all $x \in (0, \sigma)$ by Fubini's theorem. The proofs of the other equalities are similar.

(b) We now pass to the definition of the derivatives of arbitrary order and some of their main properties.

Assume α $(0 < \alpha < +\infty)$ to be a given number and the integer $p \ge 1$ to be defined by the inequalities

$$p - 1 < \alpha \le p. \tag{14}$$

Assume, in addition, that $f(x) \in L_1(0, \sigma)$ and that

$$\omega(x) = D^{-(p-\alpha)} f(x), \quad x \in (0, \sigma). \tag{15}$$

Then, according to proposition 1°,

$$\omega(x) \in L_1(0, \sigma), \tag{15'}$$

and, if $\alpha = p \ge 1$ is an integer, definition (12) should be used. Now formally we introduce the function

$$D^\alpha f(x) \equiv \frac{d^p}{dx^p} \omega(x) = \frac{d^p}{dx^p} \left\{ D^{-(p-\alpha)} f(x) \right\} \tag{16}$$

which is called *the derivative of order $\alpha > 0$ of $f(x)$ with the origin at the point $x = 0$. According to (12),*

$$D^\alpha f(x) = D^p f(x) \equiv \frac{d^p}{dx^p} f(x) \tag{17}$$

when $\alpha = p$ is an integer, so $D^p f(x)$ in this case simply coincides with the ordinary derivative of $f(x)$ of order $\alpha = p \geq 1$. From (16) it also follows that

$$D^\alpha f(x) = \frac{d}{dx}\left\{ D^{-(1-\alpha)} f(x) \right\} \quad (0 < \alpha \leq 1). \tag{16$_1$}$$

This formula may be extended formally to the value $\alpha = 0$, setting

$$D^{+0} f(x) = \frac{d}{dx}\left\{ D^{-1} f(x) \right\} = f(x), \quad x \in (0,\sigma). \tag{16$_2$}$$

Further, it follows from (16$_1$) and (12) that for almost all $x \in (0,\sigma)$

$$D^1 f(x) = \frac{d}{dx}\left\{ D^{-0} f(x) \right\} = f'(x). \tag{16$_3$}$$

Obviously the derivative D^α_σ of order $\alpha > 0$ of a function $f(x)$, with the end at the point $x = \sigma$, may be defined in a similar way. Specifically, it should again be supposed that $p - 1 < \alpha \leq p \ (p \geq 1)$ and taken

$$D^\alpha_\sigma f(x) \equiv \frac{d^p}{d(\sigma - x)^p}\left\{ D^{-(p-\alpha)}_\sigma f(x) \right\}, \quad x \in (0,\sigma). \tag{18}$$

Then, evidently,

$$D^p_\sigma f(x) = \frac{d^p}{d(\sigma - x)^p} f(x) = (-1)^p \frac{d^p}{dx^p} f(x) \quad (p \geq 1), \tag{18$_1$}$$

$$D^\alpha_\sigma f(x) = \frac{d}{d(\sigma - x)}\left\{ D^{-(1-\alpha)}_\sigma f(x) \right\} \quad (0 < \alpha \leq 1), \tag{18$_2$}$$

$$D^{+0}_\sigma f(x) = f(x), \quad D^1_\sigma f(x) = -f'(x). \tag{18$_3$}$$

In later sections formulas (7) will play a significant role. Therefore, we have to be convinced that these formulas remain true when α is replaced by $-\alpha$. Indeed, if we suppose that $p - 1 < \alpha \leq p$ and $\gamma \in (-1,+\infty)$, then according to the definitions of derivatives D^α and D^α_σ,

$$D^\alpha\left\{ \frac{x^\gamma}{\Gamma(1+\gamma)} \right\} = \frac{d^p}{dx^p}\left\{ D^{-(p-\alpha)}\left(\frac{x^\gamma}{\Gamma(1+\gamma)} \right) \right\}, \quad x \in (0,\sigma)$$

and

$$D^{\alpha}_{\sigma}\left\{\frac{(\sigma-x)^{\gamma}}{\Gamma(1+\gamma)}\right\} = \frac{d^p}{d(\sigma-x)^p}\left\{D^{-(p-\alpha)}_{\sigma}\left(\frac{(\sigma-x)^{\gamma}}{\Gamma(1+\gamma)}\right)\right\}, \quad x\in(0,\sigma).$$

Therefore, the use of (7) gives

$$D^{\alpha}\left\{\frac{x^{\gamma}}{\Gamma(1+\gamma)}\right\} = \frac{d^p}{dx^p}\left\{\frac{x^{\gamma+p-\alpha}}{\Gamma(1+\gamma+p-\alpha)}\right\} = \frac{x^{\gamma-\alpha}}{\Gamma(1+\gamma-\alpha)}, \quad x\in(0,\sigma)$$

and

$$D^{\alpha}_{\sigma}\left\{\frac{(\sigma-x)^{\gamma}}{\Gamma(1+\gamma)}\right\} = \frac{d^p}{d(\sigma-x)^p}\left\{\frac{(\sigma-x)^{\gamma+p-\alpha}}{\Gamma(1+\gamma+p-\alpha)}\right\} = \frac{(\sigma-x)^{\gamma-\alpha}}{\Gamma(1+\gamma-\alpha)}, \quad x\in(0,\sigma).$$

Thus the following general formulas are true for any $\gamma\in(-1,+\infty)$ *and* $\alpha\in(-\infty,+\infty)$:

$$D^{\alpha}\left\{\frac{x^{\gamma}}{\Gamma(1+\gamma)}\right\} = \frac{x^{\gamma-\alpha}}{\Gamma(1+\gamma-\alpha)}, \quad x\in(0,\sigma), \tag{19_1}$$

$$D^{\alpha}_{\sigma}\left\{\frac{(\sigma-x)^{\gamma}}{\Gamma(1+\gamma)}\right\} = \frac{(\sigma-x)^{\gamma-\alpha}}{\Gamma(1+\gamma-\alpha)}, \quad x\in(0,\sigma). \tag{19_2}$$

Finally, note that the functions $\Gamma^{-1}(1+\gamma)x^{\gamma}$ *and* $\Gamma^{-1}(1+\gamma)(\sigma-x)^{\gamma}$ *are of class* $L_1(0,\sigma)$, *but their transformations* (19_1) *and* (19_2) *are of the same class only when* $\alpha<1+\gamma$.

(c) *A natural question arises: is it possible to find additional conditions ensuring the inclusion* $D^{\alpha}f(x)\in L_1(0,\sigma)$ $(\alpha>0)$ *for a function* $f(x)\in L_1(0,\sigma)$? *It appears that such conditions can be found by use of the well-known functions which are absolutely continuous in the segment* $\Omega=[a,b]$ $(-\infty<a<b<+\infty)$ *(or, briefly, which are of class* $AC(\Omega)$). *Remember that a function* $f(x)$ *is said to be absolutely continuous in* Ω, *if for any* $\epsilon>0$ *there exists a* $\delta\equiv\delta(\epsilon)>0$, *such that the inequality*

$$\sum_{k=1}^{n}|f(b_k)-f(a_k)|<\epsilon$$

is true for any finite system $\{[a_k,b_k]\}_{k=1}^{n}\subset\Omega$ *of disjoint segments for which*

$$\sum_{k=1}^{n}(b_k-a_k)<\delta.$$

It is well known that the class $AC(\Omega)$ *coincides with the set of primitives of functions which are summable on* Ω *in the Lebesgue sense, i.e.,* $AC(\Omega)$ *coincides with the set of functions representable in the form*

$$f(x)=c+\int_a^x \varphi(\tau)d\tau, \quad \varphi(\tau)\in L_1(a,b). \tag{20}$$

We shall also use the following definition: $AC^n(\Omega)(n \geq 1)$ is the set of functions $f(x)$ continuously differentiable in Ω up to the order $n-1$ and such that $f^{(n-1)}(x) \in AC(\Omega)$. It is obvious that $AC^1(\Omega) = AC(\Omega)$ and also that, similar to (20), $AC^n(\Omega)$ coincides with the set of functions representable in the form

$$f(x) = \sum_{k=0}^{n-1} c_k(x-a)^k + \frac{1}{\Gamma(n)} \int_a^x (x-\tau)^{n-1}\varphi(\tau)d\tau, \quad x \in \Omega, \tag{21}$$

where $\{c_k\}_0^{n-1}$ are arbitrary constants and $\varphi(\tau) \in L_1(a,b)$ is an arbitrary function. This assertion is an immediate consequence of representation (20) of the function $f^{(n-1)}(x) \in AC(\Omega)$ and of formula (4). Observe that in (21)

$$\varphi(\tau) = f^{(n)}(\tau) \in L_1(a,b) \ \text{ and } \ c_k = f^{(k)}(a)/k! \quad (0 \leq k \leq n-1). \tag{22}$$

Also observe that (21) and (22) yield Maclaurin's formula

$$f(x) = \sum_{k=0}^{n-1} f^{(k)}(0)\frac{x^k}{\Gamma(1+k)} + D^{-n}\left\{f^{(n)}(x)\right\}, \quad x \in [0,\sigma] \tag{21_1}$$

for any function $f(x) \in AC^n[0,\sigma]$. Hence, if we apply the operator $D^{-\alpha}(0 \leq \alpha < +\infty)$ to both sides of (21_1), then, in view of (7) and (13_1), we arrive at the representation

$$D^{-\alpha}f(x) = \sum_{k=0}^{n-1} f^{(k)}(0)\frac{x^{k+\alpha}}{\Gamma(1+\alpha+k)} + D^{-(n+\alpha)}\left\{f^{(n)}(x)\right\}, \quad x \in [0,\sigma]. \tag{23}$$

Taking $n = 1$, we arrive at the following proposition.
$4°$. If $f(x) \in AC[0,\sigma]$, then for any $\alpha \in [0,+\infty)$

$$D^{-\alpha}f(x) = f(0)\frac{x^\alpha}{\Gamma(1+\alpha)} + D^{-(1+\alpha)}f'(x), \quad x \in [0,\sigma], \tag{24}$$

and

$$D^{-\alpha}f(x) \in AC[0,\sigma]. \tag{25}$$

Using the classes $AC^n(\Omega)$ we find the desired additional conditions, which are given in the next proposition.
$5°$. If the function $f(x) \in L_1(0,\sigma)$ satisfies the condition

$$\omega(x) \equiv D^{-(p-\alpha)}f(x) \in AC^p[0,\sigma] \quad (p-1 < \alpha \leq p, p \geq 1), \tag{26}$$

then the derivative

$$D^\alpha f(x) = \frac{d^p}{dx^p}\omega(x) \tag{27}$$

exists almost everywhere in $(0, \sigma)$ *and is of class* $L_1(0, \sigma)$.

Remark 1. Let $p-1 < \alpha \le p(p \ge 1)$ and let the function $f(x)$ satisfy the conditions

$$f(x) \in L_1(0, \sigma) \ \text{ and } \ \omega(x) = D^{-(p-\alpha)} f(x) \in AC^p[0, \sigma].$$

Then $D^\alpha f(x) \in L_1(0, \sigma)$ and also

$$D^{\alpha-k} f(x) \in L_1(0, \sigma) \quad (1 \le k \le p-1). \tag{28}$$

Besides,

$$\begin{aligned}
D^{\alpha-k} f(x) &= \frac{d^{p-k}}{dx^{p-k}} \left\{ D^{-(p-k-(\alpha-k))} f(x) \right\} \\
&= \frac{d^{p-k}}{dx^{p-k}} \left\{ D^{-(p-\alpha)} f(x) \right\} = \frac{d^{p-k}}{dx^{p-k}} \omega(x) \in AC^k[0, \sigma] \quad (1 \le k \le p-1),
\end{aligned} \tag{29}$$

since $p - k - 1 < \alpha - k \le p - k$ when $1 \le k \le p - 1$. Now we move on to some propositions concerning compositions of integro-differential operators.

6°. If $f(x) \in L_1(0, \sigma)$, then

$$D^\alpha D^{-\alpha} f(x) = D_\sigma^\alpha D_\sigma^{-\alpha} f(x) = f(x) \tag{30}$$

for any $\alpha \in [0, +\infty)$ *and for almost all* $x \in (0, \sigma)$.

Note that this proposition is trivial in the case $\alpha = 0$ in view of (12), (16_2) and (18_3). If $p - 1 < \alpha \le p(p \ge 1)$, then, by proposition 3°, $D^{-p} f(x) = D^{-(p-\alpha)} D^{-\alpha} f(x)$ for almost all $x \in (0, \sigma)$. Hence

$$D^\alpha D^{-\alpha} f(x) = \frac{d^p}{dx^p} \left\{ D^{-(p-\alpha)} D^{-\alpha} f(x) \right\} = \frac{d^p}{dx^p} \left\{ D^{-p} f(x) \right\} = f(x)$$

almost everywhere in $(0, \sigma)$, and the first of equalities (30) is proved. The proof of the second equality of (30) is similar.

7°. *Let* $p - 1 < \alpha \le p(p \ge 1)$ *and let the function* $f(x) \in L_1(0, \sigma)$ *satisfy the additional condition*

$$\omega(x) = D^{-(p-\alpha)} f(x) \in AC^p[0, \sigma].$$

Then the following formula is true almost everywhere in $(0, \sigma)$:

$$\begin{aligned}
D^{-\alpha} D^\alpha f(x) &= f(x) - \sum_{k=1}^{p} \left\{ D^{\alpha-k} f(\tau) \right\} \Big|_{\tau=0} \frac{x^{\alpha-k}}{\Gamma(1 + \alpha - k)} \\
&= f(x) - \sum_{k=1}^{p} \left\{ \omega^{(p-k)}(\tau) \right\} \Big|_{\tau=0} \frac{x^{\alpha-k}}{\Gamma(1 + \alpha - k)}.
\end{aligned} \tag{31}$$

Indeed,

$$D^{-\alpha}D^{\alpha}f(x) = \frac{d}{dx}\left\{D^{-(1+\alpha)}D^{\alpha}f(x)\right\}$$
$$= \frac{d}{dx}\left\{D^{-(1+\alpha)}\left(\omega^{(p)}(x)\right)\right\} = \frac{d}{dx}\left\{D^{-(p+\alpha-p+1)}\left(\omega^{(p)}(x)\right)\right\}. \tag{32}$$

And, if we replace n by p, α by $\alpha - p + 1$ and $f(x)$ by $\omega(x)$ in (23), then we obtain

$$D^{-(\alpha-p+1)}\omega(x) = \sum_{k=0}^{p-1}\frac{\omega^{(k)}(0)}{\Gamma(2+\alpha-p+k)}x^{k+\alpha-p+1} + D^{-(p+\alpha-p+1)}\left\{\omega^{(p)}(x)\right\}$$

for any $x \in [0, \sigma]$. Thus (32) yields

$$D^{-\alpha}D^{\alpha}f(x) = \frac{d}{dx}\left\{D^{-(\alpha-p+1)}\omega(x) - \sum_{k=0}^{p-1}\frac{\omega^{(k)}(0)}{\Gamma(2+\alpha-p+k)}x^{k+\alpha-p+1}\right\}$$
$$= \frac{d}{dx}\left\{D^{-1}f(x)\right\} - \sum_{k=0}^{p-1}\frac{\omega^{(k)}(0)}{\Gamma(1+\alpha-p+k)}x^{k+\alpha-p}$$
$$= f(x) - \sum_{k=1}^{p}\frac{\omega^{(p-k)}(0)}{\Gamma(1+\alpha-k)}x^{\alpha-k},$$

if we use notations (26) and equalities (13_1).

Particularly, if $p = 1$ and $0 < \alpha \leq 1$, this proposition takes the following form.

8°. *Let $f(x) \in L_1(0,\sigma)$ and let $D^{-(1-\alpha)}f(x) \in AC[0,\sigma]$ for some $\alpha \in (0,1]$ then almost everywhere in $(0,\sigma)$*

$$D^{-\alpha}D^{\alpha}f(x) = f(x) - \left\{D^{-(1-\alpha)}f(\tau)\right\}\Big|_{\tau=0}\frac{x^{\alpha-1}}{\Gamma(\alpha)}. \tag{33}$$

The following generalizations of the previous assertions are also useful.

9°. *If $f(x) \in L_1(0,\sigma)$, then, almost everywhere in $0,\sigma)$,*

$$D^{\alpha}D^{-\beta}f(x) = D^{-(\beta-\alpha)}f(x), \beta \geq \alpha \geq 0 \tag{34}$$

and

$$D^{\alpha}D^{-\beta}f(x) = D^{\alpha-\beta}f(x), \alpha > \beta \geq 0 \tag{35}$$

if the derivative $D^{\alpha-\beta}f(x)$ exists almost everywhere in $(0,\sigma)$.

Indeed, if $\beta \geq \alpha \geq 0$, then propositions 3° and 6° yield $D^{\alpha}D^{-\beta}f(x) = D^{\alpha}D^{-\alpha}D^{-(\beta-\alpha)}f(x) = D^{-(\beta-\alpha)}f(x)$. On the other hand, if $\alpha > \beta \geq 0$, then we assume $p - 1 < \alpha \le p (p \ge 1)$ and $q - 1 < \alpha - \beta \le q (q \ge 1)$. Hence $q \le p$, and therefore, by proposition 3°, almost everywhere in $(0, \sigma)$

$$D^\alpha D^{-\beta} f(x) = \frac{d^p}{dx^p} \left\{ D^{-(p-\alpha)} D^{-\beta} f(x) \right\} = \frac{d^p}{dx^p} \left\{ D^{-(p-\alpha+\beta)} f(x) \right\}$$

$$= \frac{d^q}{dx^q} \frac{d^{p-q}}{dx^{p-q}} \left\{ D^{-(p-q)} D^{-(q-\alpha+\beta)} f(x) \right\}$$

$$= \frac{d^q}{dx^q} \left\{ D^{-(q-\alpha+\beta)} f(x) \right\} = D^{\alpha-\beta} f(x).$$

10°. *If $f(x) \in L_1(0, \sigma)$ and, in addition,*

$$D^{-(p-\beta)} f(x) \in AC^p[0, \sigma] \quad (p - 1 < \beta \le p, p \ge 1),$$

then the equality

$$D^{-\alpha} D^\beta f(x) = D^{\beta-\alpha} f(x) - \sum_{k=1}^{p} \left\{ D^{\beta-k} f(\tau) \right\} \bigg|_{\tau=0} \frac{x^{\alpha-k}}{\Gamma(1 + \alpha - k)} \tag{36}$$

is true almost everywhere in $(0, \sigma)$ for any $\alpha > 0$.

Indeed, the use of proposition 3° (when $\beta \le \alpha$), or proposition 9° (when $\beta \ge \alpha$), gives $D^{-\alpha} D^\beta f(x) = D^{\beta-\alpha} D^{-\beta} D^\beta f(x)$. Hence, by (31),

$$D^{-\alpha} D^\beta f(x) = D^{\beta-\alpha} \left\{ f(x) - \sum_{k=1}^{p} \left\{ D^{\beta-k} f(\tau) \right\} \bigg|_{\tau=0} \frac{x^{\beta-k}}{\Gamma(1 + \beta - k)} \right\},$$

and formula (36) follows, if (19_1) is taken into account.

Now we shall give a simple condition sufficient for the existence of the derivative $D^\alpha f(x) (\alpha > 0)$ and also a representation formula for this derivative.

11°. *Let $f(x) \in AC^q[0, \sigma] (q \ge 1)$. Then the derivative $D^\alpha f(x)$ exists almost everywhere in $(0, \sigma)$ for any $\alpha \in (0, q]$. And if $p - 1 < \alpha \le p$, then*

$$D^\alpha f(x) - \sum_{k=0}^{p-1} \frac{f^{(k)}(0)}{\Gamma(1 + k - \alpha)} x^{k-\alpha} = \frac{1}{\Gamma(p - \alpha)} \int_0^x (x - t)^{p-\alpha-1} f^{(p)}(t) dt \tag{37}$$

$$= D^{-(p-\alpha)} f^{(p)}(x)$$

almost everywhere in $(0, \sigma)$.

To prove these assertions, we observe that $f(x) \in AC^p[0, \sigma]$ since $p \le q$. The replacement of n by p and α by $p - \alpha$ in formula (23) leads to the representation

$$\omega(x) = D^{-(p-\alpha)} f(x) = \sum_{k=0}^{p-1} \frac{f^{(k)}(0)}{\Gamma(1 + p - \alpha + k)} x^{p-\alpha+k} + D^{-(2p-\alpha)} f^{(p)}(x). \tag{38}$$

Using this representation we can easily verify the existence of the derivative $D^\alpha f(x) = (d^p/dx^p)(\omega(x))$ and the validity of (37).

Remark 2. Representation (37) shows that, if $f(x) \in AC^q[0, \sigma]$ ($q \geq 1, 0 < \alpha \leq q$), then the derivative $D^\alpha f(x)$ exists almost everywhere in $(0, \sigma)$. But in the same conditions, $D^\alpha f(x)$ is not necessarily of class $L_1(0, \sigma)$.

The next helpful proposition follows from the previous one.

12°. *If* $f(x) \in AC[0, \sigma]$, *then* $D^\alpha f(x) \in L_1(0, \sigma)$ *for any* $\alpha \in (0, 1]$ *and*

$$D^\alpha f(x) = \frac{f(0)}{\Gamma(1-\alpha)} x^{-\alpha} + D^{-(1-\alpha)} f'(x), \quad \alpha \in (0, 1] \tag{39}$$

almost everywhere in $(0, \sigma)$.

The following proposition will also be used later.

13°. *If* $f(x) \in L_1(0, \sigma)$ *and* $D^{-(1-\alpha)} f(x) \in AC[0, \sigma]$ *for a given* $\alpha \in (0, 1]$ *(and, consequently,* $D^\alpha f(x) \in L_1(0, \sigma)$*), then* $D^\beta f(x) \in L_1(0, \sigma)$ *for any* $\beta \in (0, \alpha)$.

Indeed, by the given definition,

$$D^\beta f(x) = \frac{d}{dx} \left\{ D^{-(1-\beta)} f(x) \right\}, \tag{40}$$

and in addition

$$D^{-(1-\beta)} f(x) = D^{-(\alpha-\beta)} \left\{ D^{-(1-\alpha)} f(x) \right\}. \tag{41}$$

But $D^{-(1-\alpha)} f(x) \in AC[0, \sigma]$, so $D^{-(1-\beta)} f(x) \in AC[0, \sigma]$ by proposition 4°, and the desired inclusion follows from (40).

(d) We shall conclude this section with a theorem relating to the solvability of Abel's integral equation.

Theorem 10.2-1. *Let* $f(x) \in L_1(0, \sigma)$. *Then Abel's integral equation*

$$f(x) = D^{-\alpha} g(x) = \frac{1}{\Gamma(\alpha)} \int_0^x (x-t)^{\alpha-1} g(t) dt \quad (\alpha > 0) \tag{42}$$

has a solution $g(x) \in L_1(0, \sigma)$ *if and only if the following pair of conditions, where* $p \geq 1(p-1 < \alpha \leq p)$ *is an integer, is satisfied:*

$$\omega(x) \equiv D^{-(p-\alpha)} f(x) \in AC^p[0, \sigma] \ \left(i.e., (d^{p-1}/dx^{p-1})\omega(x) \in AC[0, \sigma]\right), \tag{i}$$

$$\omega(0) = \omega'(0) = \ldots = \omega^{(p-1)}(0) = 0. \tag{ii}$$

If these conditions are satisfied, then the solution $g(x) \in L_1(0, \sigma)$ *of equation (42) is unique and*

$$g(x) = D^\alpha f(x) = \frac{d^p}{dx^p} \left\{ D^{-(p-\alpha)} f(x) \right\} \tag{43}$$

almost everywhere in $(0, \sigma)$.

Proof. If condition (i) is satisfied, then the derivative $D^\alpha f(x) = \omega^{(p)}(x)$ obviously exists almost everywhere in $(0, \sigma)$ and is of class $L_1(0, \sigma)$. Besides, it proves to be a solution of equation (42). Indeed, according to proposition $7°$,

$$D^{-\alpha} g(x) = D^{-\alpha} D^\alpha f(x) = f(x) - \sum_{k=1}^{p} \frac{\omega^{(p-k)}(0)}{\Gamma(1 + \alpha - k)} x^{\alpha - k}$$

almost everywhere in $(0, \sigma)$. Therefore, condition (ii) implies that $g(x)$ is a solution of (42). This solution is unique since proposition $6°$ shows that, if $D^{-\alpha} g(x) = 0$ almost everywhere in $(0, \sigma)$, then we also have $D^\alpha D^{-\alpha} g(x) = g(x) = 0$ almost everywhere in $(0, \sigma)$. Thus it remains to prove that conditions (i) and (ii) are necessary for solvability of equation (42). To this end we suppose that there exists a function $g(x) \in L_1(0, \sigma)$ such that $D^{-\alpha} g(x) = f(x)$ almost everywhere in $(0, \sigma)$. Then, using proposition $3°$, we obtain

$$\omega(x) = D^{-(p-\alpha)} f(x) = D^{-p} g(x) = \frac{1}{\Gamma(p)} \int_0^x (x - t)^{p-1} g(t) dt, \quad x \in (0, \sigma).$$

Hence

$$\omega^{(k)}(x) = \frac{1}{\Gamma(p - k)} \int_0^x (x - t)^{p-k-1} g(t) dt \quad (k = 0, 1, \dots, p - 1),$$

and evidently $\omega(x)$ satisfies conditions (i) and (ii).

10.3 A Cauchy type problem

(a) Let the set of parameters

$$\{\gamma\} \equiv \{\gamma_j\}_0^3 \tag{1}$$

satisfy the conditions

$$1/2 < \gamma_0, \ \gamma_3 \le 1, \ 0 \le \gamma_1, \ \gamma_2 \le 1 \tag{2}$$

and

$$\sum_{j=0}^{3} \gamma_j = 3. \tag{3}$$

Then, obviously,

$$1 < \gamma_0 + \gamma_3 \le 2 \text{ and } 1 \le \gamma_1 + \gamma_2 < 2. \tag{4}$$

Therefore only one of the parameters γ_1 and γ_2 may be equal to zero. And if it is so, then evidently

$$\gamma_0 = \gamma_2 = \gamma_3 = 1 \text{ when } \gamma_1 = 0,$$
$$\text{or } \gamma_0 = \gamma_1 = \gamma_3 = 1 \text{ when } \gamma_2 = 0. \tag{5}$$

We shall consider these as special cases and instead of (2) we shall suppose henceforth

$$1/2 < \gamma_0, \ \gamma_3 \leq 1 \text{ and } 0 < \gamma_1, \gamma_2 \leq 1. \tag{2_1}$$

Observe that, together with (3), these inequalities also imply inequalities (4).

We associate a given set of parameters (1) with a set $\{L_j\}_0^3$ of integro-differential operators of fractional orders setting

$$L_0 y(x) = D^{-(1-\gamma_0)} y(x), \tag{5_1}$$

$$L_1 y(x) = D^{-(1-\gamma_1)} \frac{d}{dx} L_0 y(x) = D^{-(1-\gamma_1)} D^{\gamma_0} y(x), \tag{5_2}$$

$$L_2 y(x) = D^{-(1-\gamma_2)} \frac{d}{dx} L_1 y(x) = D^{-(1-\gamma_2)} D^{\gamma_1} D^{\gamma_0} y(x), \tag{5_3}$$

$$\mathbb{L}_{1/2} y(x) \equiv L_3 y(x)$$
$$= D^{-(1-\gamma_3)} \frac{d}{dx} L_2 y(x) = D^{-(1-\gamma_3)} D^{\gamma_2} D^{\gamma_1} D^{\gamma_0} y(x). \tag{6}$$

It is necessary to define correctly the domain of definition of the operators $\{L_j\}_0^3$ so that it is large enough to suit our later purposes.

Definition 1. Denote by $AC_{\{\gamma\}}[0,\sigma](0 < \sigma < +\infty)$ the set of functions $y(x)$ satisfying the conditions

$$\begin{aligned} 1) & \quad y(x) \in L_1(0,\sigma) \\ 2) & \quad L_j y(x) \in AC[0,\sigma] \quad (j = 0, 1, 2). \end{aligned} \tag{7}$$

Thus, if $y(x) \in AC_{\{\gamma\}}[0,\sigma]$, then obviously the functions $L_j y(x)(j = 0, 1, 2)$ are continuous in $[0,\sigma]$, and the function $\mathbb{L}_{1/2} y(x) \equiv L_3 y(x)$ is of class $L_1(0,\sigma)$ since it is the fractional integral of the function $(d/dx)L_2 y(x) \in L_1(0,\sigma)$ of order $1 - \gamma_3$.

(b) Let $y(x) \in AC_{\{\gamma\}}[0,\sigma]$ be a given function. We introduce the following constants:

$$m_k(y) = \{L_k y(x)\}\big|_{x=0} \quad (k = 0, 1, 2). \tag{8}$$

Further, we denote

$$\mu_k = \sum_{j=0}^{k} \gamma_j \quad (k = 0, 1, 2) \tag{9}$$

and observe that

$$\begin{aligned} \mu_0 &= \gamma_0 \in (1/2, 1], \\ \mu_1 &= \gamma_0 + \gamma_2 \in (\gamma_0, 2], \\ \mu_2 &= \gamma_0 + \gamma_1 + \gamma_2 = 3 - \gamma_3 \in [2, 5/2) \end{aligned} \tag{10}$$

according to (2_1) and (3). Besides, conditions (2_1) imply

$$\gamma_{j_1} + \gamma_{j_2} \leq 2 \quad (j_1 \neq j_2) \tag{11}$$

for any pair of numbers of (1).

Now we are ready to establish three useful lemmas.

Lemma 10.3-1. *If $y(x) \in AC_{\{\gamma\}}[0, \sigma]$, then*

$$D^{-2}\mathbb{L}_{1/2}y(x) = y(x) - \sum_{k=0}^{2} m_k(y)\frac{x^{\mu_k-1}}{\Gamma(\mu_k)} \tag{12}$$

almost everywhere in $(0, \sigma)$.

Proof. Since $\mathbb{L}_{1/2}y(x) \in L_1(0, \sigma)$, using proposition $3°$ of Section 10.2 and formula 10.2(7) we arrive at the following equalities which are true almost everywhere in $(0, \sigma)$:

$$\begin{aligned}
D^{-2}\mathbb{L}_{1/2}y(x) &= D^{-2}D^{-(1-\gamma_3)}\frac{d}{dx}L_2y(x) = D^{-(2-\gamma_3)}\left\{D^{-1}\frac{d}{dx}L_2y(x)\right\} \\
&= D^{-(2-\gamma_3)}\left\{L_2y(x) - m_2(y)\right\} = D^{-(2-\gamma_3)}L_2y(x) - m_2(y)\frac{x^{2-\gamma_3}}{\Gamma(3-\gamma_3)}.
\end{aligned} \tag{12_1}$$

Further, we obtain

$$\begin{aligned}
D^{-(2-\gamma_3)}L_2y(x) &= D^{-(2-\gamma_3)}\left\{D^{-(1-\gamma_2)}\frac{d}{dx}L_1y(x)\right\} \\
&= D^{-(2-\gamma_2-\gamma_3)}\left\{D^{-1}\frac{d}{dx}L_1y(x)\right\} = D^{-(2-\gamma_2-\gamma_3)}\left\{L_1y(x) - m_1(y)\right\} \\
&= D^{-(2-\gamma_2-\gamma_3)}L_1y(x) - m_1(y)\frac{x^{2-\gamma_2-\gamma_3}}{\Gamma(3-\gamma_2-\gamma_3)}.
\end{aligned} \tag{12_2}$$

Similarly, using relation (3) we obtain

$$\begin{aligned}
D^{-(2-\gamma_2-\gamma_3)}L_1y(x) &= D^{-\gamma_0}\frac{d}{dx}L_0y(x) \\
&= D^{-\gamma_0}\frac{d}{dx}\left\{D^{-(1-\gamma_0)}y(x)\right\} = D^{-\gamma_0}D^{\gamma_0}y(x).
\end{aligned}$$

According to proposition $8°$ of Section 10.2,

$$D^{-\gamma_0}D^{\gamma_0}y(x) = y(x) - \left\{D^{-(1-\gamma_0)}y(\tau)\right\}\Big|_{\tau=0}\frac{x^{\gamma_0-1}}{\Gamma(\gamma_0)}$$

almost everywhere in $(0, \sigma)$, and therefore

$$D^{-(2-\gamma_2-\gamma_3)}L_1y(x) = y(x) - m_0(y)\frac{x^{\gamma_0-1}}{\Gamma(\gamma_0)} \tag{12_3}$$

almost everywhere in $(0, \sigma)$. This formula, together with (12_1) and (12_2) implies the desired representation (12), if notations (9) and relation (3) are used.

Lemma 10.3-2. *If $y(x) \in AC_{\{\gamma\}}[0,\sigma]$, then almost everywhere in $(0,\sigma)$ there exists the derivative $y''(x)$ and*

$$\mathbb{L}_{1/2}y(x) = y''(x) - \sum_{k=0}^{2} m_k(y)\frac{x^{\mu_k-3}}{\Gamma(\mu_k-2)}. \tag{13}$$

Proof.. Applying the operation D^2 to both sides of identity (12) and using proposition $6°$ of Section 10.2 we arrive at formula (13).

Consider now the Cauchy type problem

$$\mathbb{L}_{1/2}y(x) + \lambda y(x) = 0, \quad x \in (0,\sigma), \tag{14}$$

$$L_k y(x)\big|_{x=0} = a_k \quad (k=0,1,2), \tag{15}$$

where $\{a_k\}_0^2$ and λ are arbitrary complex numbers in general.

Lemma 10.3-3. *The Cauchy type problem (14)-(15) may have only a unique solution in the class $AC_{\{\gamma\}}[0,\sigma]$.*

Proof. If there exist two solutions $y_j(x) \in AC_{\{\gamma\}}[0,\sigma](j=1,2)$, then their difference $y(x) = y_1(x) - y_2(x)$ is of the same class, and it is a solution of the homogeneous Cauchy type problem

$$\mathbb{L}_{1/2}y(x) + \lambda y(x) = 0, \quad x \in (0,\sigma), \ L_k y(x)\big|_{x=0} = m_k(y) = 0, \quad k=0,1,2. \tag{16}$$

By (12) and (16), $D^{-2}\mathbb{L}_{1/2}y(x) = y(x) = -\lambda D^{-2}y(x), x \in (0,\sigma)$. In other words, the function $y(x) \in L_1(0,\sigma)$ is a solution of the homogeneous integral equation of Volterra type

$$y(x) = -\lambda \int_0^x (x-t)y(t)dt, \quad x \in (0,\sigma).$$

Hence directly, or by use of the main apparatus of the theory of such integral equations, we obtain $y(x) = 0$(i.e., $y_1(x) = y_2(x)$) almost everywhere in $(0,\sigma)$.

(c) The next lemma deals with the function

$$y_\mu(x;\lambda) \equiv E_{1/2}\left(-\lambda x^2;\mu\right) x^{\mu-1} = \sum_{k=0}^{\infty} \frac{(-\lambda)^k x^{2k+\mu-1}}{\Gamma(\mu+2k)} \tag{17}$$

which obviously satisfies the conditions

$$y_\mu(x;\lambda) \in \begin{cases} L_1(0,\sigma) \text{ when } \mu \geq 0, \\ L_2(0,\sigma) \text{ when } \mu > 1/2. \end{cases} \tag{17_1}$$

Lemma 10.3-4. *If $r = 0, 1, 2$ is arbitrary and*

$$\mu_r = \sum_{j=0}^{r} \gamma_j, \tag{18}$$

then the function $y(x) = y_{\mu_r}(x; \lambda) \in L_2(0, \sigma)$ is a solution of the Cauchy type problem

$$\mathbb{L}_{1/2} y(x) + \lambda y(x) = 0, \quad x \in (0, \sigma), \tag{19_1}$$

$$(\mathcal{L}_r) \qquad L_k y(x)|_{x=0} = \delta_{k,r} \equiv \begin{cases} 1 \text{ when } k = r \\ 0 \text{ when } k \neq r \end{cases} \quad (k = 0, 1, 2). \tag{19_2}$$

And this solution is unique in the class $AC_{\{\gamma\}}[0, \sigma]$.

Proof. We start from results of some simple calculations related to $y_\mu(x; \lambda)$, in which we frequently use the general formula $10.2(19_1)$ and notation (9). We consider three cases.

$1°$. If $\mu = \mu_0 = \gamma_0$, then

$$L_0 y_{\mu_0}(x; \lambda) = \sum_{k=0}^{\infty} (-\lambda)^k \frac{x^{2k}}{\Gamma(1 + 2k)}, \tag{20_1}$$

$$L_1 y_{\mu_0}(x; \lambda) = -\lambda \sum_{k=0}^{\infty} (-\lambda)^k \frac{x^{2k+2-\gamma_1}}{\Gamma(3 - \gamma_1 + 2k)}, \tag{20_2}$$

$$L_2 y_{\mu_0}(x; \lambda) = -\lambda \sum_{k=0}^{\infty} (-\lambda)^k \frac{x^{2k+2-\gamma_1-\gamma_2}}{\Gamma(3 - \gamma_1 - \gamma_2 + 2k)}, \tag{20_3}$$

and, in addition, (4) yields $\gamma_1 + \gamma_2 < 2$. Finally,

$$L_3 y_{\mu_0}(x; \lambda) \equiv \mathbb{L}_{1/2} y_{\mu_0}(x; \lambda) = -\lambda \sum_{k=0}^{\infty} (-\lambda)^k \frac{x^{2k+2-\gamma_1-\gamma_2-\gamma_3}}{\Gamma(3 - \gamma_1 - \gamma_2 - \gamma_3 + 2k)},$$

and, since $3 - \gamma_1 - \gamma_2 - \gamma_3 = \gamma_0 = \mu_0$ by (3), this identity becomes

$$\mathbb{L}_{1/2} y_{\mu_0}(x; \lambda) + \lambda y_{\mu_0}(x; \lambda) = 0, \quad x \in (0, \sigma). \tag{20_4}$$

Formulas $(20_1) - (20_4)$ imply that the function $y_{\mu_0}(x; \lambda) \in AC_{\{\gamma\}}[0, \sigma]$ is a solution of equation (19_1), and they imply also that this function satisfies the initial conditions (19_2) for $r = 0$.

$2°$. If $\mu = \mu_1 = \gamma_0 + \gamma_1$, then

$$L_0 y_{\mu_1}(x; \lambda) = \sum_{k=0}^{\infty} (-\lambda)^k \frac{x^{2k+\gamma_1}}{\Gamma(1 + \gamma_1 + 2k)}, \tag{21_1}$$

$$L_1 y_{\mu_1}(x; \lambda) = \sum_{k=0}^{\infty} (-\lambda)^k \frac{x^{2k}}{\Gamma(1 + 2k)}, \tag{21_2}$$

$$L_2 y_{\mu_1}(x; \lambda) = -\lambda \sum_{k=0}^{\infty} (-\lambda)^k \frac{x^{2k-\gamma_2+2}}{\Gamma(3 - \gamma_2 + 2k)}, \tag{21_3}$$

$$L_3 y_{\mu_1}(x; \lambda) \equiv \mathbb{L}_{1/2} y_{\mu_1}(x; \lambda) = -\lambda \sum_{k=0}^{\infty} (-\lambda)^k \frac{x^{2k+2-\gamma_2-\gamma_3}}{\Gamma(3 - \gamma_2 - \gamma_3 + 2k)}.$$

But in the considered case, $3 - \gamma_2 - \gamma_3 = \gamma_0 + \gamma_1 = \mu_1$ according to (3). Therefore the last identity may be written in the form

$$\mathbb{L}_{1/2} y_{\mu_1}(x; \lambda) + \lambda y_{\mu_1}(x; \lambda) = 0, x \in (0, \sigma). \tag{21_4}$$

By formulas $(21_1) - (21_4)$, $y(x) = y_{\mu_1}(x; \lambda) \in AC_{\{\gamma\}}[0, \sigma]$ and, in addition, this function satisfies (19_1) with initial conditions (19_2) for $r = 1$.

$3°$. If $\mu = \mu_2 = \gamma_0 + \gamma_1 + \gamma_2$, then

$$L_0 y_{\mu_2}(x; \lambda) = \sum_{k=0}^{\infty} (-\lambda)^k \frac{x^{2k+\gamma_1+\gamma_2}}{\Gamma(1 + \gamma_1 + \gamma_2 + 2k)}, \tag{22_1}$$

$$L_1 y_{\mu_2}(x; \lambda) = \sum_{k=0}^{\infty} (-\lambda)^k \frac{x^{2k+\gamma_2}}{\Gamma(1 + \gamma_2 + 2k)}, \tag{22_2}$$

$$L_2 y_{\mu_2}(x; \lambda) \equiv \sum_{k=0}^{\infty} (-\lambda)^k \frac{x^{2k}}{\Gamma(1 + 2k)}, \tag{22_3}$$

$$L_3 y_{\mu_2}(x; \lambda) = \mathbb{L}_{1/2} y_{\mu_2}(x; \lambda) = -\lambda \sum_{k=0}^{\infty} (-\lambda)^k \frac{x^{2k+2-\gamma_3}}{\Gamma(3 - \gamma_3 + 2k)}.$$

But $\mu_2 = 3 - \gamma_3$ according to (3), and the last identity can be written in the form

$$\mathbb{L}_{1/2} y_{\mu_2}(x; \lambda) + \lambda y_{\mu_2}(x; \lambda) = 0, \qquad x \in (0, \sigma). \tag{22_4}$$

Formulas $(22_1) - (22_4)$ yield $y(x) = y_{\mu_2}(x; \lambda) \in AC_{\{\gamma\}}[0, \sigma]$. Using these formulas it is easy to verify the validity of (19_1) with initial conditions (19_2) for $r = 2$.

In the three cases considered above we have actually proved all the assertions we desired to prove except the assertion on the uniqueness of the solutions of the corresponding Cauchy type problems, proved earlier in Lemma 10.3-3.

Now we are ready to prove the following theorem.

Theorem 10.3-1. *Let $\sigma > 0$ be an arbitrary number. Then the function*

$$Y(x; \lambda) \equiv \sum_{j=0}^{2} a_j y_{\mu_j}(x; \lambda) \in L_2(0, \sigma) \tag{23}$$

is the unique solution of the Cauchy type problem (14)-(15) in the class $AC_{\{\gamma\}}[0, \sigma]$.

Proof. By Lemma 10.3-4,

$$\mathbb{L}_{1/2} Y(x; \lambda) + \lambda Y(x; \lambda) = \sum_{j=0}^{2} a_j \left\{ \mathbb{L}_{1/2} y_{\mu_j}(x; \lambda) + \lambda y_{\mu_j}(x; \lambda) \right\} = 0, \quad x \in (0, \sigma)$$

and

$$L_k Y(x; \lambda)\big|_{x=0} = \sum_{j=0}^{2} a_j L_k y_{\mu_j}(x; \lambda)\bigg|_{x=0} = \sum_{j=0}^{2} a_j \delta_{k,j} = a_k \quad (k = 0, 1, 2).$$

The uniqueness of the solution is proved in Lemma 10.3-3.

(d) Finally, we present the forms taken by the operators $\{L_j\}_0^3$, the corresponding Cauchy type problems and their solutions in the special cases (5).

In the case when $\gamma_1 = 0$ and $\gamma_0 = \gamma_2 = \gamma_3 = 1$,

$$L_0 y(x) = y(x), \quad L_1 y(x) = D^{-1} D^1 y(x) = y(x) - y(0),$$
$$L_2 y(x) = y'(x), \quad \mathbb{L}_{1/2} y(x) = L_3 y(x) = y''(x),$$

and we have also $\mu_0 = 1, \mu_1 = 1, \mu_2 = 2$.

In the case when $\gamma_2 = 0$ and $\gamma_0 = \gamma_1 = \gamma_3 = 1$,

$$L_0 y(x) = y(x), \quad L_1 y(x) = y'(x),$$
$$L_2 y(x) = D^{-1} y''(x) = y'(x) - y'(0), \quad \mathbb{L}_{1/2} y(x) = L_3 y(x) = y''(x),$$

and we have also $\mu_0 = 1, \mu_1 = 2, \mu_2 = 2$.

In both these cases we arrive at the same well-known Cauchy problem:

$$y''(x) + \lambda y(x) = 0, \quad x \in (0, \sigma), \quad y(0) = \alpha, \quad y'(0) = \beta, \tag{24}$$

whose solution is

$$Y(x; \lambda) = \alpha E_{1/2}\left(-\lambda x^2; 1\right) + \beta E_{1/2}\left(-\lambda x^2; 2\right) x$$
$$= \alpha \cos \sqrt{\lambda} x + \beta \frac{\sin \sqrt{\lambda} x}{\sqrt{\lambda}}. \tag{25}$$

10.4 The associated Cauchy type problem and the analog of Lagrange formula.

In this section we suppose that the parameters $\{\gamma\} = \{\gamma_j\}_0^3$ satisfy the same conditions as in 10.3, i.e.,

$$1/2 < \gamma_0, \gamma_3 \le 1, \quad 0 < \gamma_1, \gamma_2 \le 1, \quad \sum_{j=0}^{3} \gamma_j = 3 \tag{1}$$

which yield

$$1 < \gamma_0 + \gamma_3 \le 2, \quad 1 \le \gamma_1 + \gamma_2 < 2. \tag{2}$$

(a) In the same way as we introduced $\{L_j\}_0^3$ we introduce the operators $\{L_j^*\}_0^3$ of integro-differentiation, but now with the end at the point $x = \sigma$ $(0 < \sigma < +\infty)$ setting

$$L_0^* z(x) = D_\sigma^{-(1-\gamma_3)} z(x), \tag{3_1}$$

$$L_1^* z(x) = D_\sigma^{-(1-\gamma_2)} \left\{ \frac{d}{d(\sigma - x)} L_0^* z(x) \right\} = D_\sigma^{-(1-\gamma_2)} D_\sigma^{\gamma_3} z(x), \tag{3_2}$$

$$L_2^* z(x) = D_\sigma^{-(1-\gamma_1)} \left\{ \frac{d}{d(\sigma - x)} L_1^* z(x) \right\} = D_\sigma^{-(1-\gamma_1)} D_\sigma^{\gamma_2} D_\sigma^{\gamma_3} z(x), \tag{3_3}$$

$$\mathbb{L}_{1/2}^* z(x) \equiv L_3^* z(x) = D_\sigma^{-(1-\gamma_0)} \left\{ \frac{d}{d(\sigma - x)} L_2^* z(x) \right\}$$
$$= D_\sigma^{-(1-\gamma_0)} D_\sigma^{\gamma_1} D_\sigma^{\gamma_2} D_\sigma^{\gamma_3} z(x). \tag{4}$$

Now we define a domain of definition of these operators.

Definition 2. Denote by $AC_{\{\gamma\}}^*[0, \sigma](0 < \sigma < +\infty)$ the set of functions $z(x)$ satisfying the conditions

$$z(x) \in L_1(0, \sigma),$$
$$L_j^* z(x) \in AC[0, \sigma] \quad (j = 0, 1, 2). \tag{5}$$

It is obvious that, if $z(x) \in AC_{\{\gamma\}}^*[0, \sigma]$, then the functions $L_j^* z(x)(j = 0, 1, 2)$ are continuous in $[0, \sigma]$ and the function $\mathbb{L}_{1/2}^* z(x)$, as a fractional integral of $(d/d(\sigma - x))L_2^* z(x) \in L_1(0, \sigma)$ of order $1 - \gamma_0$, is of class $L_1(0, \sigma)$.

Further, assuming $z(x) \in AC_{\{\gamma\}}^*[0, \sigma]$, we denote

$$\{L_k^* z(x)\}|_{x=\sigma} = m_k^*(z) \quad (k = 0, 1, 2). \tag{6}$$

We denote also

$$\mu_k^* = \sum_{j=0}^{k} \gamma_{3-j} \quad (k = 0, 1, 2) \tag{7}$$

and observe that

$$\mu_0^* = \gamma_3 \in (1/2, 1], \quad \mu_1^* = \gamma_3 + \gamma_2 \in (\gamma_3, 2],$$
$$\mu_2^* = \gamma_3 + \gamma_2 + \gamma_1 = 3 - \gamma_0 \in [2, 5/2). \tag{8}$$

One may prove the following four lemmas in the same way as the similar ones in Section 10.3.

Lemma 10.4-1. *If $z(x) \in AC^*_{\{\gamma\}}[0, \sigma]$, then the equality*

$$D_\sigma^{-2} \mathbb{L}^*_{1/2} z(x) = z(x) - \sum_{k=0}^{2} m_k^*(z) \frac{(\sigma - x)^{\mu_k^* - 1}}{\Gamma(\mu_k^*)} \tag{9}$$

is true almost everywhere in $(0, \sigma)$.

Lemma 10.4-2. *If $z(x) \in AC^*_{\{\gamma\}}[0, \sigma]$, then the derivative $z''(x)$ exists almost everywhere in $(0, \sigma)$ and*

$$\mathbb{L}^*_{1/2} z(x) = z''(x) - \sum_{k=0}^{2} m_k^*(z) \frac{(\sigma - x)^{\mu_k^* - 3}}{\Gamma(\mu_k^* - 2)}. \tag{10}$$

almost everywhere in $(0, \sigma)$.

Lemma 10.4-3. *The Cauchy type problem*

$$\mathbb{L}^*_{1/2} z(x) + \lambda^* z(x) = 0, x \in (0, \sigma),$$
$$L_j^* z(x)\big|_{x=\sigma} = b_j \quad (j = 0, 1, 2),$$

where $\{b_j\}_0^2$ and λ^* are arbitrary complex numbers, has a unique solution in $AC^*_{\{\gamma\}}[0, \sigma]$.

To state the fourth lemma, we introduce the function

$$z_{\mu^*}(x; \lambda^*) = y_{\mu^*}(\sigma - x; \lambda^*) = E_{1/2}\left(-\lambda^*(\sigma - x)^2; \mu^*\right)(\sigma - x)^{\mu^* - 1}$$
$$= \sum_{k=0}^{\infty} (-\lambda^*)^k \frac{(\sigma - x)^{2k + \mu^* - 1}}{\Gamma(\mu^* + 2k)} \tag{11}$$

and note that

$$z_{\mu^*}(x; \lambda^*) \in \begin{cases} L_1(0, \sigma), & \text{when } \mu^* \geq 0, \\ L_2(0, \sigma), & \text{when } \mu^* > 1/2. \end{cases} \tag{12}$$

Lemma 10.4-4. *If $r = 0$, 1, 2 is an arbitrary number and*

$$\mu_r^* = \sum_{j=0}^{r} \gamma_{3-j},$$

then the Cauchy type problem

$$\mathbb{L}_{1/2}^* z(x) + \lambda^* z(x) = 0, \quad x \in (0, \sigma), \tag{13}$$

$$(\mathcal{L}_r^*) \qquad L_k^* z(x)\big|_{x=\sigma} = \delta_{k,r} = \begin{cases} 1, & \text{when } k = r \\ 0, & \text{when } k \neq r \end{cases} \quad (k = 0, 1, 2) \tag{14}$$

has a unique solution in $AC_{\{\gamma\}}^*[0, \sigma]$. This solution is $z(x) = z_{\mu_r^*}(x; \lambda^*) \in L_2(0, \sigma)$.

The following theorem is a consequence of the last lemma.

Theorem 10.4-1. *The Cauchy type problem*

$$\mathbb{L}_{1/2}^* z(x) + \lambda^* z(x) = 0, \quad x \in (0, \sigma),$$
$$L_j^* z(x)\big|_{x=\sigma} = b_j, \quad (j = 0, 1, 2) \tag{15}$$

has a unique solution in $AC_{\{\gamma\}}^*[0, \sigma]$. This solution is

$$Z(x; \lambda^*) \equiv \sum_{j=0}^{2} b_j z_{\mu_j^*}(x; \lambda^*) \in L_2(0, \sigma). \tag{16}$$

(b) The operators $\{L_j^*\}_0^3$ and the corresponding Cauchy type problems take more simple forms in the special cases 10.3(5). Specifically, one may easily verify that in the case when $\gamma_1 = 0$ and $\gamma_0 = \gamma_2 = \gamma_3 = 1$

$$L_0^* z(x) = z(x), \quad L_1^* z(x) = D_\sigma^1 z(x) = -z'(x),$$
$$L_2^* z(x) = D_\sigma^{-1} z''(x) = z'(\sigma) - z'(x), \quad \mathbb{L}_{1/2}^* z(x) = L_3^* z(x) = z''(x),$$

and in the case when $\gamma_2 = 0$ and $\gamma_0 = \gamma_1 = \gamma_3 = 1$

$$L_0^* z(x) = z(x), \quad L_1^* z(x) = z(x) - z(\sigma),$$
$$L_2^* z(x) = -z'(x), \quad \mathbb{L}_{1/2}^* z(x) = L_3^* z(x) = z''(x).$$

Thus, the Cauchy type problem is the same in these two cases:

$$z''(x) + \lambda^* z(x) = 0, \quad x \in (0, \sigma),$$
$$z(\sigma) = \alpha, \quad -z'(\sigma) = \beta. \tag{17}$$

The solution of this problem can be expressed in the form

$$Z(x; \lambda^*) = \alpha E_{1/2}\left(-\lambda^*(\sigma - x)^2; 1\right) + \beta E_{1/2}\left(-\lambda^*(\sigma - x)^2; 2\right)(\sigma - x)$$

$$= \alpha \cos \sqrt{\lambda^*}(\sigma - x) + \beta \frac{\sin \sqrt{\lambda^*}(\sigma - x)}{\sqrt{\lambda^*}}. \tag{18}$$

(c) Remember that, if the parameters $\{\gamma_j\}_0^3$ are given, then we have three different ways of choosing μ and μ^*:

$$1°. \quad \mu = \mu_0 = \gamma_0, \quad \mu^* = \mu_0^* = \gamma_3. \tag{19_1}$$

$$2°. \quad \mu = \mu_1 = \gamma_0 + \gamma_1, \quad \mu^* = \mu_1^* = \gamma_3 + \gamma_2. \tag{19_2}$$

$$3°. \quad \mu = \mu_2 = \gamma_0 + \gamma_1 + \gamma_2, \quad \mu^* = \mu_2^* = \gamma_3 + \gamma_2 + \gamma_1. \tag{19_3}$$

Summarizing the assertions of Lemmas 10.3-4 and 10.4-4, we arrive at the following conclusion. If $r = 0,\ 1,\ 2$ is arbitrary, then $y_{\mu_r}(x; \lambda)$ and $z_{\mu_r^*}(x; \lambda^*)$ are the corresponding unique solutions of Cauchy type problems $(\mathcal{L}_r)$ and $(\mathcal{L}_r^*)$ in the classes $AC_{\{\gamma\}}[0, \sigma]$ and $AC_{\{\gamma\}}^*[0, \sigma]$. It is useful to recall these problems:

$$(\mathcal{L}_r) \qquad \mathbb{L}_{1/2}y(x) + \lambda y(x) = 0, \quad x \in (0, \sigma), \tag{20_1}$$

$$L_k y(x)|_{x=0} = L_k y(0) = \delta_{k,r} \quad (k = 0, 1, 2). \tag{20_2}$$

$$(\mathcal{L}_r^*) \qquad \mathbb{L}_{1/2}^* z(x) + \lambda^* z(x) = 0, \quad x \in (0, \sigma), \tag{21_1}$$

$$L_k^* z(x)|_{x=\sigma} = L_k^* z(0) = \delta_{k,r} \quad (k = 0, 1, 2). \tag{21_2}$$

Now note that, as follows from the definition of $y_\mu(x; \lambda)$ and from formulas 10.3 (20_k), 10.3 (21_k), 10.3 (22_k) $(k = 1, 2, 3, 4)$, for any way of choosing μ and μ^*, all the functions

$$y_\mu(x; \lambda),\ L_k y_\mu(x; \lambda)\ (0 \le k \le 3),\ \frac{d}{dx} L_k y_\mu(x; \lambda)\ (0 \le k \le 2) \tag{22}$$

are continuous in $[0, \sigma]$, except at the point $x = 0$, where they may have integrable singularities. Similarly, the functions

$$z_{\mu^*}(x; \lambda^*),\ L_k^* z_{\mu^*}(x; \lambda^*)\ (0 \le k \le 3),\ \frac{d}{d(\sigma - x)} L_k^* z_{\mu^*}(x; \lambda^*)\ (0 \le k \le 2) \tag{23}$$

are also continuous in $[0, \sigma]$, except at the point $x = \sigma$, where they may have integrable singularities. Therefore the product of any two functions of (22) and (23) is of class $L_1(0, \sigma)$.

Further, note that, if $y(x) \in AC_{\{\gamma\}}[0, \sigma]$ and $z(x) \in AC_{\{\gamma\}}^*[0, \sigma]$ are any solutions of equations (20_1) and (21_1)(which do not necessarily satisfy the initial conditions (20_2) and (21_2)), and if we denote formally

$$P(\lambda; \lambda^*) \equiv \int_0^\sigma \left\{ z(x)\mathbb{L}_{1/2}y(x) - y(x)\mathbb{L}_{1/2}^* z(x) \right\} dx, \tag{24}$$

then obviously

$$P(\lambda; \lambda^*) = (\lambda^* - \lambda)U(y; z), \tag{25}$$

where

$$U(y; z) = \int_0^\sigma y(x)z(x)dx. \tag{26}$$

The following lemma formally establishes an important integral identity which may be considered as a special analog for operators $\mathbb{L}_{1/2}$ and $\mathbb{L}_{1/2}^*$ of the classical Lagrange formula. Of course, the mentioned identity is undoubtedly valid if some additional conditions are satisfied.

Lemma 10.4-5. *Let $y(x) \in AC_{\{\gamma\}}[0, \sigma]$ and $z(x) \in AC_{\{\gamma\}}^*[0, \sigma]$ be any solutions of equations (20_1) and (21_1), then*

$$P(\lambda; \lambda^*) = \sum_{k=0}^{2} \left\{ L_k y(\sigma) L_{2-k}^* z(\sigma) - L_k y(0) L_{2-k}^* z(0) \right\}. \tag{27}$$

Proof. Using the definitions of corresponding operators and the operations of ordinary and fractional integration by parts, we obtain the following equalities:

$$\int_0^\sigma z(x)\mathbb{L}_{1/2}y(x)dx$$

$$= \int_0^\sigma z(x)D^{-(1-\gamma_3)}\left\{\frac{d}{dx}L_2 y(x)\right\} dx = \int_0^\sigma D_\sigma^{-(1-\gamma_3)}z(x)\left\{\frac{d}{dx}L_2 y(x)\right\} dx$$

$$= L_0^* z(x)L_2 y(x)\big|_0^\sigma + \int_0^\sigma \left\{\frac{d}{d(\sigma-x)}L_0^* z(x)\right\} D^{-(1-\gamma_2)}\left\{\frac{d}{dx}L_1 y(x)\right\} dx$$

$$= L_0^* z(x)L_2 y(x)\big|_0^\sigma + \int_0^\sigma L_1^* z(x)\left\{\frac{d}{dx}L_1 y(x)\right\} dx.$$

Repeating this argument we arrive at the equality

$$\int_0^\sigma z(x)\mathbb{L}_{1/2}y(x)dx = \sum_{k=0}^{2} L_k y(x)L_{2-k}^* z(x)\bigg|_0^\sigma + \int_0^\sigma y(x)\mathbb{L}_{1/2}^* z(x)dx$$

which coincides with formula (27).

If $y(x) \in AC_{\{\gamma\}}[0, \sigma]$ and $z(x) \in AC_{\{\gamma\}}^*[0, \sigma]$ are not only solutions of equations (20_1) and (21_1), but they also satisfy the initial conditions (20_2) and (21_2), i.e. $y(x) = y_{\mu_r}(x; \lambda)$ and $z(x) = z_{\mu_r^*}(x; \lambda^*)$ $(r = 0, 1, 2)$, then the formal representation (27) of $P(\lambda; \lambda^*)$ is indeed true. Moreover, this representation takes a simple form given in the following lemma.

Lemma 10.4-6. *The following representations are true when we choose μ and μ^* in three different ways.*
1°. *In the case* (19_1), *when* $\mu = \mu_0 = \gamma_0$ *and* $\mu^* = \mu_0^* = \gamma_3$,

$$P(\lambda; \lambda^*) = - \left\{ \lambda E_{1/2}\left(-\sigma^2\lambda; \mu + \mu^*\right) - \lambda^* E_{1/2}\left(-\sigma^2\lambda^*; \mu + \mu^*\right) \right\} \sigma^{\mu + \mu^* - 1}. \quad (28)$$

2°, 3°. *In both cases* (19_2) *and* (19_3) *when respectively,* $\mu = \mu_1 = \gamma_0 + \gamma_1, \mu^* = \mu_1^* = \gamma_3 + \gamma_2$ *and* $\mu = \mu_2 = \gamma_0 + \gamma_1 + \gamma_2, \mu^* = \mu_2^* = \gamma_3 + \gamma_2 + \gamma_1$,

$$P(\lambda; \lambda^*) = \left\{ E_{1/2}\left(-\sigma^2\lambda; \mu + \mu^* - 2\right) - E_{1/2}\left(-\sigma^2\lambda^*; \mu + \mu^* - 2\right) \right\} \sigma^{\mu + \mu^* - 3}. \quad (29)$$

Proof. Here we use formulas obtained in the proof of Lemma 10.3-4.
1°. From formulas $10.3(20_k)(k = 1,\ 2,\ 3)$ it follows that

$$L_0 y(0) = 1, \quad L_0 y(\sigma) = E_{1/2}\left(-\sigma^2\lambda; 1\right),$$

and, similarly,

$$L_0^* z(0) = E_{1/2}\left(-\sigma^2\lambda^*; 1\right), \quad L_0^* z(\sigma) = 1.$$

Further, we obtain

$$L_1 y(0) = 0, \quad L_1 y(\sigma) = -\lambda E_{1/2}\left(-\sigma^2\lambda; 3 - \gamma_1\right) \sigma^{2 - \gamma_1},$$

and, similarly,

$$L_1^* z(0) = -\lambda^* E_{1/2}\left(-\sigma^2\lambda^*; 3 - \gamma_2\right) \sigma^{2 - \gamma_2}, \quad L_1^* z(\sigma) = 0.$$

Also, we obtain

$$L_2 y(0) = 0, \quad L_2 y(\sigma) = -\lambda E_{1/2}\left(-\sigma^2\lambda; 3 - \gamma_1 - \gamma_2\right) \sigma^{2 - \gamma_1 - \gamma_2}$$

and

$$L_2^* z(0) = -\lambda^* E_{1/2}\left(-\sigma^2\lambda^*; 3 - \gamma_1 - \gamma_2\right) \sigma^{2 - \gamma_1 - \gamma_2}, \quad L_2^* z(\sigma) = 0.$$

We insert these quantities into the right-hand side of (27) and, taking into account that

$$\mu + \mu^* = \gamma_0 + \gamma_3 = 3 - \gamma_1 - \gamma_2 \quad (30)$$

in the considered case, we arrive at formula (28).
2°. From formulas $10.3(21_k)(k = 1,\ 2,\ 3)$ it follows that

$$L_0 y(0) = 0, \quad L_0 y(\sigma) = E_{1/2}\left(-\sigma^2\lambda; 1 + \gamma_1\right) \sigma^{\gamma_1},$$

and

$$L_0^* z(0) = E_{1/2}\left(-\sigma^2\lambda^*; 1 + \gamma_2\right) \sigma^{\gamma_2}, \quad L_0^* z(\sigma) = 0.$$

Further, we obtain

$$L_1 y(0) = 1, \quad L_1 y(\sigma) = E_{1/2}\left(-\sigma^2\lambda; 1\right),$$
$$L_1^* z(0) = E_{1/2}\left(-\sigma^2\lambda^*; 1\right), \quad L_1^* z(\sigma) = 1.$$

Finally, we obtain

$$L_2 y(0) = 0, \quad L_2 y(\sigma) = -\lambda E_{1/2}\left(-\sigma^2\lambda; 3 - \gamma_2\right)\sigma^{2-\gamma_2},$$
$$L_2^* z(0) = -\lambda^* E_{1/2}\left(-\sigma^2\lambda^*; 3 - \gamma_1\right)\sigma^{2-\gamma_1}, \quad L_2^* z(\sigma) = 0$$

and insert these quantities into the right-hand side of (27). Then, taking into account that

$$\mu + \mu^* = 3 \tag{31}$$

in the considered case, we arrive at formula (29).

$3°$. From formulas $10.3(22_k)(k = 1, 2, 3)$ it follows that

$$L_0 y(0) = 0, \quad L_0 y(\sigma) = E_{1/2}\left(-\sigma^2\lambda; 1 + \gamma_1 + \gamma_2\right)\sigma^{\gamma_1+\gamma_2},$$
$$L_0^* z(0) = E_{1/2}\left(-\sigma^2\lambda^*; 1 + \gamma_1 + \gamma_2\right)\sigma^{\gamma_1+\gamma_2}, \quad L_0^* z(\sigma) = 0.$$

Further, we obtain

$$L_1 y(0) = 0, \quad L_1 y(\sigma) = E_{1/2}\left(-\sigma^2\lambda; 1 + \gamma_2\right)\sigma^{\gamma_2},$$
$$L_1^* z(0) = E_{1/2}\left(-\sigma^2\lambda^*; 1 + \gamma_1\right)\sigma^{\gamma_1}, \quad L_1^* z(\sigma) = 0$$

and

$$L_2 y(0) = 1, \quad L_2 y(\sigma) = E_{1/2}\left(-\sigma^2\lambda; 1\right),$$
$$L_2^* z(0) = E_{1/2}\left(-\sigma^2\lambda^*; 1\right), \quad L_2^* z(\sigma) = 1.$$

We put these quantities into the right-hand side of (27) and, taking into account that $\mu + \mu^* = 3 + \gamma_1 + \gamma_2$ in the considered case, we arrive again at formula (29).

Remark. From the preceding lemma it follows that

$$P(\lambda; \lambda^*) = \sigma^{\mu+\mu^*-3}\left\{E_{1/2}\left(-\sigma^2\lambda; \mu + \mu^* - 2\right) - E_{1/2}\left(-\sigma^2\lambda^*; \mu + \mu^* - 2\right)\right\}$$

for $y(x) = y_\mu(x; \lambda)$, $z(x) = z_{\mu^*}(x; \lambda^*)$ and for any of three ways of choosing μ and μ^*. This formula can be deduced also from $1.2(10)$ on the basis of (25)-(26).

The following theorem is a consequence of Lemma 10.4-6.

Theorem 10.4-2. *Let $y_\mu(x; \lambda)$ and $z_{\mu^*}(x; \lambda^*)$ be the solutions of the Cauchy type problems $(20_1) - (20_2)$ and $(21_1) - (21_2)$, respectively. Then the following integral formulas are true:*

if $\mu = \mu_0$, $\mu^* = \mu_0^*$, then

$$U(y_\mu; z_{\mu^*}) = \frac{\lambda E_{1/2}\left(-\sigma^2\lambda; 1+\nu\right) - \lambda^* E_{1/2}\left(-\sigma^2\lambda^*; 1+\nu\right)}{\lambda - \lambda^*}\sigma^\nu, \qquad (32_1)$$

where

$$1 + \nu = \mu + \mu^* = \mu_0 + \mu_0^* = \gamma_0 + \gamma_3, \quad 0 < \nu \le 1, \qquad (32_2)$$

and, if $\mu = \mu_1$, $\mu^* = \mu_1^*$ or $\mu = \mu_2$, $\mu^* = \mu_2^*$, then

$$U(y_\mu; z_{\mu^*}) = -\frac{E_{1/2}\left(-\sigma^2\lambda; 1+\nu\right) - E_{1/2}\left(-\sigma^2\lambda^*; 1+\nu\right)}{\lambda - \lambda^*}\sigma^\nu, \qquad (33_1)$$

where

$$\begin{aligned}
1 + \nu &= \mu + \mu^* - 2 = \mu_1 + \mu_1^* - 2 = 1, \quad \nu = 0, \text{ or} \\
1 + \nu &= \mu + \mu^* - 2 = \mu_2 + \mu_2^* - 2 = 1 + \gamma_1 + \gamma_2, \quad 1 \le \nu < 2.
\end{aligned} \qquad (33_2)$$

(d) It must be mentioned that the statements of Theorem 10.4-2 are true for arbitrary parameters $\mu \ge 0$ and $\mu^* \ge 0$. Although this general assertion follows immediately from identities 10.2(10), if we put $\rho = 1/2, \alpha = \mu, \beta = \mu^*, z = -\lambda$ and $\lambda = -\lambda^*$, nevertheless, we preferr to give a proof of Theorem 10.4-2 based essentially on the properties of functions $y_\mu(x; \lambda)$ and $z_{\mu^*}(x; \lambda^*)$ arising from the fact that they are solutions of Cauchy type problems formulated in terms of integro-differential operators $\mathbb{L}_{1/2}$ and $\mathbb{L}_{1/2}^*$ of fractional order. The given proof is based also on an analog of the classical Lagrange formula.

10.5 Boundary value problems and eigenfunction expansions

Later on we frequently use the notations introduced in previous sections without references.

(a) First we shall state a lemma which follows immediately from the expansions of functions $y_\mu(x; \lambda)$ and $z_{\mu^*}(x; \lambda^*)$, if the simple formulas 10.2$(19_1) - (19_2)$ are used.

Lemma 10.5-1. *Each of the following formulas is true if we choose appropriate μ and μ^*.*

$$1°. \text{ Let } \mu = \mu_0 = \gamma_0, \quad \mu^* = \mu_0^* = \gamma_3 \text{ and let}$$

$$\nu_0 = \mu_0 + \mu_0^* - 1 = \gamma_0 + \gamma_3 - 1 \quad (0 < \nu_0 \le 1). \qquad (1)$$

Then

$$\begin{aligned}
D^{1-\nu_0} D^{\mu_0} y_{\mu_0}(x; \lambda) &= D^{2-\gamma_0-\gamma_3} D^{\gamma_0} y_{\mu_0}(x; \lambda) \\
&= -\lambda E_{1/2}\left(-\lambda x^2; 1+\nu_0\right) x^{\nu_0},
\end{aligned} \qquad (2_1)$$

$$D_\sigma^{1-\nu_0} D_\sigma^{\mu_0^*} z_{\mu_0^*}(x;\lambda^*) = D_\sigma^{2-\gamma_0-\gamma_3} D_\sigma^{\gamma_3} z_{\mu_0^*}(x;\lambda^*)$$
$$= -\lambda^* E_{1/2}\left(-\lambda^*(\sigma-x)^2; 1+\nu_0\right)(\sigma-x)^{\nu_0}. \tag{2_2}$$

$2°$, $3°$. Let

$$\mu = \mu_1 = \gamma_0 + \gamma_1, \quad \mu^* = \mu_1^* = \gamma_3 + \gamma_2 \tag{3_1}$$

or

$$\mu = \mu_2 = \gamma_0 + \gamma_1 + \gamma_2, \quad \mu^* = \mu_2^* = \gamma_3 + \gamma_2 + \gamma_1, \tag{3_2}$$

and let

$$\nu_r = \mu_r + \mu_r^* - 3 = \begin{cases} 0 & \text{when } r = 1 \quad (\nu_1 = 0), \tag{4_1} \\ \gamma_1 + \gamma_2 & \text{when } r = 2 \quad (1 \le \nu_2 < 2). \end{cases}$$
$$\tag{4_2}$$

Then

$$D^{\mu_r - \nu_r - 1} y_{\mu_r}(x;\lambda) = \begin{cases} D^{\gamma_0+\gamma_1-1} y_{\mu_1}(x;\lambda) = E_{1/2}\left(-\lambda x^2; 1+\nu_1\right) \\ \qquad\qquad\qquad \text{when } r = 1, \tag{5_1} \\[2mm] D^{-(1-\gamma_0)} y_{\mu_2}(x;\lambda) = E_{1/2}\left(-\lambda x^2; 1+\nu_2\right) x^{\nu_2} \\ \qquad\qquad\qquad \text{when } r = 2, \tag{5_2} \end{cases}$$

and also

$$D_\sigma^{\mu_r^*-\nu_r-1} z_{\mu_r^*}(x;\lambda^*) = \begin{cases} D_\sigma^{\gamma_2+\gamma_3-1} z_{\mu_1^*}(x;\lambda^*) \\ \quad = E_{1/2}\left(-\lambda^*(\sigma-x)^2; 1+\nu_1\right) \\ \qquad\qquad \text{when } r = 1, \tag{6_1} \\[2mm] D_\sigma^{-(1-\gamma_3)} z_{\mu_2^*}(x;\lambda^*) \\ \quad = E_{1/2}\left(-\lambda^*(\sigma-x)^2; 1+\nu_2\right)(\sigma-x)^{\nu_2} \\ \qquad\qquad \text{when } r = 2. \tag{6_2} \end{cases}$$

(b) Now we state four boundary value problems relating to the ways of choosing μ and μ^*.

Problem I *consists in finding those values of λ for which the solution $y_{\mu_0}(x;\lambda)$ of the Cauchy type problem $(\mathcal{L}_0)$ satisfies the boundary condition*

$$D^{1-\nu_0} D^{\mu_0} y_{\mu_0}(x;\lambda)\big|_{x=\sigma} = 0. \tag{7_1}$$

Problem I* *consists in finding those values of λ^* for which the solution $z_{\mu_0^*}(x;\lambda^*)$ of the Cauchy type problem $(\mathcal{L}_0^*)$ satisfies the boundary condition*

$$D_\sigma^{1-\nu_0} D_\sigma^{\mu_0^*} z_{\mu_0^*}(x;\lambda^*)\Big|_{x=0} = 0. \tag{7_2}$$

Problem II *consists in finding those values of λ for which the solution $y_{\mu_r}(x;\lambda)$ of the Cauchy type problem $(\mathcal{L}_r)$ satisfies the boundary condition*

$$D^{\mu_r-\nu_r-1} y_{\mu_r}(x;\lambda)\big|_{x=\sigma} = 0 \quad (r = 1,2). \tag{8_1}$$

Problem II* *consists in finding those values of λ^* for which the solution $z_{\mu_r^*}(x;\lambda^*)$ of the Cauchy type problem $(\mathcal{L}_r^*)$ satisfies the boundary condition*

$$D_\sigma^{\mu_r^*-\nu_r-1} z_{\mu_r^*}(x;\lambda^*)\Big|_{x=0} = 0 \quad (r=1,2). \tag{8_2}$$

It is natural to call the desired values of λ and λ^* the eigenvalues of the boundary value problems I–I^* and II–II^*. Similarly, it is natural to call the corresponding solutions $y_\mu(x;\lambda)$ and $z_{\mu^*}(x;\lambda^*)$ the eigenfunctions of these problems. The existence of the solutions of the considered boundary value problems is established in the following two theorems.

Theorem 10.5-1. $1°$. *The set of eigenvalues of problems I and I* coincides with the sequence $\{\lambda_k\}_0^\infty (\lambda_0 = 0)$ of zeros of the function*

$$z\mathcal{E}_\sigma(z;\nu) = zE_{1/2}\left(-\sigma^2 z; 1+\nu\right) \quad (\nu = \nu_0 \in (0,1]). \tag{9_1}$$

$2°$. *The set of eigenvalues of problems II and II* coincides with the sequence $\{\lambda_k\}_1^\infty$ of zeros of the function*

$$\mathcal{E}_\sigma(z;\nu) = E_{1/2}\left(-\sigma^2 z; 1+\nu\right), \tag{9_2}$$

where $\nu = \nu_1 = 0$ when $r = 1$ and $\nu = \nu_2 \in [1,2)$ when $r = 2$.

Proof. According to Theorem 1.4-3, all zeros $\{\lambda_k = \lambda_k(\sigma;\nu)\}_1^\infty (0 < \lambda_k < \lambda_{k+1}, 1 \leq k < +\infty)$ of the function $\mathcal{E}_\sigma(z;\nu)(0 \leq \nu < 2)$ are simple and positive. Therefore assertions $1°$ and $2°$ follow immediately from conditions $(7_1) - (7_2)$ and $(8_1) - (8_2)$ of boundary value problems $I - I^*$ and $II - II^*$ by formulas $(2_1) - (2_2), (5_1) - (5_2)$ and $(6_1) - (6_2)$ of Lemma 10.5-1.

The next theorem easily follows from the preceding one.

Theorem 10.5-2. $1°$. *The boundary value problems I and I* have correspondingly the following systems of eigenfunctions:*

$$\{y_{\mu_0}(x;\lambda_k)\}_0^\infty \equiv \left\{\frac{x^{\mu_0-1}}{\Gamma(\mu_0)}, \left\{E_{1/2}\left(-\lambda_k x^2; \mu_0\right) x^{\mu_0-1}\right\}_1^\infty\right\} \tag{10_1}$$

and

$$\{z_{\mu_0^*}(x;\lambda_k)\}_0^\infty \equiv \left\{\frac{(\sigma-x)^{\mu_0^*-1}}{\Gamma(\mu_0^*)}, \left\{E_{1/2}\left(-\lambda_k(\sigma-x)^2; \mu_0^*\right) (\sigma-x)^{\mu_0^*-1}\right\}_1^\infty\right\}. \tag{10_2}$$

$2°$. *The boundary value problems II and II* have correspondingly the eigenfunctions*

$$\{y_{\mu_r}(x;\lambda_k)\}_1^\infty \equiv \left\{E_{1/2}\left(-\lambda_k x^2; \mu_r\right) x^{\mu_r-1}\right\}_1^\infty \quad (r=1,2) \tag{11_1}$$

and

$$\{z_{\mu_r^*}(x;\lambda_k)\}_1^\infty \equiv \left\{E_{1/2}\left(-\lambda_k(\sigma-x)^2; \mu_r^*\right) (\sigma-x)^{\mu_r^*-1}\right\}_1^\infty \quad (r=1,2). \tag{11_2}$$

(c) Now we need to pass from systems $(10_1)-(10_2)$ and $(11_1)-(11_2)$ to the systems introduced in Chapter 5. This passage makes it possible to establish at the end of this chapter the important fact that Theorems 5.3-1 and 5.3-2 actually state expansions in terms of eigenfunctions of the non-ordinary boundary value problems stated above. To this end, we need to recall some definitions and notations from Chapter 5. Assuming $\nu \in [0,2)$, we denoted the sequence of zeros of the function $\mathcal{E}_\sigma(z;\nu) \equiv E_{1/2}(-\sigma^2 z; 1+\nu)$ by $\{\lambda_k\}_1^\infty$ $(0 < \lambda_k < \lambda_{k+1}, 1 \le k < +\infty)$. The systems of functions

$$\{\varphi_k(\tau)\}_1^\infty \equiv \{\varphi_k(\tau;\mu)\}_1^\infty \text{ and } \{\varphi_k^*(\tau)\}_0^\infty \equiv \{\varphi_k^*(\tau;\mu)\}_0^\infty \tag{12}$$

were defined for $\tau \in (0,\sigma)$ as follows:

$$\varphi_k(\tau;\mu) = -\frac{\sigma^{-\nu}}{\mathcal{E}_\sigma'(\lambda_k;\nu)} E_{1/2}\left(-\lambda_k(\sigma-\tau)^2; 3+\nu-\mu\right)(\sigma-\tau)^{\nu-\mu+2} \quad (k \ge 1),$$
$$\tag{13}$$

where $0 \le \mu < 3+\nu$;

$$\varphi_0^*(\tau;\mu) = \sigma^{-\nu}\frac{\Gamma(1+\nu)}{\Gamma(1+\nu-\mu)}(\sigma-\tau)^{\nu-\mu},$$
$$\tag{14}$$
$$\varphi_k^*(\tau;\mu) = \frac{\sigma^{-\nu}}{\lambda_k \mathcal{E}_\sigma'(\lambda_k;\nu)} E_{1/2}\left(-\lambda_k(\sigma-\tau)^2; 1+\nu-\mu\right)(\sigma-\tau)^{\nu-\mu} \quad (k \ge 1),$$

where $0 \le \mu < 1+\nu$. We also defined the systems of functions

$$\{e_k(\tau;\mu)\}_1^\infty \text{ and } \{e_k(\tau;\mu)\}_0^\infty (\mu \ge 0, \tau \in (0,\sigma)), \tag{15}$$

where

$$e_0(\tau;\mu) = \frac{\tau^{\mu-1}}{\Gamma(\mu)}, \quad e_k(\tau;\mu) = E_{1/2}\left(-\lambda_k\tau^2;\mu\right)\tau^{\mu-1} \quad (k \ge 1). \tag{16}$$

Using these definitions we can easily obtain several relations between systems $(10_1)-(10_2), (11_1)-(11_2)$ and $(12), (15)$, if we take into account that

$$\nu_0 = \mu_0 + \mu_0^* - 1, \quad 0 \le \mu_0 < 1+\nu_0,$$
$$\nu_r = \mu_r + \mu_r^* - 3, \quad 0 \le \mu_r < 3+\nu_r \quad (r=1,2). \tag{17}$$

These relations are the following:

$$e_k(\tau;\mu_0) = y_{\mu_0}(\tau;\lambda_k) \quad (k \ge 0), \quad \nu = \nu_0, \tag{18_1}$$
$$\varphi_k^*(\tau;\mu_0) = c_k^* z_{\mu_0^*}(\tau;\lambda_k) \quad (k \ge 0), \quad \nu = \nu_0, \tag{18_2}$$

where

$$c_0^* = \sigma^{-\nu_0}\Gamma(1+\nu_0), \quad c_k^* = \frac{\sigma^{-\nu_0}}{\lambda_k \mathcal{E}_\sigma'(\lambda_k;\nu_0)} \quad (k \ge 1), \tag{19}$$

and also

$$e_k(\tau;\mu_r) = y_{\mu_r}(\tau;\lambda_k) \quad (k \geq 1), \quad \nu = \nu_r \quad (r = 1,2), \tag{20_1}$$

$$\varphi_k(\tau;\mu_r) = c_k^r z_{\mu_r^*}(\tau;\lambda_k) \quad (k \geq 1), \quad \nu = \nu_r \quad (r = 1,2), \tag{20_2}$$

where

$$c_k^r = -\frac{\sigma^{-\nu_r}}{\mathcal{E}_\sigma'(\lambda_k;\nu_r)} \quad (k \geq 1,\ r = 1,2). \tag{21}$$

The following theorem is a consequence of the mentioned relations and of Lemma 5.2-2.

Theorem 10.5-3. *The systems* (10_1), (10_2) *and* (11_1), (11_2) *are biorthogonal, i.e.,*

$$\int_0^\sigma y_{\mu_0}(\tau;\lambda_k) z_{\mu_0^*}(\tau;\lambda_k)d\tau = \delta_{k,n}/c_k^* \quad (k,n = 0,1,2,\dots),$$
$$\int_0^\sigma y_{\mu_r}(\tau;\lambda_k) z_{\mu_r^*}(\tau;\lambda_k)d\tau = \delta_{k,n}/c_k^r \quad (k,n = 1,2,\dots), \quad r = 1,2. \tag{22}$$

(d) Using formulas $(18_1) - (18_2)$ and $(20_1) - (20_2)$ we can reformulate the main Theorems 5.3-1 and 5.3-2 in terms of systems $(10_1) - (10_2)$ and $(11_1) - (11_2)$ and thus to obtain a theorem on expansions in terms of eigenfunctions of our boundary value problems. First it is useful to recall those conditions which the parameters ω, μ, κ and ν satisfy in the mentioned theorems.

In Theorem 5.3-1 we assumed

$$-1 < \omega < 1, \mu = 3/2 + \omega, \quad \kappa = 1 + 2\omega \text{ and } \nu \in \Delta(\kappa,2) = (\omega, \omega + 1) \cap [0,2). \tag{23}$$

In Theorem 5.3-2 we assumed

$$-1 < \omega < 1, \quad \mu = 3/2 + \omega, \quad \kappa = 1 + 2\omega \text{ and } \nu \in \Delta^*(\kappa,2) = (\omega + 1, \omega + 2) \cap [0,2). \tag{24}$$

In view of these assumptions, it becomes necessary to discuss each of three ways of choosing μ and μ^* separately.

1°. If $\mu_0 = \gamma_0$, $\mu_0^* = \gamma_3$ and $\nu_0 = \mu_0 + \mu_0^* - 1 = \gamma_0 + \gamma_3 - 1$, then we have $\nu_0 \in (0,1]$ since $1/2 < \gamma_0, \gamma_3 \leq 1$. So, if we put $\mu_0 = 3/2 + \omega_0$ $(-1 < \omega_0 \leq -1/2)$, then we shall obtain

$$\nu_0 = 3/2 + \omega_0 + \gamma_3 - 1 = 1/2 + \omega_0 + \gamma_3 \in (\omega_0 + 1, \omega_0 + 3/2] \subset (\omega_0 + 1, \omega_0 + 2)$$
$$= (\omega_0 + 1, \omega_0 + 2) \cap [0,2) = \Delta^*(\kappa_0, 2),$$

i.e., (24) is satisfied.

2°. Let $\mu_1 = \gamma_0 + \gamma_1$, $\mu_1^* = \gamma_3 + \gamma_2, \nu_1 = \mu_1 + \mu_1^* - 3 = 0$ and let $\mu_1 = 3/2 + \omega_1 (\omega_1 = \gamma_0 + \gamma_1 - 3/2)$. Then, since $1/2 < \gamma_0 \leq 1$ and $0 < \gamma_1 \leq 1$, we have $-1 < \omega_1 \le 1/2$. And it remains to observe that, provided $\gamma_0 + \gamma_1 < 3/2$, we have $\nu_1 = 0 \in (\omega_1, \omega_1 + 1)$ which implies $\nu_1 = 0 \in (\omega_1, \omega_1 + 1) \cap [0, 2) = \Delta(\kappa_1, 2)$. Thus, (23) is satisfied if it is assumed in addition that $\gamma_0 + \gamma_1 < 3/2$.

$3°$. If $\mu_2 = \gamma_0 + \gamma_1 + \gamma_2$, $\quad \mu_2^* = \gamma_3 + \gamma_2 + \gamma_1$ and $\nu_2 = \mu_2 + \mu_2^* - 3 = \gamma_1 + \gamma_2 \in [1, 2)$, then we put $\mu_2 = 3/2 + \omega_2 (\omega_2 = 3/2 - \gamma_3 \in [1/2, 1))$. In this case it is easy to verify that $\quad \nu_2 \in (\omega_2, \omega_2 + 1) = (3/2 - \gamma_3, 5/2 - \gamma_3) \subset [0, 2)$. Hence (23) is satisfied.

Now Theorems 5.3-1 and 5.3-2 can be easily reformulated in the following way.

Theorem 10.5-4. *If it is assumed $\mu_1 = \gamma_0 + \gamma_1 < 3/2$ when $r = 1$ and the boundary value problems II and II* are considered, then:*

$1°$. Any function $\varphi(x) \in L_2(0, \sigma)$ can be expanded in the series of eigenfunctions of both problems I and II :

$$\varphi(x) = \sum_{k=0}^{\infty} b_k^* y_{\mu_0}(x; \lambda_k), \quad b_k^* = c_k^* \int_0^\sigma \varphi(\tau) z_{\mu_0^*}(\tau; \lambda_k) d\tau \quad (k \ge 0), \qquad (25_1)$$

$$\varphi(x) = \sum_{k=1}^{\infty} b_k^r y_{\mu_r}(x; \lambda_k), \quad b_k^r = c_k^r \int_0^\sigma \varphi(\tau) z_{\mu_r^*}(\tau; \lambda_k) d\tau \quad (k \ge 1, r = 1, 2). \qquad (26_1)$$

These series converge to $\varphi(x)$ in the norm of $L_2(0, \sigma)$. Besides,

$$\|\varphi\|_2 \asymp \|\{b_k^*\}_0^\infty\|_{2, -\kappa_0} = \left\{ \sum_{k=0}^{\infty} |b_k^*|^2 (1 + k)^{-\kappa_0} \right\}^{1/2}, \qquad (27_1)$$

$$\|\varphi\|_2 \asymp \|\{b_k^r\}_1^\infty\|_{2, -\kappa_r} = \left\{ \sum_{k=1}^{\infty} |b_k^r|^2 (1 + k)^{-\kappa_r} \right\}^{1/2} \quad (r = 1, 2), \qquad (28_1)$$

where $\kappa_r = -2(1 - \mu_r) \quad (r = 0, 1, 2)$.

$2°$. Any function $\varphi(x) \in L_2(0, \sigma)$ can be expanded in the series of eigenfunctions of both boundary value problems I* and II*:

$$\varphi(x) = \sum_{k=0}^{\infty} a_k^* z_{\mu_0^*}(x; \lambda_k), \quad a_k^* = c_k^* \int_0^\sigma \varphi(\tau) y_{\mu_0}(\tau; \lambda_k) d\tau \quad (k \ge 0), \qquad (25_2)$$

$$\varphi(x) = \sum_{k=1}^{\infty} a_k^r z_{\mu_r^*}(x; \lambda_k), \quad a_k^r = c_k^r \int_0^\sigma \varphi(\tau) y_{\mu_r}(\tau; \lambda_k) d\tau \quad (k \ge 1, \ r = 1, 2). \qquad (26_2)$$

These series converge to $\varphi(x)$ in the norm of $L_2(0, \sigma)$. Besides,

$$\|\varphi\|_2 \asymp \|\{a_k^*\}_0^\infty\|_{2, -\widetilde{\kappa}_0} = \left\{ \sum_{k=0}^{\infty} |a_k^*|^2 (1 + k)^{-\widetilde{\kappa}_0} \right\}^{1/2}, \qquad (27_2)$$

$$\|\varphi\|_2 \asymp \|\{a_k^r\}_1^\infty\|_{2, -\widetilde{\kappa}_r} = \left\{ \sum_{k=1}^{\infty} |a_k^r|^2 (1 + k)^{-\widetilde{\kappa}_r} \right\}^{1/2}, \quad (r = 1, 2), \qquad (28_2)$$

where $\widetilde{\kappa}_r = -2(1 - \mu_r^*) \ (r = 0, 1, 2)$.

Note that Lemma 4.2-5 and formulas (19) and (21) are also used to prove the preceding reformulation of Theorems 5.3-1 and 5.3-2.

(e) The operators $\mathbb{L}_{1/2}$ and $\mathbb{L}_{1/2}^*$ and the corresponding Cauchy type problems may be written in simpler forms. Namely, if we introduce the notations

$$q_r(x) = \frac{x^{\mu_r-3}}{\Gamma(\mu_r-2)}, \quad q_r^*(x) = \frac{(\sigma-x)^{\mu_r^*-3}}{\Gamma(\mu_r^*-2)} \quad (r = 0,\ 1,\ 2) \tag{29}$$

then, using Lemmas 10.3-2, 10.3-4, 10.4-2 and 10.4-4, we arrive at the following theorem.

Theorem 10.5-5. 1°. *If $r = 0,\ 1,\ 2$ is arbitrary, then:*
1°. *The function*

$$y_{\mu_r}(x;\lambda) = E_{1/2}\left(-\lambda x^2; \mu_r\right) x^{\mu_r-1}$$

is the unique solution of the Cauchy type problem

$$\begin{aligned}
y''(x) - q_r(x) + \lambda y(x) &= 0, \quad x \in (0,\sigma), \\
L_k y(0) &= \delta_{k,r} \quad (k = 0,1,2)
\end{aligned} \tag{30}$$

in the class $AC_{\{\gamma\}}[0,\sigma]$.
2°. *The function*

$$z_{\mu_r^*}(x;\lambda^*) = E_{1/2}\left(-\lambda^*(\sigma-x)^2; \mu_r^*\right)(\sigma-x)^{\mu_r^*-1}$$

is the unique solution of the Cauchy type problem

$$\begin{aligned}
z''(x) - q_r^*(x) + \lambda^* z(x) &= 0, \quad x \in (0,\sigma), \\
L_k^* z(\sigma) &= \delta_{k,r} \quad (k = 0,1,2)
\end{aligned} \tag{31}$$

in the class $AC_{\{\gamma\}}^*[0,\sigma]$.

The boundary value problems $I-I^*$ and II- II* also can be written in simpler forms. Namely,

$$\begin{aligned}
I_1 \quad & y''(x) - q_0(x) + \lambda y(x) = 0, \quad x \in (0,\sigma), \\
& L_k y(0) = \delta_{k,0} \quad (k = 0,1,2), \\
& D^{1-\nu_0} D^{\mu_0} y(\sigma) = 0 \quad (\nu_0 = \gamma_0 + \gamma_3 - 1,\ \mu_0 = \gamma_0);
\end{aligned} \tag{32_1}$$

$$\begin{aligned}
I_1^* \quad & z''(x) - q_0^*(x) + \lambda^* z(x) = 0, \quad x \in (0,\sigma), \\
& L_k^* z(\sigma) = \delta_{k,0} \quad (k = 0,1,2), \\
& D_\sigma^{1-\nu_0} D_\sigma^{\mu_0^*} z(0) = 0 \quad (\nu_0 = \gamma_0 + \gamma_3 - 1,\ \mu_0^* = \gamma_3);
\end{aligned} \tag{32_2}$$

$$\begin{aligned}
II_1 \quad & y''(x) - q_r(x) + \lambda y(x) = 0, \quad x \in (0,\sigma) \quad (r = 1,2), \\
& L_k y(0) = \delta_{k,r} \quad (k = 0,1,2;\ r = 1,2), \\
& D^{\mu_r-\nu_r-1} y(\sigma) = 0 \quad (r = 1,2;\ \mu_1 = \gamma_0 + \gamma_1, \\
& \qquad \mu_2 = \gamma_0 + \gamma_1 + \gamma_2,\ \nu_1 = 0,\ \nu_2 = \gamma_1 + \gamma_2);
\end{aligned} \tag{33_1}$$

$$II_1^* \quad \begin{aligned} &z''(x) - q_r^*(x) + \lambda^* z(x) = 0, \quad x \in (0,\sigma) \quad (r = 1,2), \\ &L_k^* z(\sigma) = \delta_{k,r} \quad (k = 0,1,2; \ r = 1,2), \\ &D_\sigma^{\mu_r^* - \nu_r - 1} z(0) = 0 \quad (r = 1,2; \ \mu_1^* = \gamma_3 + \gamma_2, \\ &\qquad \mu_2^* = \gamma_3 + \gamma_2 + \gamma_1, \ \nu_1 = 0, \ \nu_2 = \gamma_1 + \gamma_2). \end{aligned} \tag{33_2}$$

As a result, we obtain the following theorem.

Theorem 10.5-6. *The assertions of Theorems 10.5-1, 10.5.2 and 10.5-4 remain true for the boundary value problems $I_1 - I_1^*$ and $II_1 - II_1^*$.*

(f) We conclude this section with two particular cases of our boundary value problems where we have solutions very close to the classical ones.

First we assume that in the boundary value problems I_1 and I_1^* $\gamma_0 = 1, \gamma_3 \in (1/2, 1], 0 < \gamma_1, \gamma_2 \leq 1$ and $\gamma_0 + \gamma_1 + \gamma_2 + \gamma_3 = 3$. Then $\mu_0 = \gamma_0 = 1, \mu_0^* = \gamma_3$ and $\nu_0 = \gamma_0 + \gamma_3 - 1 = \gamma_3$. Therefore, by Theorem 10.5-6, the set of eigenvalues $\{\lambda_k\}_0^\infty (0 = \lambda_0 < \lambda_1 < \lambda_2 < \ldots)$ of both problems I_1 and I_1^* coincides with the set of zeros of the entire function

$$\lambda \mathcal{E}_\sigma(\lambda; \nu_0) = \lambda E_{1/2}\left(-\sigma^2 \lambda; 1 + \nu_0\right), \nu_0 = \gamma_3.$$

The eigenfunctions of these problems are correspondingly

$$\left\{\cos \sqrt{\lambda_k} x\right\}_0^\infty \quad \text{and} \quad \left\{E_{1/2}\left(-\lambda_k(\sigma - x)^2; \gamma_3\right)(\sigma - x)^{\gamma_3 - 1}\right\}_0^\infty. \tag{34}$$

In addition, if we put $\gamma_3 = 1$, then $\gamma_1 + \gamma_2 = 1, 0 < \gamma_1, \gamma_2 \leq 1$ and $\mu_0 = \mu_0^* = 1, \nu_0 = 1$. Therefore in this case $\lambda_k = (\pi k/\sigma)^2 (k = 0, 1, 2 \ldots)$ and the systems (34) become

$$\left\{\cos \frac{\pi k}{\sigma} x\right\}_0^\infty \quad \text{and} \quad \left\{\cos \frac{\pi k}{\sigma}(\sigma - x)\right\}_0^\infty. \tag{35}$$

Actually these are the same systems which, after suitable normalization, pass to the classical Fourier system

$$\left\{\frac{1}{\sqrt{\sigma}}, \left\{\sqrt{\frac{2}{\sigma}} \cos \frac{\pi k}{\sigma} x\right\}_1^\infty\right\}. \tag{36}$$

Next we assume $\gamma_3 = 1, \ \gamma_0 \in (1/2, 1], \quad 0 < \gamma_1, \gamma_2 \leq 1$ and $\gamma_0 + \gamma_1 + \gamma_2 + \gamma_3 = 3$ in the boundary value problems II_1 and II_1^*, where $r = 2$. Then $\mu_2 = \gamma_0 + \gamma_1 + \gamma_2 = 2, \ \mu_2^* = \gamma_3 + \gamma_2 + \gamma_1 = 1 + \gamma_2 + \gamma_1$ and $\nu_2 = \gamma_1 + \gamma_2$. Therefore, by Theorem 10.5-6, the set of eigenvalues $\{\lambda_k\}_1^\infty (0 < \lambda_1 < \lambda_2 < \ldots)$ of both problems II_1 and II_1^* (when $r = 2$) coincides with the set of zeros of the entire function

$$\mathcal{E}_\sigma(\lambda; \nu_2) = E_{1/2}\left(-\sigma^2 \lambda; 1 + \nu_2\right), \quad \nu_2 = \gamma_1 + \gamma_2.$$

The eigenfunctions of these problems are correspondingly

$$\left\{ \frac{\sin x\sqrt{\lambda_k}}{\sqrt{\lambda_k}} \right\}_1^\infty \quad \text{and} \quad \left\{ E_{1/2}\left(-\lambda_k(\sigma - x)^2; 1 + \gamma_1 + \gamma_2\right)(\sigma - x)^{\gamma_1 + \gamma_2} \right\}_1^\infty. \tag{37}$$

In addition, if we put $\gamma_0 = 1$, then $\gamma_1 + \gamma_2 = 1$, $\quad 0 < \gamma_1, \gamma_2 \leq 1$, $\quad \mu_2 = \mu_2^* = 1$ and $\nu_2 = 1$. In this case $\lambda_k = (\pi k/\sigma)^2 (k = 1, 2 \ldots)$, and the systems (37) take the forms

$$\left\{ \left(\frac{\pi k}{\sigma}\right)^{-1} \sin \frac{\pi k}{\sigma} x \right\}_1^\infty \quad \text{and} \quad \left\{ \left(\frac{\pi k}{\sigma}\right)^{-1} \sin \frac{\pi k}{\sigma}(\sigma - x) \right\}_1^\infty. \tag{38}$$

Actually, these are the same systems which, after suitable normalization, pass to the classical Fourier system

$$\left\{ \sqrt{\frac{2}{\sigma}} \sin \frac{\pi k}{\sigma} x \right\}_1^\infty. \tag{39}$$

10.6 Notes

10.2 The basis of this section lies on the paper of Hille-Tamarkin [1]. Some additional properties of the operators $D^{-\alpha}(-\infty < \alpha < +\infty)$ were obtained in collaboration with A. B. Nersesian and published in M.M. Djrbashian [5, Chapter 9, §1]. A very detailed account of the theory of fractional integrodifferentiation is given in the book of S.G. Samko-A.A. Kilbas-O.I. Marichev [1].

10.3–10.5 These sections contain an extended and improved account of the results of the paper of M.M. Djrbashian-S.G. Raphaelian [5]. We must mention also the papers of M.M. Djrbashian-A.B. Nersesian [1-4], as they are close to the considered problems. We already mentioned (see Notes 5.4) that [1] contains the proofs of biorthogonal expansions in $[0, l]$ in terms of linear combinations of pairs of Mittag-Leffler type functions. In [2, 3] these expansions were connected with definite boundary value problems generated by integrodifferential operators of fractional orders. The boundary value problems considered in this chapter are essentially different from them. Moreover, in contrast to [2, 3], the solution of the boundary value problems considered here reaches its logical completion, i.e., we obtain theorems on expansions in $L_2(0, \sigma)$ in terms of eigenfunctions. Finally, the Cauchy type problems with an arbitrary index n which are considered in [4] are similar to the Cauchy type problems of this chapter in the case when $n = 3$.

11 Cauchy type problems and boundary value problems in the complex domain (the case of odd segments)

11.1 Introduction

In this chapter we state and solve some Cauchy type problems in the complex domain formulated in terms of associated integro-differential operators $\mathbb{L}_{s+1/2}$ and $\mathbb{L}^*_{s+1/2}$ $(s = 1, 2, \dots)$ of fractional order. We represent explicitly the solutions of these problems by the Mittag-Leffler type function $E_{s+1/2}(z; \mu)$, and we prove an analog of the classical Lagrange formula for these solutions. Then we state some special boundary value problems in the complex domain by means of the mentioned operators. Namely, we assume that the solutions of the mentioned Cauchy type problems satisfy some boundary conditions at the ends of the sum of odd segments

$$\gamma_{2s+1}(\sigma) = \bigcup_{h=-s}^{s} \{z = r \exp\left(i\pi(h + 1/2)\right),\ 0 \le r \le \sigma\}$$

situated in the Riemann surface G^∞ of $Ln\ z$.

Using the results of Chapters 6 and 7, we prove the main Theorem 11.4-1 on the basis property of definite systems of functions in the space $L_2\{\gamma_{2s+1}(\sigma)\}$. The use of this theorem gives an exhaustive solution of the considered boundary value problems, i.e., we arrive at Theorem 11.4-2 on expansion in terms of eigenfunctions and adjoint functions converging in the norm of $L_2\{\gamma_{2s+1}(\sigma)\}$. On the other hand, Theorem 11.4-1 immediately implies Theorem 11.4-3, which gives an explicit apparatus of basic Fourier type systems of entire functions for some weighted spaces L_2 over the sum of odd segments of the complex plain

$$\Gamma_{2s+1}(\sigma) = \bigcup_{h=-s}^{s} \left\{z = r \exp i\pi \left[\frac{h + 1/2}{s + 1/2}\right],\ 0 \le r \le \sigma^{1/(s+1/2)}\right\}.$$

These theorems are essential generalizations of the main results of Chapter 10.

11.2 Preliminaries

(a) An auxiliary parameter μ exists in Theorem 2.4-1 on parametric representation of the class $W^{2,\omega}_{s+1/2,\sigma}$ (where $s \ge 1$ is an integer and $-1 < \omega < 1$) of entire functions of order $\rho = s + 1/2$ and of type $\le \sigma$. This parameter was defined by the formula

$$\mu = \frac{3/2 + s + \omega}{2s + 1}. \tag{1}$$

Therefore,

$$\mu \in \left(\frac{1}{2}, \frac{1}{2} + \frac{1}{s + 1/2}\right) = \left(\frac{1}{2}, \frac{1}{2} + \frac{1}{\rho}\right). \tag{2}$$

Two intervals of variation of parameter $\nu \in [0, 2)$ were defined by formulas 7.2(9). If we replace ω in these formulas by its expression which follows from (1), then the mentioned intervals can be written as follows:

$$\Delta_s(1^0) = \left(\mu + \frac{s - 3/2}{2s + 1}, \mu + \frac{s - 1/2}{2s + 1} \right) \subset [0, 2), \tag{3}$$

$$\Delta_s(2^0) = \left(\mu + \frac{s - 1/2}{2s + 1}, \mu + \frac{s + 1/2}{2s + 1} \right) \subset [0, 2). \tag{4}$$

(b) Assuming that $s \geq 1$ is any integer, we denote

$$\alpha_s = \exp\left\{ i \frac{\pi}{s + 1/2} \right\}, \tag{5}$$

as was done in Sections 6.2 and 7.2. Next, we put for any $\sigma \in (0, +\infty)$

$$\zeta_{\sigma,h}^{1/(s+1/2)} = \sigma^{1/(s+1/2)} \alpha_s^{h+1/2} \quad (-s \leq h \leq s), \tag{6}$$

or, which is the same,

$$\zeta_{\sigma,h} = \sigma \alpha_s^{(h+1/2)(s+1/2)} = \sigma e^{i\pi(h+1/2)} \quad (-s \leq h \leq s). \tag{7}$$

Further, we introduce the sum of $2s + 1$ segments of length σ situated in the Riemann surface G^∞:

$$\begin{aligned}
\gamma_{2s+1}(\sigma) &= \bigcup_{h=-s}^{s} [0, \zeta_{\sigma,h}] \\
&= \bigcup_{h=-s}^{s} \left\{ \tau \alpha_s^{(h+1/2)(s+1/2)} = \tau e^{i\pi(h+1/2)}, \quad 0 \leq \tau \leq \sigma \right\}.
\end{aligned} \tag{8}$$

These segments start from their common endpoint $\zeta = 0$ at angles $\pi(h + 1/2)$ $(-s \leq h \leq s)$, so that any pair of successive segments forms an angle of opening π. Note that, if we take

$$w = \zeta^{1/(s+1/2)}, \quad w_{\sigma,h} = \zeta_{\sigma,h}^{1/(s+1/2)} \quad (-s \leq h \leq s), \tag{9}$$

then the sum of segments $\gamma_{2s+1}(\sigma) \subset G^\infty$ changes to the sum of segments of the w-plane

$$\Gamma_{2s+1}(\sigma) = \bigcup_{h=-s}^{s} [0, w_{\sigma,h}] = \bigcup_{h=-s}^{s} \left\{ r \alpha_s^{h+1/2}, \ 0 \leq r \leq \sigma^{1/(s+1/2)} \right\}. \tag{10}$$

These segments are all of length $\sigma^{1/(s+1/2)}$ and form equal angles of opening $\pi/(s+1/2)$ with a common vertex at $w = 0$.

(c) Chapters 6 and 7 were based on some important facts related to the character of distribution of zeros of the entire function

$$\mathcal{E}_{s+1/2,\sigma}(z;\nu) = E_{1/2}\left(-\sigma^2 z^{2s+1}; 1+\nu\right) \quad (s \geq 1). \tag{11}$$

It is useful to remember particularly that if $\nu \in [0,2)$, then the zeros $\{\mu_n\}_1^\infty (0 < |\mu_n| \leq |\mu_{n+1}|, \ n \geq 1)$ of the function (11) are simple and are situated on the sum of rays

$$\Gamma_{2s+1} = \bigcup_{h=-s}^{s} \left\{ z = r \exp\left[i\frac{\pi h}{s+1/2}\right] \right\}, \quad 0 \leq r < +\infty. \tag{12}$$

Note also that we have given a precise order of numeration of the sequence $\{\mu_n\}_1^\infty \subset \Gamma_{2s+1}$.

(d) One can easily verify that, if the notations (6) are used, then the identities $7.2(7_1) - (7_2)$ can be written down as follows:

$$(2s+1)\sigma^{\frac{2s}{2s+1}} z^s E_{1/2}\left(-\sigma^2 z^{2s+1}; 1+\nu\right)$$

$$= \sum_{h=-s}^{s} \alpha_s^{-(h+1/2)s} E_{s+1/2}\left(z\zeta_{\sigma,h}^{1/(s+1/2)}; \nu + \frac{1}{2s+1}\right) \quad (s \geq 1), \tag{13}$$

$$(2s+1)\sigma^{\frac{2s+2}{2s+1}} z^{s+1} E_{1/2}\left(-\sigma^2 z^{2s+1}; 1+\nu\right)$$

$$= \sum_{h=-s}^{s} \alpha_s^{-(h+1/2)(s+1)} E_{s+1/2}\left(z\zeta_{\sigma,h}^{1/(s+1/2)}; \nu - \frac{1}{2s+1}\right) \quad (s \geq 1). \tag{14}$$

11.3 Cauchy type problems and boundary value problems containing the operators $\mathbb{L}_{s+1/2}$ and $\mathbb{L}_{s+1/2}^*$.

Let

$$G^\infty = \{\zeta : |\operatorname{Arg}\zeta| < +\infty, 0 < |\zeta| < +\infty\} \tag{1}$$

be the Riemann surface of the function $Ln\ z$. Assume $G \subset G^\infty$ is any set of points which is star-shaped with respect to the point $\zeta = 0$ of the branching of G^∞. In other words, we assume that G contains the whole interval $(0, \zeta_0)$, if it contains the point ζ_0. Now, if a function $y(\zeta)$ is given on G, and $\zeta_0 \in G$ is an arbitrary fixed point, then the operations of fractional integration and differentiation of order $\alpha \in (0, +\infty)$ can be formally defined on the interval $(0, \zeta_0)$ by the following formulas:

$$D^{-\alpha}y(\zeta) = \frac{1}{\Gamma(\alpha)} \int_0^\zeta (\zeta - w)^{\alpha-1} y(w)\,dw, \quad \zeta \in (0, \zeta_0),$$

$$D^\alpha y(\zeta) = \frac{d^p}{d\zeta^p}\left\{ D^{-(p-\alpha)}y(\zeta) \right\}, \quad \zeta \in (0, \zeta_0) \ (p-1 < \alpha \leq p,), \tag{2}$$

where the integration and differentiation operations are assumed to be performed along the interval $(0, \zeta_0) \subset G$. So, we have defined the operations of integro-differentiation *with the origin at* $\zeta = 0$. In a similar way we can also define the integro-differentiation operations of order $\alpha \in (0, +\infty)$ *with the end at* $\zeta_0 \in G$. To this end we put

$$D_{\zeta_0}^{-\alpha} y(\zeta) = \frac{1}{\Gamma(\alpha)} \int_{\zeta}^{\zeta_0} (w - \zeta)^{\alpha-1} y(w) dw, \quad \zeta \in (0, \zeta_0),$$

$$D_{\zeta_0}^{\alpha} y(\zeta) = \frac{d^p}{d(\zeta_0 - \zeta)^p} \left\{ D_{\zeta_0}^{-(p-\alpha)} y(\zeta) \right\}, \quad \zeta \in (0, \zeta_0) \ (p - 1 < \alpha \leq p)$$

$$(3)$$

assuming again that integrations and differentiations are performed along $(0, \zeta_0)$. Of course, we must assume in addition

$$D^{-0} y(\zeta) \equiv D^{+0} y(\zeta) \equiv y(\zeta), \ \zeta \in (0, \zeta_0),$$

$$D_{\zeta_0}^{-0} y(\zeta) \equiv D_{\zeta_0}^{+0} y(\zeta) \equiv y(\zeta), \ \zeta \in (0, \zeta_0).$$

It is obvious that the propositions of Section 10.2 remain true for the introduced definitions (2) and (3) of fractional integro-differentiation in the complex domain for any set of points $G \subset G^{\infty}$ star-shaped with respect to $\zeta = 0$. We shall give below a pair of formulas which simply follow from $10.2(19_1) - (19_2)$. To this end we assume $l_{\varphi} = \{\zeta = re^{i\varphi} : 0 < r < +\infty\} \ (|\varphi| < +\infty)$ to be an arbitrary ray of the Riemann surface G^{∞}. Then

$$D^{-\alpha} \left\{ \frac{\zeta^{\gamma}}{\Gamma(1+\gamma)} \right\} = \frac{\zeta^{\gamma+\alpha}}{\Gamma(1+\gamma+\alpha)}, \quad \zeta \in l_{\varphi}, \tag{4}$$

$$D_{\zeta_\sigma}^{-\alpha} \left\{ \frac{(\zeta_\sigma - \zeta)^{\gamma}}{\Gamma(1+\gamma)} \right\} = \frac{(\zeta_\sigma - \zeta)^{\gamma+\alpha}}{\Gamma(1+\gamma+\alpha)}, \quad \zeta \in (0, \zeta_\sigma) \subset l_{\varphi} \tag{5}$$

for arbitrary $\gamma \in (-1, +\infty), \alpha \in (-\infty, +\infty)$ and an arbitrary point $\zeta_\sigma = \sigma e^{i\varphi} \in l_{\varphi}$ $(0 < \sigma < +\infty)$. We shall assume $\sigma \in (0, +\infty)$ to be a given fixed number.

(a) Now we introduce some functions which play an important role.

Let $s \geq 1$ be an integer, let $0 < \mu, \mu^* < +\infty$. Also let λ and λ^* be arbitrary complex numbers. If we set

$$\mathcal{Y}_{s+1/2,\mu}(\zeta; \lambda) \equiv E_{s+1/2}\left(\lambda \zeta^{1/(s+1/2)}; \mu\right) \zeta^{\mu-1} \tag{6}$$

and

$$\mathcal{Z}_{s+1/2,\mu^*}(\zeta; \lambda^*) \equiv E_{s+1/2}\left(\lambda^*(\zeta_\sigma - \zeta)^{1/(s+1/2)}; \mu^*\right) (\zeta_\sigma - \zeta)^{\mu^*-1} \tag{7}$$

for any $\zeta \in (0, \zeta_\sigma)$, then we can observe that, since the point $\zeta_\sigma \in G^{\infty}$ is not fixed for a given $\sigma \in (0, +\infty)$, these formulas define the functions $\mathcal{Y}_{s+1/2,\mu}(\zeta; \lambda)$ and $\mathcal{Z}_{s+1/2,\mu^*}(\zeta; \lambda^*)$ in the whole domain $G_\sigma^{\infty} = \{\zeta \in G^{\infty} : |\zeta| < \sigma\}$.

(b) Suppose $s \geq 1$ is an integer, the parameters μ and μ^* satisfy the conditions

$$2/3 \leq \mu, \mu^* \leq 1, \ \text{ if } \ s = 1,$$
$$1/2 < \mu, \mu^* < 1/2 + 1/(s + 1/2), \ \text{ if } \ s \geq 2. \tag{8}$$

Further, we deal with the class $AC[0, \zeta_\sigma]$ of functions which are absolutely continuous in the segment $[0, \zeta_\sigma]$, where $\zeta_\sigma \in G^\infty$ is an arbitrary point. We shall say that the functions $y(\zeta)$ and $z(\zeta)$ are of classes $AC_\mu[0, \zeta_\sigma]$ and $AC^*_{\mu^*}[0, \zeta_\sigma]$, respectively, if the following conditions are satisfied:

$$\text{(i)} \quad y(\zeta) \in L_1(0, \zeta_\sigma) \text{ and } z(\zeta) \in L_1(0, \zeta_\sigma), \tag{9}$$

$$\text{(ii)} \quad L_0 y(\zeta) \equiv D^{-(1-\mu)} y(\zeta) \in AC[0, \zeta_\sigma],$$
$$L_0^* y(\zeta) \equiv D_{\zeta_\sigma}^{-(1-\mu^*)} z(\zeta) \in AC[0, \zeta_\sigma]. \tag{10}$$

The operators

$$\mathbb{L}_{s+1/2} y(\zeta) \equiv D^{-(\mu - 1/(s+1/2))} \left\{ \tfrac{d}{d\zeta} L_0 y(\zeta) \right\}$$
$$= D^{-(\mu - 1/(s+1/2))} D^\mu y(\zeta), \tag{11}$$

$$\mathbb{L}_{s+1/2}^* z(\zeta) \equiv D_{\zeta_\sigma}^{-(\mu^* - 1/(s+1/2))} \left\{ \tfrac{d}{d(\zeta_\sigma - \zeta)} L_0^* z(\zeta) \right\}$$
$$= D_{\zeta_\sigma}^{-(\mu^* - 1/(s+1/2))} D_{\zeta_\sigma}^{\mu^*} z(\zeta) \tag{12}$$

will be considered in the mentioned classes. Note that, as follows from (10), both quantities

$$m_0(y) = L_0 y(\zeta)|_{\zeta=0} \quad \text{and} \quad m_0^*(z) = L_0^* z(\zeta)|_{\zeta=\zeta_\sigma} \tag{13}$$

are finite for any functions $y(\zeta) \in AC_\mu[0, \zeta_\sigma]$ and $z(\zeta) \in AC^*_{\mu^*}[0, \zeta_\sigma]$.

(c) Now we pass to the proofs of the main results of this section. Henceforth we assume that $s \geq 1$ is any integer and that the parameters μ and μ^* satisfy condition (8). Note also that the lemmas established below are similar to those of Section 10.3, and therefore some fragments of their proofs will be omitted. Finally, note that we generally omit references on propositions of theory of fractional integro-differentiation given in Section 10.2.

Lemma 11.3-1. *Let* $y(\zeta) \in AC_\mu[0, \zeta_\sigma]$ *and* $z(\zeta) \in AC^*_{\mu^*}[0, \zeta_\sigma]$, *where* $\zeta_\sigma \in G^\infty$ *is an arbitrary point. Then the following representations are true almost everywhere in* $(0, \zeta_\sigma)$:

$$\mathbb{L}_{s+1/2} y(\zeta) = D^{1/(s+1/2)} y(\zeta) - m_0(y) \frac{\zeta^{\mu - 1 - 1/(s+1/2)}}{\Gamma(\mu - 1/(s + 1/2))}, \tag{14}$$

$$\mathbb{L}_{s+1/2}^* z(\zeta) = D_{\zeta_\sigma}^{1/(s+1/2)} z(\zeta) - m_0^*(z) \frac{(\zeta_\sigma - \zeta)^{\mu^* - 1 - 1/(s+1/2)}}{\Gamma(\mu^* - 1/(s + 1/2))}. \tag{15}$$

Proof. By formula (11),

$$D^{-1/(s+1/2)}\mathbb{L}_{s+1/2}y(\zeta) = D^{-1/(s+1/2)}D^{-(\mu-1/(s+1/2))}D^{\mu}y(\zeta)$$
$$= D^{-\mu}D^{\mu}y(\zeta) \tag{16}$$

almost everywhere in $(0,\zeta_\sigma)$. But, by proposition $8°$ of Section 10.2,

$$D^{-\mu}D^{\mu}y(\zeta) = y(\zeta) - m_0(y)\frac{\zeta^{\mu-1}}{\Gamma(\mu)} \tag{17}$$

almost everywhere in $(0,\zeta_\sigma)$. Hence

$$D^{-1/(s+1/2)}\mathbb{L}_{s+1/2}y(\zeta) = y(\zeta) - m_0(y)\frac{\zeta^{\mu-1}}{\Gamma(\mu)}, \quad \zeta \in (0,\zeta_\sigma) \tag{18}$$

almost everywhere. Applying the operator $D^{1/(s+1/2)}$ to both sides of this formula and using (4), we obtain representation (14). Representation (15) is obtained in a similar way.

Lemma 11.3-2. $1°$. *The Cauchy type problem*

$$\mathbb{L}_{s+1/2}y(\zeta) - \lambda y(\zeta) = 0, \qquad \zeta \in (0,\zeta_\sigma)$$
$$L_0 y(\zeta)|_{\zeta=0} = 1 \tag{19}$$

considered in the class $AC_\mu[0,\zeta_\sigma]$ *has the unique solution*

$$\mathcal{Y}_{s+1/2,\mu}(\zeta;\lambda) = E_{s+1/2}\left(\lambda\zeta^{1/(s+1/2)};\mu\right)\zeta^{\mu-1} \in L_2(0,\zeta_\sigma). \tag{20}$$

$2°$. *The Cauchy type problem*

$$\mathbb{L}^*_{s+1/2}z(\zeta) - \lambda^* z(\zeta) = 0, \qquad \zeta \in (0,\zeta_\sigma)$$
$$L_0^* z(\zeta)|_{\zeta=\zeta_\sigma} = 1 \tag{21}$$

considered in the class $AC^*_{\mu^*}[0,\zeta_\sigma]$ *has the unique solution*

$$\mathcal{Z}_{s+1/2,\mu^*}(\zeta;\lambda^*) = E_{s+1/2}\left(\lambda^*(\zeta_\sigma - \zeta)^{1/(s+1/2)};\mu^*\right)(\zeta_\sigma - \zeta)^{\mu^*-1} \in L_2(0,\zeta_\sigma). \tag{22}$$

Proof. $1°$. Obviously

$$\mathcal{Y}_{s+1/2,\mu}(\zeta;\lambda) = \sum_{k=0}^{\infty}\lambda^k\frac{\zeta^{k/(s+1/2)+\mu-1}}{\Gamma(\mu + k/(s + 1/2))}.$$

Hence, using (4) we easily obtain

$$L_0 \mathcal{Y}_{s+1/2,\mu}(\zeta;\lambda) = E_{s+1/2}\left(\lambda\zeta^{1/(s+1/2)};1\right), \quad L_0 \mathcal{Y}_{s+1/2,\mu}(\zeta;\lambda)\big|_{\zeta=0} = 1,$$

$$\mathbb{L}_{s+1/2}\mathcal{Y}_{s+1/2,\mu}(\zeta;\lambda) = D^{-(\mu-1/(s+1/2))}\left\{\frac{d}{d\zeta}E_{s+1/2}\left(\lambda\zeta^{1/(s+1/2)};1\right)\right\}$$

$$= D^{-(\mu-1/(s+1/2))}\left\{\lambda E_{s+1/2}\left(\lambda\zeta^{1/(s+1/2)};1/(s+1/2)\right)\zeta^{1/(s+1/2)-1}\right\}$$

$$= \lambda E_{s+1/2}\left(\lambda\zeta^{1/(s+1/2)};\mu\right)\zeta^{\mu-1} = \lambda\mathcal{Y}_{s+1/2,\mu}(\zeta;\lambda).$$

Thus function (20) of $AC_\mu[0,\zeta_\sigma]$ is a solution of Cauchy type problem (19). It is obvious that the uniqueness of this solution will be established as soon as we prove the uniqueness of the solution $y(\zeta) = 0$ of the homogeneous Cauchy type problem

$$\mathbb{L}_{s+1/2}y(\zeta) - \lambda y(\zeta) = 0, \qquad \zeta \in (0,\zeta_\sigma), \tag{23}$$

$$L_0 y(\zeta)\big|_{\zeta=0} = 0. \tag{24}$$

in $AC_\mu[0,\zeta_\sigma]$. To this end, suppose $y(\zeta) \in AC_\mu[0,\zeta_\sigma]$ is any solution of problem (23)-(24) and observe that condition (24) can be written in the form $m_0(y) = 0$. By (18),

$$y(\zeta) = \lambda D^{-1/(s+1/2)}y(\zeta), \qquad \zeta \in (0,\zeta_\sigma),$$

and $y(\zeta)$ is a solution of the Volterra homogeneous integral equation

$$y(\zeta) = \frac{\lambda}{\Gamma(\alpha)}\int_0^\zeta (\zeta - w)^{\alpha-1}y(w)dw, \qquad \zeta \in (0,\zeta_\sigma),$$

where $\alpha = (s+1/2)^{-1} \in (0,2/3]$. Hence $y(\zeta) \equiv 0$, $\zeta \in (0,\zeta_\sigma)$.
$2°$. This assertion can be proved in a similar way based on formula (5).

Theorem 11.3-1. *The following formula is true for the solutions* $\mathcal{Y}_{s+1/2,\mu}(\zeta;\lambda)$ *and* $\mathcal{Z}_{s+1/2,\mu^*}(\zeta;\lambda^*)$ *of Cauchy type problems (19) and (21):*

$$\int_0^\sigma \mathcal{Y}_{s+1/2,\mu}(\zeta;\lambda)\mathcal{Z}_{s+1/2,\mu^*}(\zeta;\lambda^*)d\zeta =$$

$$\frac{E_{s+1/2}\left(\lambda\zeta_\sigma^{\frac{1}{s+1/2}};\mu+\mu^*-\frac{1}{s+1/2}\right) - E_{s+1/2}\left(\lambda^*\zeta_\sigma^{\frac{1}{s+1/2}};\mu+\mu^*-\frac{1}{s+1/2}\right)}{\lambda - \lambda^*} \tag{25}$$

$$\times \zeta^{\mu+\mu^*-1/(s+1/2)-1}.$$

Proof. Using definitions (11) and (12) of operators $\mathbb{L}_{s+1/2}$ and $\mathbb{L}^*_{s+1/2}$, formulas (4) and (5) and integration by parts, we obtain

$$
\lambda \int_0^{\zeta_\sigma} \mathcal{Y}_{s+1/2,\mu}(\zeta;\lambda) \mathcal{Z}_{s+1/2,\mu^*}(\zeta;\lambda^*)d\zeta
$$

$$
= \int_0^{\zeta_\sigma} \left[\mathbb{L}_{s+1/2}\mathcal{Y}_{s+1/2,\mu}(\zeta;\lambda)\right] \mathcal{Z}_{s+1/2,\mu^*}(\zeta;\lambda^*)d\zeta
$$

$$
= \int_0^{\zeta_\sigma} \left[D^{-(\mu-1/(s+1/2))}\frac{d}{d\zeta}L_0\mathcal{Y}_{s+1/2,\mu}(\zeta;\lambda)\right] \mathcal{Z}_{s+1/2,\mu^*}(\zeta;\lambda^*)d\zeta \qquad (26)
$$

$$
= \int_0^{\zeta_\sigma} \left\{\frac{d}{d\zeta}L_0\mathcal{Y}_{s+1/2,\mu}(\zeta;\lambda)\right\} D_{\zeta_\sigma}^{-(\mu-1/(s+1/2))} \mathcal{Z}_{s+1/2,\mu^*}(\zeta;\lambda^*)d\zeta
$$

$$
= \int_0^{\zeta_\sigma} \left\{\frac{d}{d\zeta}L_0\mathcal{Y}_{s+1/2,\mu}(\zeta;\lambda)\right\} \mathcal{Z}_{s+1/2,\mu+\mu^*-1/(s+1/2)}(\zeta;\lambda^*)d\zeta.
$$

But

$$
\mathcal{Z}_{s+1/2,\mu+\mu^*-1/(s+1/2)}(\zeta;\lambda^*)
$$
$$
\equiv \frac{(\zeta_\sigma - \zeta)^{\mu+\mu^*-1-1/(s+1/2)}}{\Gamma(\mu+\mu^*-1/(s+1/2))} + \lambda^* \mathcal{Z}_{s+1/2,\mu+\mu^*}(\zeta;\lambda^*), \qquad \zeta \in (0,\zeta_\sigma), \qquad (27)
$$

and consequently

$$
\lambda \int_0^{\zeta_\sigma} \mathcal{Y}_{s+1/2,\mu}(\zeta;\lambda) \mathcal{Z}_{s+1/2,\mu^*}(\zeta;\lambda^*)d\zeta = I_1 + I_2, \qquad (28)
$$

where

$$
I_1 = \frac{1}{\Gamma\left(\mu+\mu^*-\frac{1}{s+1/2}\right)} \int_0^{\zeta_\sigma} (\zeta_\sigma - \zeta)^{\mu+\mu^*-1-\frac{1}{s+1/2}} \frac{d}{d\zeta}L_0\mathcal{Y}_{s+1/2,\mu}(\zeta;\lambda)d\zeta, \quad (29)
$$

$$
I_2 = \lambda^* \int_0^{\zeta_\sigma} \mathcal{Z}_{s+1/2,\mu+\mu^*}(\zeta;\lambda^*)\frac{d}{d\zeta}L_0\mathcal{Y}_{s+1/2,\mu}(\zeta;\lambda)d\zeta. \qquad (30)
$$

Further, obviously

$$
I_1 = \frac{1}{\Gamma\left(\mu+\mu^*-\frac{1}{s+1/2}\right)}
$$
$$
\times \int_0^{\zeta_\sigma} (\zeta_\sigma - \zeta)^{\mu+\mu^*-1-\frac{1}{s+1/2}} \lambda\mathcal{Y}_{s+1/2,1/(s+1/2)}(\zeta;\lambda)d\zeta \qquad (31)
$$
$$
= \lambda D^{\mu+\mu^*-1/(s+1/2)}\mathcal{Y}_{s+1/2,1/(s+1/2)}(\zeta;\lambda)\Big|_{\zeta=\zeta_\sigma}
$$
$$
= \lambda\mathcal{Y}_{s+1/2,\mu+\mu^*}(\zeta;\lambda)\Big|_{\zeta=\zeta_\sigma} = \lambda E_{s+1/2}\left(\lambda\zeta_\sigma^{1/(s+1/2)};\mu+\mu^*\right)\zeta_\sigma^{\mu+\mu^*-1},
$$

and

$$
\begin{aligned}
I_2 &= \lambda^* \left[L_0 \mathcal{Y}_{s+1/2,\mu}(\zeta;\lambda) \right] \mathcal{Z}_{s+1/2,\mu+\mu^*}(\zeta;\lambda^*)\Big|_0^{\zeta_\sigma} \\
&\quad + \lambda^* \int_0^{\zeta_\sigma} \left[L_0 \mathcal{Y}_{s+1/2,\mu}(\zeta;\lambda) \right] \frac{d}{d(\zeta_\sigma - \zeta)} \mathcal{Z}_{s+1/2,\mu+\mu^*}(\zeta;\lambda^*)d\zeta \\
&= \lambda^* \mathcal{Y}_{s+1/2,1}(\zeta;\lambda) \mathcal{Z}_{s+1/2,\mu+\mu^*}(\zeta;\lambda^*)\Big|_0^{\zeta_\sigma} \\
&\quad + \lambda^* \int_0^{\zeta_\sigma} \left[D^{-(1-\mu)} \mathcal{Y}_{s+1/2,\mu}(\zeta;\lambda) \right] \mathcal{Z}_{s+1/2,\mu+\mu^*-1}(\zeta;\lambda^*)d\zeta \\
&= -\lambda^* E_{s+1/2} \left(\lambda^* \zeta_\sigma^{1/(s+1/2)}; \mu+\mu^* \right) \zeta_\sigma^{\mu+\mu^*-1} \\
&\quad + \lambda^* \int_0^{\zeta_\sigma} \mathcal{Y}_{s+1/2,\mu}(\zeta;\lambda) \mathcal{Z}_{s+1/2,\mu^*}(\zeta;\lambda^*)d\zeta.
\end{aligned}
\tag{32}
$$

Combining formulas (28), (31) and (32), we obtain

$$
\begin{aligned}
(\lambda - \lambda^*) \int_0^{\zeta_\sigma} &\mathcal{Y}_{s+1/2,\mu}(\zeta;\lambda) \mathcal{Z}_{s+1/2,\mu^*}(\zeta;\lambda^*)d\zeta \\
&= \lambda E_{s+1/2} \left(\lambda \zeta_\sigma^{1/(s+1/2)}; \mu+\mu^* \right) \zeta_\sigma^{\mu+\mu^*-1} \\
&\quad - \lambda^* E_{s+1/2} \left(\lambda^* \zeta_\sigma^{1/(s+1/2)}; \mu+\mu^* \right) \zeta_\sigma^{\mu+\mu^*-1}.
\end{aligned}
\tag{33}
$$

It remains to observe that this formula coincides with (25).

Remark. Identity (25) can be directly derived from formula 1.2(10) where we assume $\rho = s+1/2$, $\alpha = \mu$ and $\beta = \mu^*$, but we preferred to give another proof based on the fact that $\mathcal{Y}_{s+1/2,\mu}(\zeta;\lambda)$ and $\mathcal{Z}_{s+1/2,\mu^*}(\zeta;\lambda^*)$ are solutions of the Cauchy type problems (19) and (21).

(d) We shall now formulate two pairs of boundary value problems assuming again that $s \geq 1$ is any integer, that the parameters μ and μ^* satisfy the conditions (8) and that $\nu \in [0,2)$.

Problem $\mathbf{I}_{s+1/2}$ *consists in finding those values of λ for which the solution $\mathcal{Y}_{s+1/2,\mu}(\zeta;\lambda)$ of the Cauchy type problem (19) satisfies the additional boundary condition*

$$
\sum_{h=-s}^{s} \alpha_s^{-\nu(h+1/2)(s+1/2)} D^{-(\nu+1/(2s+1)-\mu)} \mathcal{Y}_{s+1/2,\mu}(\zeta;\lambda)\Bigg|_{\zeta=\zeta_{\sigma,h}} = 0
\tag{35}
$$

at the endpoints $\zeta_{\sigma,h}$ $(-s \leq h \leq s)$ of the sum of segments $\gamma_{2s+1}(\sigma)$.

Problem $\mathbf{I}^*_{s+1/2}$ *consists in finding those values of λ^* for which the solution $\mathcal{Z}_{s+1/2,\mu^*}(\zeta;\lambda^*)$ of the Cauchy type problem (21) satisfies the additional boundary condition*

$$
\sum_{h=-s}^{s} \alpha_s^{-\nu(h+1/2)(s+1/2)} D_{\zeta_{\sigma,h}}^{-(\nu+1/(2s+1)-\mu^*)} \mathcal{Z}_{s+1/2,\mu^*}(\zeta;\lambda^*)\Bigg|_{\zeta=0} = 0
\tag{36}
$$

at the common endpoint $\zeta = 0$ of the segments of $\gamma_{2s+1}(\sigma)$.

The second pair of boundary value problems is formulated as follows.

Problem II$_{s+1/2}$ *consists in finding those values of λ for which the solution $\mathcal{Y}_{s+1/2,\mu}(\zeta;\lambda)$ of the Cauchy type problem (19) satisfies the additional boundary condition*

$$\sum_{h=-s}^{s} \alpha_s^{-\nu(h+1/2)(s+1/2)} D^{-(\nu-1/(2s+1)-\mu)} \mathcal{Y}_{s+1/2,\mu}(\zeta;\lambda)\Big|_{\zeta=\zeta_{\sigma,h}} = 0 \qquad (37)$$

at the endpoints $\zeta_{\sigma,h}(-s \le h \le s)$ of the segments of $\gamma_{2s+1}(\sigma)$.

Problem II$^*_{s+1/2}$ *consists in finding those values of λ^* for which the solution $\mathcal{Z}_{s+1/2,\mu^*}(\zeta;\lambda^*)$ of the Cauchy type problem (21) satisfies the additional boundary conditions*

$$\sum_{h=-s}^{s} \alpha_s^{-\nu(h+1/2)(s+1/2)} D_{\zeta_{\sigma,h}}^{-(\nu-1/(2s+1)-\mu^*)} \mathcal{Z}_{s+1/2,\mu^*}(\zeta;\lambda^*)\Big|_{\zeta=0} = 0 \qquad (38)$$

at the common endpoint $\zeta = 0$ of the segments of $\gamma_{2s+1}(\sigma)$.

Later on the desired values of λ and λ^* will be called *eigenvalues* of the formulated boundary value problems, and the corresponding solutions $\mathcal{Y}_{s+1/2,\mu}(\zeta;\lambda)$ and $\mathcal{Z}_{s+1/2,\mu^*}(\zeta;\lambda^*)$ of Cauchy type problems will be called *eigenfunctions* of these problems.

(e) Denote the sums (35) and (36) respectively by $R_{s+1/2}(\lambda)$ and $R^*_{s+1/2}(\lambda^*)$, and the sums (37) and (38) respectively by $S_{s+1/2}(\lambda)$ and $S^*_{s+1/2}(\lambda^*)$.

Lemma 11.3-3. *The following identities are true:*

$$1°. \quad R_{s+1/2}(\lambda) = (2s+1)\sigma^\nu \lambda^s E_{1/2}\left(-\sigma^2 \lambda^{2s+1}; 1+\nu\right),$$

$$R^*_{s+1/2}(\lambda^*) = R_{s+1/2}(\lambda^*). \qquad (39)$$

$$2°. \quad S_{s+1/2}(\lambda) = (2s+1)\sigma^\nu \lambda^{s+1} E_{1/2}\left(-\sigma^2 \lambda^{2s+1}; 1+\nu\right),$$

$$S^*_{s+1/2}(\lambda^*) = S_{s+1/2}(\lambda^*) \qquad (40)$$

Proof. 1°. It is easy to see that

$$D^{-(\nu+1/(2s+1)-\mu)} \mathcal{Y}_{s+1/2,\mu}(\zeta;\lambda)\Big|_{\zeta=\zeta_{\sigma,h}} = \mathcal{Y}_{s+1/2,\nu+1/(2s+1)}(\zeta;\lambda)\Big|_{\zeta=\zeta_{\sigma,h}}$$

$$= E_{s+1/2}\left(\lambda\zeta_{\sigma,h}^{1/(s+1/2)}; \nu + 1/(2s+1)\right) \zeta_{\sigma,h}^{\nu+1/(2s+1)-1} \qquad (-s \le h \le s). \qquad (41)$$

Thus, using the identity 11.2(13), we obtain

$$R_{s+1/2}(\lambda)$$

$$= \sum_{h=-s}^{s} \alpha_s^{-\nu(h+1/2)(s+1/2)} E_{s+1/2}\left(\lambda\zeta_{\sigma,h}^{1/(s+1/2)}; \nu + \frac{1}{2s+1}\right) \zeta_{\sigma,h}^{\nu+1/(2s+1)-1}$$

$$= \sigma^{\nu+1/(2s+1)-1} \sum_{h=-s}^{s} E_{s+1/2}\left(\lambda\zeta_{\sigma,h}^{1/(s+1/2)}; \nu + \frac{1}{2s+1}\right) \alpha_s^{-(h+1/2)s}$$

$$= (2s+1)\sigma^\nu \lambda^s E_{1/2}\left(-\sigma^2 \lambda^{2s+1}; 1+\nu\right).$$

Thus the first of identities (39) is proved. The proof of the second is similar.
2°. Identities (40) can be proved in the same way, but this time the proof is based on formula 11.2(14).

The following theorem is an immediate consequence of the preceding lemma and of the formulations of boundary value problems given above.

Theorem 11.3-2. *If $s \geq 1$ is any integer, μ and μ^* satisfy the conditions*

$$2/3 \leq \mu, \mu^* \leq 1 \text{ when } s = 1,$$
$$1/2 < \mu, \mu^* < 1/2 + 1/(s + 1/2) \text{ when } s \geq 2$$

and $\nu \in [0, 2)$, then:
1°. The sets of eigenvalues of both boundary value problems $I_{s+1/2}$ and $I^*_{s+1/2}$ coincide with the sequence $\{\mu_n\}_0^\infty$ ($\mu_0 = 0$) of zeros of the entire function

$$\lambda^s \mathcal{E}_{s+1/2,\sigma}(\lambda; \nu) \equiv \lambda^s E_{1/2}\left(-\sigma^2 \lambda^{2s+1}; 1 + \nu\right).$$

Thus all these eigenvalues are simple with the exception of $\mu_0 = 0$ which is of order s.
2°. The sets of eigenvalues of both boundary value problems $II_{s+1/2}$ and $II^*_{s+1/2}$ coincide with the sequence $\{\mu_n\}_0^\infty$ ($\mu_0 = 0$) of zeros of the entire function

$$\lambda^{s+1} \mathcal{E}_{s+1/2}(\lambda; \nu) \equiv \lambda^{s+1} E_{1/2}\left(-\sigma^2 \lambda^{2s+1}; 1 + \nu\right).$$

Thus all these eigenvalues are simple with the exception of $\mu_0 = 0$ which is of order $s + 1$.

(f) It is useful to give explicit representations of eigenfunctions of both pairs of considered boundary value problems, but it is necessary first to introduce some notations.

Let $s \geq 1$ be an integer, let $0 < \mu, \mu^* < +\infty$ and also let $\nu \in [0, 2)$. As in Section 11.2(c) we shall denote by $\{\mu_n\}_1^\infty$ the sequence of zeros of the entire function

$$\mathcal{E}_{s+1/2,\sigma}(z; \nu) \equiv E_{1/2}\left(-\sigma^2 z^{2s+1}; 1 + \nu\right).$$

Further, we denote

$$\begin{aligned}
\mathcal{Y}_{\nu,n}(\zeta) &\equiv \frac{\partial^{-n}}{\partial\lambda^{-n}}\left\{\mathcal{Y}_{s+1/2,\mu}(\zeta; \lambda)\right\}\big|_{\lambda=0} \\
&= \frac{\Gamma(1-n)}{\Gamma(\mu - n/(s+1/2))}\zeta^{-n/(s+1/2)+\mu-1}, \quad \zeta \in (0, \zeta_\sigma),\ n \leq 0,
\end{aligned} \tag{42_1}$$

$$\begin{aligned}
\mathcal{Y}_{\nu,n}(\zeta) &\equiv \mathcal{Y}_{s+1/2,\mu}(\zeta; \mu_n) \\
&= E_{s+1/2}\left(\mu_n \zeta^{1/(s+1/2)}; \mu\right)\zeta^{\mu-1}, \quad \zeta \in (0, \zeta_\sigma),\ n \geq 1,
\end{aligned} \tag{42_2}$$

$$\mathcal{Z}_{\nu,n}^{(1)}(\zeta) \equiv \frac{\partial^{s-1+n}}{\partial\lambda^{s-1+n}}\left\{\mathcal{Z}_{s+1/2,\mu^*}(\zeta;\lambda)\right\}\Big|_{\lambda=0}$$

$$= \frac{\Gamma(s+n)}{\Gamma\left(\mu^* + \frac{s-1+n}{s+1/2}\right)}(\zeta_\sigma - \zeta)^{\frac{s-1+n}{s+1/2}+\mu^*-1}, \quad \zeta \in (0,\zeta_\sigma), -(s-1) \le n \le 0, \tag{43_1}$$

$$\mathcal{Z}_{\nu,n}^{(2)}(\zeta) \equiv \frac{\partial^{s+n}}{\partial\lambda^{s+n}}\left\{\mathcal{Z}_{s+1/2,\mu^*}(\zeta;\lambda)\right\}\Big|_{\lambda=0}$$

$$= \frac{\Gamma(1+s+n)}{\Gamma\left(\mu^* + \frac{s+n}{s+1/2}\right)}(\zeta_\sigma - \zeta)^{\frac{s+n}{s+1/2}+\mu^*-1}, \quad \zeta \in (0,\zeta_\sigma), -s \le n \le 0, \tag{43_2}$$

$$\mathcal{Z}_{\nu,n}^{(1)}(\zeta) \equiv \mathcal{Z}_{\nu,n}^{(2)}(\zeta) \equiv \mathcal{Z}_{s+1/2,\mu^*}(\zeta;\mu_n)$$

$$= E_{s+1/2}\left(\mu_n(\zeta_\sigma - \zeta)^{1/(s+1/2)};\mu^*\right)(\zeta_\sigma - \zeta)^{\mu^*-1}, \quad \zeta \in (0,\zeta_\sigma), n \ge 1. \tag{43_3}$$

Obviously these formulas define the introduced functions in the whole domain $G_\sigma^\infty = \{\zeta \in G^\infty : |\zeta| < \sigma\}$.

Theorem 11.3-3. *If $s \ge 1$ is an integer, μ and μ^* satisfy the conditions*

$$2/3 \le \mu, \mu^* \le 1 \quad when \quad s = 1$$
$$1/2 < \mu, \mu^* < 1/2 + 1/(s+1/2) \quad when \quad s \ge 2$$

and $\nu \in [0, 2)$, then:
1°. The sequence of functions

$$\{\mathcal{Y}_{\nu,n}(\zeta)\}_{-(s-1)}^{+\infty} \tag{44}$$

is a system of eigenfunctions (when $n \ge 0$) and adjoint functions (when $s \ge 2$ and $-(s-1) \le n \le -1$) of the boundary value problem $I_{s+1/2}$. On the other hand, the sequence of functions

$$\left\{\mathcal{Z}_{\nu,n}^{(1)}(\zeta)\right\}_{-(s-1)}^{+\infty} \tag{45}$$

is a system of eigenfunctions (when $n > 0$ and $n = -(s-1)$) and adjoint functions (when $s \ge 2$ and $-(s-1) < n \le 0$) of the boundary value problem $I_{s+1/2}^*$.
2°. The sequence of functions

$$\{\mathcal{Y}_{\nu,n}(\zeta)\}_{-s}^{+\infty} \tag{46}$$

is a system of eigenfunctions (when $n \ge 0$) and adjoint functions (when $-s \le n \le -1$) of the boundary value problem $II_{s+1/2}$. On the other hand, the sequence of functions

$$\left\{\mathcal{Z}_{\nu,n}^{(2)}(\zeta)\right\}_{-s}^{+\infty} \tag{47}$$

is a system of eigenfunctions (when $n > 0$ and $n = -s$) and adjoint functions (when $-s < n \le 0$) of the boundary value problem $II_{s+1/2}^*$.

Proof. If Theorem 11.3-2 and relations $(42_1) - (42_2)$, $(43_1) - (43_3)$ are taken into account, then the desired assertions easily follow from the definitions of eigenvalues and eigenfunctions and from the ordinary method of definition of adjoint functions. As we shall see later, this method is stipulated also by the explicit formulas for biorthogonal systems of $2s + 1$-dimensional vector functions of $L_2^{2s+1}(0, \sigma)$ established in Chapter 7.

11.4 Expansions in $L_2\{\gamma_{2s+1}(\sigma)\}$ in terms of Riesz bases

It is necessary to give beforehand a summary of some notations and results of Chapter 7.

(a) In Chapter 7 it was assumed

$$\mu = \frac{3/2 + s + \omega}{2s + 1} \in \left(\frac{1}{2}, \frac{1}{2} + \frac{1}{s + 1/2}\right) \tag{1}$$

for any integer $s \geq 1$ and for any $\omega \in (-1, 1)$. Besides, it was assumed there

$$\kappa_{-s} = \frac{1 + 2(\omega - s)}{2s + 1} = 2(\mu - 1), \tag{2}$$

and that the condition $\nu \in [0, 2)$ is satisfied whenever the parameter ν is present. In addition $\{\mu_n\}_1^\infty$ was assumed to be the sequence of zeros of the entire function

$$\mathcal{E}_{s+1/2, \sigma}(z; \nu) \equiv E_{1/2}\left(-\sigma^2 z^{2s+1}; 1 + \nu\right). \tag{3}$$

The following non-intersecting intervals, in which ν varies, were also introduced in Chapter 7:

$$\Delta_s(1^0) = \left(\mu + \frac{s - 3/2}{2s + 1}, \mu + \frac{s - 1/2}{2s + 1}\right) \subset [0, 2),$$

$$\Delta_s(2^0) = \left(\mu + \frac{s - 1/2}{2s + 1}, \mu + \frac{s + 1/2}{2s + 1}\right) \subset [0, 2).$$

(b) When $\nu \in [0, 2)$, the sequence of vector functions

$$\{\kappa_n(\tau)\}_{-\infty}^{+\infty} \equiv \left\{\{\kappa_{n,h}(\tau)\}_{h=-s}^{s}\right\}_{-\infty}^{+\infty}, \tau \in (0, \sigma) \tag{4}$$

was defined as follows:

$$\kappa_{n,h}(\tau) \equiv \alpha_s^{-(h+1/2)n} \frac{\Gamma(1 - n)}{\Gamma\left(\mu - \frac{n}{s+1/2}\right)} \tau^{-\frac{n}{s+1/2} + \mu - 1} \quad (-s \leq h \leq s, n \leq 0), \tag{5_1}$$

$$\kappa_{n,h}(\tau) \equiv E_{s+1/2}\left(\alpha_s^{h+1/2} \mu_n \tau^{1/(s+1/2)}; \mu\right) \tau^{\mu - 1} \quad (-s \leq h \leq s, n \geq 1). \tag{5_2}$$

(c) When $\nu \in \overline{\Delta}_s(1^0)$, the parameter

$$\eta_{2,0} = \nu + \frac{3/2 - s - \omega}{2s + 1} = \nu + \frac{3}{2s + 1} - \mu \tag{6}$$

was introduced. Using this parameter, the sequence of vector functions

$$\left\{\Omega_m^{(1)}(\tau; s)\right\}_{-(s-1)}^{+\infty} \equiv \left\{\{\Omega_{m,h}^{(1)}(\tau; s)\}_{h=-s}^{s}\right\}_{-(s-1)}^{+\infty}, \quad \tau \in (0, \sigma) \tag{7}$$

was defined as follows:

$$\Omega_{m,h}^{(1)}(\tau; s) = \alpha_s^{(h+1/2)m} \frac{\sigma^{-\nu}\Gamma(1+\nu)}{(2s+1)\Gamma(1-m)} \frac{(\sigma - \tau)^{\frac{s-1+m}{s+1/2} + \eta_{2,0} - 1}}{\Gamma\left(\frac{s-1+m}{s+1/2} + \eta_{2,0}\right)}, \tag{8_1}$$

when $-(s - 1) \le m \le 0,\ -s \le h \le s$ and

$$\Omega_{m,h}^{(1)}(\tau; s) = \frac{\alpha_s^{-(h+1/2)(s-1)}\sigma^{-\nu}}{(2s+1)\mu_m^s \mathcal{E}'_{s+1/2,\sigma}(\mu_m; \nu)} \tag{8_2}$$
$$\times E_{s+1/2}\left(\alpha_s^{h+1/2}\mu_m(\sigma - \tau)^{1/(s+1/2)}; \eta_{2,0}\right)(\sigma - \tau)^{\eta_{2,0}-1}$$

when $1 \le m < +\infty$ and $-s \le h \le s$. Besides, in Chapter 7 we moved from system (7) to the system

$$\{\omega_m^{(1)}(\tau)\}_{-(s-1)}^{+\infty} \equiv \left\{\{\omega_{m,h}^{(1)}(\tau)\}_{h=-s}^{s}\right\}_{-(s-1)}^{+\infty}, \quad \tau \in (0, \sigma), \tag{9}$$

where

$$\omega_{m,h}^{(1)}(\tau) \equiv \overline{\Omega_{m,h}^{(1)}(\tau; s)}, \quad \tau \in (0, \sigma), \quad -s \le h \le s, -(s - 1) \le m < +\infty. \tag{10}$$

It was also proved that, when $\nu \in \Delta_s(1^0)$, the systems of vector functions

$$\{\kappa_n(\tau)\}_{-(s-1)}^{+\infty} \quad \text{and} \quad \{\omega_m^{(1)}(\tau)\}_{-(s-1)}^{+\infty} \tag{11}$$

are biorthogonal in $L_2^{2s+1}(0, \sigma)$, i.e., for any n and m $(-(s - 1) \le n, m < +\infty)$

$$\left\{\kappa_n; \omega_m^{(1)}\right\} = \sum_{h=-s}^{s} \int_0^{\sigma} \kappa_{n,h}(\tau)\overline{\omega_{m,h}^{(1)}(\tau)}d\tau = \delta_{n,m}. \tag{12}$$

(d) When $\nu \in \Delta_s(2^0)$, we have introduced the parameter

$$\eta_{2,1} = \nu - \frac{1/2 + s + \omega}{2s + 1} = \nu + \frac{1}{2s + 1} - \mu, \tag{13}$$

which was used in the definition of the vector functions

$$\{\Omega_m^{(2)}(\tau;s)\}_{-s}^{+\infty} \equiv \left\{\left\{\Omega_{m,h}^{(2)}(\tau;s)\right\}_{h=-s}^{s}\right\}_{-s}^{+\infty}, \qquad \tau \in (0,\sigma). \tag{14}$$

These vector functions were defined as follows:

$$\Omega_{m,h}^{(2)}(\tau;s) = \alpha_s^{(h+1/2)m}\, \frac{\sigma^{-\nu}\Gamma(1+\nu)}{(2s+1)\Gamma(1-m)}\, \frac{(\sigma-\tau)^{\frac{s+m}{s+1/2}+\eta_{2,1}-1}}{\Gamma\left(\frac{s+m}{s+1/2}+\eta_{2,1}\right)} \tag{15_1}$$

when $-s \le m \le 0,\ -s \le h \le s$ and

$$\begin{aligned}
\Omega_{m,h}^{(2)}(\tau;s) &= \frac{\alpha_s^{-(h+1/2)s}\sigma^{-\nu}}{(2s+1)\mu_m^{s+1}\mathcal{E}'_{s+1/2,\sigma}(\mu_m;\nu)} \\
&\quad \times E_{s+1/2}\left(\alpha_s^{h+1/2}\mu_m(\sigma-\tau)^{1/(s+1/2)};\eta_{2,1}\right)(\sigma-\tau)^{\eta_{2,1}-1}
\end{aligned} \tag{15_2}$$

when $1 \le m < +\infty$ and $-s \le h \le s$. As in the preceding case, we moved from (14) to the system

$$\left\{\omega_m^{(2)}(\tau)\right\}_{-s}^{+\infty} \equiv \left\{\left\{\omega_{m,h}^{(2)}(\tau)\right\}_{h=-s}^{s}\right\}_{-s}^{+\infty}, \qquad \tau \in (0,\sigma), \tag{16}$$

where

$$\omega_{m,h}^{(2)}(\tau) \equiv \overline{\Omega_{m,h}^{(2)}(\tau;s)} \quad (\tau \in (0,\sigma), -s \le h \le s, m \ge -s). \tag{17}$$

In Chapter 7 it was proved that, when $\nu \in \Delta_s(2^0)$, the systems of vector functions

$$\{\kappa_n(\tau)\}_{-s}^{+\infty} \quad \text{and} \quad \{\omega_m^{(2)}(\tau)\}_{-s}^{+\infty} \tag{18}$$

are biorthogonal in $L_2^{2s+1}(0,\sigma)$, i.e.,

$$\{\kappa_n;\omega_m^{(2)}\} = \sum_{h=-s}^{s} \int_0^{\sigma} \kappa_{n,h}(\tau)\overline{\omega_{m,h}^{(2)}(\tau)}d\tau = \delta_{n,m} \quad (-s \le n, m < +\infty). \tag{19}$$

(e) Since $\kappa_{-s} = 2(\mu-1)$ by (2), the assertions of Theorem 7.4-4 can be formulated also in the following way:

1°. *If $\nu \in \Delta_s(1^0)$, then each of the biorthogonal systems of vector functions*

$$\left\{(1+|n|)^{\mu-1}\kappa_n(\tau)\right\}_{-(s-1)}^{+\infty} \quad and \quad \left\{(1+|m|)^{1-\mu}\omega_m^{(1)}(\tau)\right\}_{-(s-1)}^{+\infty} \tag{20}$$

is a Riesz basis of $L_2^{2s+1}(0,\sigma)$.

$2°$. *If $\nu \in \Delta_s(2^0)$, then each of the biorthogonal systems of vector functions*

$$\left\{(1+|n|)^{\mu-1}\kappa_n(\tau)\right\}_{-s}^{+\infty} \quad and \quad \left\{(1+|m|)^{1-\mu}\omega_m^{(2)}(\tau)\right\}_{-s}^{+\infty} \tag{21}$$

is a Riesz basis of $L_2^{2s+1}(0,\sigma)$.

(f) Remember, that we had defined the following sum of segments of the Riemann surface G^∞:

$$\gamma_{2s+1}(\sigma) = \bigcup_{h=-s}^{s} [0,\zeta_{\sigma,h}]$$

$$= \bigcup_{h=-s}^{s} \left\{\tau\alpha_s^{(h+1/2)(s+1/2)} = \tau e^{i\pi(h+1/2)} : \tau \in [0,\sigma]\right\},$$

where $\zeta_{\sigma,h} = \sigma\alpha_s^{(h+1/2)(s+1/2)} = \sigma e^{i\pi(h+1/2)}$ $(-s \le h \le s)$. Now we introduce the Hilbert space $L_2\{\gamma_{2s+1}(\sigma)\}$ of functions $\Phi(\zeta)$ having the finite norms

$$\|\Phi\| \equiv \left\{\int_{\gamma_{2s+1}(\sigma)} |\Phi(\zeta)|^2|d\zeta|\right\}^{1/2} = \left\{\sum_{h=-s}^{s}\int_0^{\zeta_{\sigma,h}} |\Phi(\zeta)|^2|d\zeta|\right\}^{1/2}$$

$$= \left\{\sum_{h=-s}^{s}\int_0^{\sigma} \left|\Phi\left(\tau\alpha_s^{(h+1/2)(s+1/2)}\right)\right|^2 d\tau\right\}^{1/2}. \tag{22}$$

Then, evidently, the inner product of any two functions $\Phi(\zeta)$ and $\Psi(\zeta)$ of $L_2\{\gamma_{2s+1}(\sigma)\}$ is

$$[\Phi, \Psi] \equiv \int_{\gamma_{2s+1}(\sigma)} \Phi(\zeta)\overline{\Phi(\zeta)}|d\zeta| \tag{23}$$

and

$$\|\Phi\| = [\Phi, \Phi]^{1/2}. \tag{24}$$

(g) Remember that, assuming $s \ge 1$ is an integer, $0 < \mu, \mu^* < +\infty$ and $\nu \in [0,2)$, we had introduced by formulas $(42_1) - (42_2)$ and $(43_1) - (43_3)$ of Section 11.3(f) the systems of functions

$$\{\mathcal{Y}_{\nu,n}(\zeta)\}_{-\infty}^{+\infty}, \quad \{\mathcal{Z}_{\nu,n}^{(1)}(\zeta)\}_{-(s-1)}^{+\infty}, \quad \{\mathcal{Z}_{\nu,n}^{(2)}(\zeta)\}_{-s}^{+\infty}. \tag{25}$$

It appears that functions of these systems, if considered only on $\gamma_{2s+1}(\sigma)$, are closely connected with vector functions of the systems

$$\{\kappa_n(\tau)\}_{-\infty}^{+\infty}, \quad \{\Omega_m^{(1)}(\tau;s)\}_{-(s-1)}^{+\infty}, \quad \{\Omega_m^{(2)}(\tau;s)\}_{-s}^{+\infty} \tag{26}$$

which were considered above. Namely, the following equalities may easily be verified.

1) Let $1/2 < \mu < 1/2 + 1/(s + 1/2)$ and $\nu \in [0, 2)$. Then for any $\tau \in (0, \sigma)$, $h(-s \le h \le s)$ and $n(-\infty < n < +\infty)$

$$\mathcal{Y}_{\nu,n}(\zeta)\big|_{\zeta \in [0, \zeta_{\sigma,h}]} = \mathcal{Y}_{\nu,n}\left(\tau \alpha_s^{(h+1/2)(s+1/2)}\right)$$

$$= \alpha_s^{(\mu-1)(h+1/2)(s+1/2)} \kappa_{n,h}(\tau). \tag{27}$$

2) Let $1/2 < \mu < 1/2 + 1/(s + 1/2)$, $\nu \in \Delta_s(1^0)$, and also let μ^* be determined by the relation $\mu + \mu^* = \nu + 3/(2s + 1)$. Then $1/2 + 1/2(s + 1/2) < \mu^* < 1/2 + 1/(s + 1/2)$ and, in addition, $\eta_{2,0} = \mu^*$ as it follows from (6). Therefore, for any $\tau \in (0, \sigma)$, $h(-s \le h \le s)$ and $m(-(s - 1) \le m < +\infty)$

$$\mathcal{Z}_{\nu,m}^{(1)}(\zeta)\Big|_{\zeta \in [0, \zeta_{\sigma,h}]} = \mathcal{Z}_{\nu,m}^{(1)}\left(\tau \alpha_s^{(h+1/2)(s+1/2)}\right)$$

$$= \alpha_s^{(\mu^*-1)(h+1/2)(s+1/2)} \alpha_s^{(h+1/2)(s-1)} C_{s,\nu,m}^{(1)} \Omega_{m,h}^{(1)}(\tau; s), \tag{28_1}$$

where

$$C_{s,\nu,m}^{(1)} = \begin{cases} (2s + 1)\sigma^\nu \Gamma(s + m)\Gamma(1 - m)\Gamma^{-1}(1 + \nu) & \text{when } -(s - 1) \le m \le 0, \\ (2s + 1)\sigma^\nu \mu_m^s \mathcal{E}'_{s+1/2,\sigma}(\mu_m; \nu) & \text{when } 1 \le m < +\infty. \end{cases} \tag{28_2}$$

3) Let $1/2 < \mu < 1/2 + 1/(s + 1/2)$, $\nu \in \Delta_s(2^0)$ and also let μ^* be determined by the relation $\mu + \mu^* = \nu + 1/(2s + 1)$. Then $1/2 < \mu^* < 1/2 + 1/2(s + 1/2)$ and, by (13), $\eta_{2,1} = \mu^*$. Therefore, for any $\tau \in (0, \sigma)$, $h(-s \le h \le s)$ and $m(-s \le m < +\infty)$

$$\mathcal{Z}_{\nu,m}^{(2)}(\zeta)\Big|_{\zeta \in [0, \zeta_{\sigma,h}]} = \mathcal{Z}_{\nu,m}^{(2)}\left(\tau \alpha_s^{(h+1/2)(s+1/2)}\right)$$

$$= \alpha_s^{(\mu^*-1)(h+1/2)(s+1/2)} \alpha_s^{(h+1/2)s} C_{s,\nu,m}^{(2)} \Omega_{m,h}^{(2)}(\tau; s), \tag{29_1}$$

where

$$C_{s,\nu,m}^{(2)} = \begin{cases} (2s + 1)\sigma^\nu \Gamma(1 + s + m)\Gamma(1 - m)\Gamma^{-1}(1 + \nu), & -s \le m \le 0, \\ (2s + 1)\sigma^\nu \mu_m^{s+1} \mathcal{E}'_{s+1/2,\sigma}(\mu_m; \nu), & 1 \le m < +\infty. \end{cases} \tag{29_2}$$

To prove the next lemma, we have to introduce two new systems of functions:

$$\left\{\tilde{\mathcal{Z}}_{\nu,m}^{(1)}(\zeta)\right\}_{-(s-1)}^{+\infty} \quad \text{and} \quad \left\{\tilde{\mathcal{Z}}_{\nu,m}^{(2)}(\zeta)\right\}_{-s}^{+\infty}, \tag{30_1}$$

where

$$\tilde{\mathcal{Z}}_{\nu,m}^{(j)}(\zeta) = \overline{\left(\frac{\zeta}{|\zeta|}\right)^{1-\nu} \frac{\mathcal{Z}_{\nu,m}^{(j)}(\zeta)}{C_{s,\nu,m}^{(j)}}} \qquad (j = 1, 2; -(s + j - 2) \le m < +\infty). \tag{30_2}$$

For any $\zeta \in [0, \zeta_{\sigma,h}]$ $(-s \le h \le s)$ and $m(-(s + j - 2) \le m < +\infty)$ these equalities can obviously also be written in the form

$$\tilde{\mathcal{Z}}_{\nu,m}^{(j)}(\zeta) = \alpha_s^{(1-\nu)(h+1/2)(s+1/2)} \overline{\frac{\mathcal{Z}_{\nu,m}^{(j)}(\zeta)}{C_{s,\nu,m}^{(j)}}} \qquad (j = 1, 2). \tag{31}$$

Lemma 11.4-1. 1°. *If* $1/2 < \mu < 1/2 + 1/(s + 1/2)$, $\nu \in \Delta_s(1^0)$ *and* $\mu^* = \nu + 3/(2s + 1) - \mu$, *then the systems of functions*

$$\{\mathcal{Y}_{\nu,n}(\zeta)\}^{+\infty}_{-(s-1)} \quad \text{and} \quad \left\{\tilde{\mathcal{Z}}^{(1)}_{\nu,m}(\zeta)\right\}^{+\infty}_{-(s-1)} \tag{32_1}$$

are biorthogonal in $L_2\{\gamma_{2s+1}(\sigma)\}$, *i.e.,*

$$[\mathcal{Y}_{\nu,n}; \tilde{\mathcal{Z}}^{(1)}_{\nu,m}] = \delta_{n,m} \qquad (-(s-1) \leq n, m < +\infty). \tag{33_1}$$

2°. *If* $1/2 < \mu < 1/2 + 1/(s + 1/2)$, $\nu \in \Delta_s(2^0)$ *and* $\mu^* = \nu + 1/(2s + 1) - \mu$, *then the systems of functions*

$$\{\mathcal{Y}_{\nu,n}(\zeta)\}^{+\infty}_{-s} \quad \text{and} \quad \left\{\tilde{\mathcal{Z}}^{(2)}_{\nu,m}(\zeta)\right\}^{+\infty}_{-s} \tag{32_2}$$

are biothogonal in $L_2\{\gamma_{2s+1}(\sigma)\}$, *i.e.*

$$[\mathcal{Y}_{\nu,n}; \tilde{\mathcal{Z}}^{(2)}_{\nu,m}] = \delta_{n,m} \qquad (-s \leq n, m < +\infty). \tag{33_2}$$

Proof. If $j = 1, 2$ and $-(s + j - 2) \leq n, m < +\infty$, then obviously

$$[\mathcal{Y}_{\nu,n}; \tilde{\mathcal{Z}}^{(j)}_{\nu,m}] = \sum_{h=-s}^{s} \int_0^{\zeta_{\sigma,h}} \mathcal{Y}_{\nu,n}(\zeta)\overline{\tilde{\mathcal{Z}}^{(j)}_{\nu,m}(\zeta)}|d\zeta|$$

$$= \frac{1}{C^{(j)}_{s,\nu,m}} \sum_{h=-s}^{s} \alpha_s^{(1-\nu)(h+1/2)(s+1/2)} \int_0^{\zeta_{\sigma,h}} \mathcal{Y}_{\nu,n}(\zeta)\mathcal{Z}^{(j)}_{\nu,m}(\zeta)|d\zeta|. \tag{34}$$

If we use formulas (27), (28_1), (29_1) and take into account the simple relation

$$(1 - \nu)(h + 1/2)(s + 1/2) + (\mu - 1)(h + 1/2)(s + 1/2)$$
$$+ (\mu^* - 1)(h + 1/2)(s + 1/2) + (h + 1/2)(s + j - 2) = 0 \quad (j = 1, 2), \tag{35}$$

then it will follow from (34) that

$$[\mathcal{Y}_{\nu,n}; \tilde{\mathcal{Z}}^{(j)}_{\nu,m}] = \sum_{h=-s}^{s} \int_0^{\sigma} \kappa_{n,h}(\tau)\Omega^{(j)}_{m,h}(\tau; s)d\tau$$

$$= \sum_{h=-s}^{s} \int_0^{\sigma} \kappa_{n,h}(\tau)\overline{\omega^{(j)}_{m,h}(\tau)}d\tau = \{\kappa_n; \omega^{(j)}_m\} = \delta_{n,m}.$$

(h) Finally, we move to the proof of the first main theorem of this chapter.

Theorem 11.4-1. 1°. *Let* $1/2 < \mu < 1/2 + 1/(s+1/2)$, $\nu \in \Delta_s(1^0)$, *and also let* $\mu^* = \nu + 3/(2s+1) - \mu$. *Then any function* $\Phi(\zeta) \in L_2\{\gamma_{2s+1}(\sigma)\}$ *can be expanded in the series*

$$\Phi(\zeta) = \sum_{n=-(s-1)}^{+\infty} [\Phi; \tilde{\mathcal{Z}}_{\nu,n}^{(1)}]\mathcal{Y}_{\nu,n}(\zeta), \qquad \zeta \in \gamma_{2s+1}(\sigma), \tag{36}$$

which is convergent in the norm of $L_2\{\gamma_{2s+1}(\sigma)\}$. *Besides,*

$$\|\Phi\|^2 \asymp \sum_{n=-(s-1)}^{+\infty} \left|[\Phi; \tilde{\mathcal{Z}}_{\nu,n}^{(1)}]\right|^2 (1 + |n|)^{2(1-\mu)}. \tag{37}$$

2°. *Let* $1/2 < \mu < 1/2 + 1/(s+1/2)$, $\nu \in \Delta_s(2^0)$, *and also let* $\mu^* = \nu + 1/(2s+1) - \mu$. *Then any function* $\Phi(\zeta) \in L_2\{\gamma_{2s+1}(\sigma)\}$ *can be expanded in the series*

$$\Phi(\zeta) = \sum_{n=-s}^{+\infty} [\Phi; \tilde{\mathcal{Z}}_{\nu,n}^{(2)}]\mathcal{Y}_{\nu,n}(\zeta), \qquad \zeta \in \gamma_{2s+1}(\sigma), \tag{38}$$

which is convergent in the norm of $L_2\{\gamma_{2s+1}(\sigma)\}$. *Besides,*

$$\|\Phi\|^2 \asymp \sum_{n=-s}^{+\infty} \left|[\Phi; \tilde{\mathcal{Z}}_{\nu,n}^{(2)}]\right|^2 (1 + |n|)^{2(1-\mu)}. \tag{39}$$

Proof. We shall prove assertions 1° and 2° simultaneously.
Let $\Phi(\zeta) \in L_2\{\gamma_{2s+1}(\sigma)\}$ be any function. We associate with it the vector function

$$\varphi(\tau) = \{\varphi_h(\tau)\}_{h=-s}^s \in L_2^{2s+1}(0,\sigma), \tag{40}$$

where

$$\varphi_h(\tau) = \alpha_s^{(1-\mu)(h+1/2)(s+1/2)} \Phi_h(\tau), \tau \in (0,\sigma), \tag{41_1}$$

and

$$\Phi_h(\tau) = \Phi(\zeta)|_{\zeta \in [0, \zeta_{\sigma,h}]} = \Phi\left(\tau \alpha_s^{(h+1/2)(s+1/2)}\right), \qquad \tau \in (0,\sigma). \tag{41_2}$$

Then, obviously,

$$\|\Phi\| = \|\varphi\|, \tag{42}$$

i.e., the norm of Φ in $L_2\{\gamma_{2s+1}(\sigma)\}$ is equal to the norm of φ in $L_2^{2s+1}(0,\sigma)$. On the other hand, the equalities

$$[\Phi; \tilde{\mathcal{Z}}_{\nu,m}^{(j)}] = \{\varphi; \omega_m^{(j)}\} \quad (j = 1, 2; -(s+j-2) \leq m < +\infty) \tag{43}$$

follow from formulas (31), (28_1), (29_1) and relations (35), $(41_1) - (41_2)$. Further, the systems of vector functions (20) and (21) are Riesz bases of $L_2^{2s+1}(\sigma)$ when suitable conditions are satisfied (see 11.4(e)). Hence the expansions

$$
\begin{aligned}
\varphi_h(\tau) &= \sum_{n=-(s+j-2)}^{+\infty} \left\{ \varphi; (1+|n|)^{1-\mu}\omega_n^{(j)} \right\} (1+|n|)^{\mu-1}\kappa_{n,h}(\tau) \\
&= \sum_{n=-(s+j-2)}^{+\infty} \left\{ \varphi; \omega_n^{(j)} \right\} \kappa_{n,h}(\tau) \qquad (\tau \in (0,\sigma), -s \le h \le s)
\end{aligned}
\tag{44}
$$

and the two-sided estimate

$$
\|\varphi\|^2 \asymp \sum_{n=-(s+j-2)}^{+\infty} \left| \left\{ \varphi; \omega_n^{(j)} \right\} \right|^2 (1+|n|)^{2(1-\mu)}
\tag{45}
$$

are true for any $j = 1, 2$. Using relations (27), $(41_1) - (41_2)$ and (43) we transform (44) into the following expansions in spaces $L_2[0, \zeta_{\sigma,h}]$ $(-s \le h \le s)$:

$$
\Phi(\zeta)|_{\zeta \in [0,\zeta_{\sigma,h}]} = \sum_{n=-(s+j-2)}^{+\infty} [\Phi; \tilde{\mathcal{Z}}_{\nu,n}^{(j)}]\mathcal{Y}_{\nu,n}(\zeta)\Bigg|_{\zeta \in [0,\zeta_{\sigma,h}]} \qquad (j = 1, 2).
\tag{46}
$$

It remains to observe that these expansions coincide with the desired expansions (36) and (38). As to estimates (37) and (39), they follow from (42), (43) and (45).

Remark. If stated otherwise, the preceding theorem says: if μ, μ^* and ν satisfy the suitable conditions, then for any $j = 1, 2$ the systems of functions

$$
\left\{ (1+|n|)^{\mu-1}\mathcal{Y}_{\nu,n}(\zeta) \right\}_{-(s+j-2)}^{+\infty} \quad \text{and} \quad \left\{ (1+|m|)^{1-\mu}\tilde{\mathcal{Z}}_{\nu,m}^{(j)}(\zeta) \right\}_{-(s+j-2)}^{+\infty}
\tag{47}
$$

are biorthogonal Riesz bases of $L_2\{\gamma_{2s+1}(\sigma)\}$.

(i) The following theorem, relating to expansions in terms of eigenfunctions of the boundary value problems $I_{s+1/2}$ and $II_{s+1/2}$ $(s \ge 1)$, is an immediate consequence of Theorem 11.4-1.

Theorem 11.4-2. *Let $s \ge 1$ be an integer and let μ satisfy the condition*

$$
\begin{aligned}
&2/3 \le \mu \le 1 \quad \text{when} \quad s = 1, \\
&1/2 < \mu < 1/2 + 1/(s+1/2) \quad \text{when} \quad s \ge 2.
\end{aligned}
$$

Then the following assertions are true:

1°. If $\nu \in \Delta_s(1^0)$ and $\mu^* = \nu + 3\,(2s+1) - \mu$, then any function $\Phi(\zeta) \in L_2\{\gamma_{2s+1}(\sigma)\}$ can be expanded in terms of the sequence $\{\mathcal{Y}_{\nu,n}(\zeta)\}_{-(s-1)}^{+\infty}$ of eigenfunctions and adjoint functions of the boundary value problem $I_{s+1/2}$:

$$\Phi(\zeta) = \sum_{n=-(s-1)}^{+\infty} [\Phi; \tilde{\mathcal{Z}}_{\nu,n}^{(1)}]\mathcal{Y}_{\nu,n}(\zeta), \qquad \zeta \in \gamma_{2s+1}(\sigma). \tag{48_1}$$

In addition,

$$\|\Phi\|^2 \asymp \sum_{n=-(s-1)}^{+\infty} \left|[\Phi; \tilde{\mathcal{Z}}_{\nu,n}^{(1)}]\right|^2 (1 + |n|)^{2(1-\mu)}. \tag{49_1}$$

2°. If $\nu \in \Delta_s(2^0)$ and $\mu^* = \nu + 1/(2s + 1) - \mu$, then any function $\Phi(\zeta) \in L_2\{\gamma_{2s+1}(\sigma)\}$ can be expanded in terms of the sequence $\{\mathcal{Y}_{\nu,n}(\zeta)\}_{-s}^{+\infty}$ of eigenfunctions and adjoint functions of the boundary value problem $II_{s+1/2}$:

$$\Phi(\zeta) = \sum_{n=-s}^{+\infty} [\Phi; \tilde{\mathcal{Z}}_{\nu,n}^{(2)}]\mathcal{Y}_{\nu,n}(\zeta), \qquad \zeta \in \gamma_{2s+1}(\sigma). \tag{48_2}$$

In addition,

$$\|\Phi\|^2 \asymp \sum_{n=-s}^{+\infty} \left|[\Phi; \tilde{\mathcal{Z}}_{\nu,n}^{(2)}]\right|^2 (1 + |n|)^{2(1-\mu)}. \tag{49_2}$$

(j) As was mentioned in Section 11.2, $w = \zeta^{1/(s+1/2)}$ maps the sum of segments $\gamma_{2s+1}(\sigma) \subset G^\infty$ onto the sum of segments $\Gamma_{2s+1}(\sigma)$ of the length $\sigma^{1/(s+1/2)}$, situated in the w-plane and having a common endpoint at $w = 0$. These segments are at angles $\pi(h + 1/2)/(s + 1/2)$ $(-s \leq h \leq s)$, so they form pairwise equal angles of the opening $\pi/(s + 1/2)$. Further, we assume $\omega \in (-1,1)$ and $\mu = (3/2 + s + \omega)/(2s + 1)$, as everywhere. We denote by $L_{2,\omega}\{\Gamma_{2s+1}(\sigma)\}$ the space of functions $\Psi(\omega)$ measurable on $\Gamma_{2s+1}(\sigma)$ and having finite norms

$$\|\Psi\|_\omega \equiv \left\{ \int_{\Gamma_{2s+1}(\sigma)} |\Psi(w)|^2 |w|^\omega |dw| \right\}^{1/2}. \tag{50}$$

In addition, we define the inner product of any two functions Ψ_1, Ψ_2 of $L_2\{\Gamma_{2s+1}(\sigma)\}$ in the following way:

$$[\Psi_1; \Psi_2]_\omega \equiv \int_{\Gamma_{2s+1}(\sigma)} \Psi_1(w)\overline{\Psi_2(w)}|w|^\omega |dw|. \tag{51}$$

Now we can observe that

$$\Psi(w) \equiv \sqrt{s + 1/2}\,\Phi\left(w^{s+1/2}\right) w^{(s+1/2)(1-\mu)} \tag{52}$$

is a transformation of a function $\Phi(\zeta)$ defined on $\gamma_{2s+1}(\sigma)$ into a function $\Psi(w)$ defined on $\Gamma_{2s+1}(\sigma)$. Moreover, a simple calculation shows that (52) represents an isometric isomorphism between the Hilbert spaces $L_2\{\gamma_{2s+1}(\sigma)\}$ and $L_{2,\omega}\{\Gamma_{2s+1}(\sigma)\}$, so it does not change the norms or the inner products. Using (52) one can easily move from the main Theorem 11.4-1 to the following theorem relating to the expansions of functions of $L_{2,\omega}\{\Gamma_{2s+1}(\sigma)\}$ in terms of definite systems of entire functions.

Theorem 11.4-3. *Let $s \geq 1$ be an integer, let $-1 < \omega < 1$ and let the parameter $\mu \in (1/2, 1/2 + 1/(s + 1/2))$ be defined by the equality*

$$\mu = \frac{3/2 + s + \omega}{2s + 1}.$$

Then the following assertions are true:
$1°$. If $\nu \in \Delta_s(1^0)$ and $\mu^* = \nu + 3/(2s + 1) - \mu$, then any function $\Psi(w) \in L_{2,\omega}\{\Gamma_{2s+1}(\sigma)\}$ can be expanded in the series

$$\Psi(w) = \sum_{n=-(s-1)}^{0} C_n^{(1)} \frac{\Gamma(1 - n)}{\Gamma(\mu - n/(s + 1/2))} w^{-n}$$

$$+ \sum_{n=1}^{+\infty} C_n^{(1)} E_{s+1/2}\left(\mu_n w; \mu\right), \ w \in \Gamma_{2s+1}(\sigma),$$

(53_1)

which converges in the norm of $L_{2,\omega}\{\Gamma_{2s+1}(\sigma)\}$, and

$$\|\Psi\|_\omega^2 \asymp \sum_{n=-(s-1)}^{+\infty} |C_n^{(1)}|^2 (1 + |n|)^{2(1-\mu)}. \tag{54_1}$$

Besides, the coefficients of expansion (53_1) are defined by the formula

$$C_n^{(1)} = \left[\Psi(w); (s + 1/2)\tilde{Z}_{\nu,n}^{(1)}\left(w^{s+1/2}\right) w^{(s+1/2)(1-\mu)}\right]_\omega \ (-(s - 1) \leq n < +\infty).$$

(55_1)

$2°$. If $\nu \in \Delta_s(2^0)$ and $\mu^* = \nu + 1/(2s + 1) - \mu$, then any function $\Psi(w) \in L_{2,\omega}\{\Gamma_{2s+1}(\sigma)\}$ can be expanded in the series

$$\Psi(w) = \sum_{n=-s}^{0} C_n^{(2)} \frac{\Gamma(1 - n)}{\Gamma(\mu - n/(s + 1/2))} w^{-n}$$

$$+ \sum_{n=1}^{+\infty} C_n^{(2)} E_{s+1/2}\left(\mu_n w; \mu\right), \quad w \in \Gamma_{2s+1}(\sigma),$$

(53_2)

which converges in the norm of $L_{2,\omega}\{\Gamma_{2s+1}(\sigma)\}$, and

$$\|\Psi\|_\omega^2 \asymp \sum_{n=-s}^{+\infty} |C_n^{(2)}|^2 (1+|n|)^{2(1-\mu)}. \tag{54_2}$$

Besides, the coefficients of the expansion (53_2) are defined by the formula

$$C_n^{(2)} = \left[\Psi(w); (s+1/2)\tilde{\mathcal{Z}}_{\nu,n}^{(2)}\left(w^{s+1/2}\right) w^{(s+1/2)(1-\mu)}\right]_\omega \quad (-s \leq n < +\infty). \tag{55_2}$$

Remark. The expansions of the preceding theorem essentially generalize the classical sin- and cos-expansions in $[0,\sigma]$ by extending them to the case when there are an arbitrary odd number of segments in the complex plane.

11.5 Notes

The results of this chapter in somewhat different forms were published in the papers of M.M. Djrbashian [10,11] where the simplest case $s = 1$ was considered.

12 Cauchy type problems and boundary value problems in the complex domain (the case of even segments)

12.1 Introduction

The results of this chapter are similar to those obtained in Chapter 11, but the problems considered here are essentially different. Namely, the Cauchy type problems considered here are formulated in terms of another pair of associated integrodifferential operators — $\mathbb{L}_s$ and $\mathbb{L}_s^*$ (where $s \geq 1$ is any integer), and the corresponding boundary conditions are assumed to be satisfied at the endpoints of the sum of even $(2s)$ segments

$$\gamma_{2s}(\sigma) = \bigcup_{h=0}^{2s-1} \{z = r \exp[i\pi(h + 1/2)] : 0 \leq r \leq \sigma\}$$

in the Riemann surface G^∞ of Lnz. Using the results of Chapters 8 and 9 we prove the main Theorem 12.4-1 on the basis property of certain systems of functions in $L_2\{\gamma_{2s}(\sigma)\}$. The consequence of this theorem is the important Theorem 12.4-2 on expansions of functions of $L_2\{\gamma_{2s}(\sigma)\}$ in terms of the systems of eigenfunctions and adjoint functions of the first of two boundary value problems considered here. Another consequence of Theorem 12.4-1 is Theorem 12.4-3 containing the construction of some systems of entire functions which are bases of weighted spaces L_2 over the sum of segments

$$\Gamma_{2s}(\sigma) = \bigcup_{h=0}^{2s-1} \left\{z = r \exp[i\pi(h + 1/2)/s] : 0 \leq r \leq \sigma^{1/s}\right\} \subset \mathbb{C}.$$

These systems of entire functions are similar to the Fourier system $\{e^{inz}\}_{-\infty}^{+\infty}$ on $[-\sigma, \sigma]$.

12.2 Preliminaries

(a) Theorem 2.4-2 on the parametric representation of classes $W_{s,\sigma}^{2,\omega}$ (where $s \geq 1$ is an integer and $-1 < \omega < 1$) of entire functions of order $\rho = s$ and of type $\leq \sigma$ contains the parameter

$$\mu = \frac{1 + s + \omega}{2s} \in \left(\frac{1}{2}, \frac{1}{2} + \frac{1}{s}\right) = \left(\frac{1}{2}, \frac{1}{2} + \frac{1}{\rho}\right) \tag{1}$$

which is used in this chapter. Here we also use definition 9.2(8) of the interval $\Delta_s \subset [0, 2)$ of variation of the parameter ν. If we take (1) into account, then this interval can be written in the form

$$\Delta_s = \left(\mu + \frac{1}{2} - \frac{1}{s}, \mu + \frac{1}{2}\right) \qquad (s \geq 1). \tag{2}$$

(b) According to the notations of Sections 8.2 and 9.2, we put

$$\beta_s = \exp\{i\pi/s\} \qquad (s \geq 1). \tag{3}$$

Further, assuming $\sigma \in (0, +\infty)$, we introduce the notation

$$\zeta_{\sigma,h}^{1/s} = \sigma^{1/s}\beta_s^{h+1/2}(0 \leq h \leq 2s-1), \tag{4}$$

or, in other words,

$$\zeta_{\sigma,h} = \sigma\beta_s^{(h+1/2)s} = \sigma e^{i\pi(h+1/2)} \qquad (0 \leq h \leq 2s-1). \tag{5}$$

Now, on the Riemann surface G^∞ of Lnz, we introduce the following sum of segments of length σ:

$$\gamma_{2s}(\sigma) = \bigcup_{h=0}^{2s-1} [0, \zeta_{\sigma,h}] = \bigcup_{h=0}^{2s-1} \left\{\tau\beta_s^{(h+1/2)s} = \tau e^{i\pi(h+1/2)} : 0 \leq \tau \leq \sigma\right\}. \tag{6}$$

Obviously these segments have a common endpoint at $\zeta = 0$, and any pair of successive segments forms an angle of opening π and, when $s = 1$, we have

$$\gamma_{2s}(\sigma) \equiv \gamma_2(\sigma) = [0, i\sigma] \cup [0, -i\sigma]. \tag{6'}$$

Note also that the mapping

$$w = \zeta^{1/s}, \quad w_{\sigma,h} = \zeta_{\sigma,h}^{1/s} \qquad (0 \leq h \leq 2s-1) \tag{7}$$

transforms $\gamma_{2s}(\sigma)$ into the following sum of $2s$ segments of the length $\sigma^{1/s}$:

$$\Gamma_{2s}(\sigma) = \bigcup_{h=0}^{2s-1} [0, w_{\sigma,h}] = \bigcup_{h=0}^{2s-1} \left\{w = r\beta_s^{h+1/2} : 0 \leq r \leq \sigma^{1/s}\right\}. \tag{8}$$

These new segments are all in the w-plane where they form a system of angles of opening π/s with a common vertex at the origin.

(c) The results of Chapters 8 and 9 were obtained using the distribution of zeros of the entire function

$$\mathcal{E}_{s,\sigma}(z; \nu) \equiv E_{1/2}\left(-\sigma^2 z^{2s}; 1 + \nu\right) \qquad (s \geq 1 \text{ is an integer}). \tag{9}$$

Remember that, if $\nu \in [0, 2)$, then the zeros $\{\mu_n\}_1^\infty$ $(0 < |\mu_n| \leq |\mu_{n+1}|, n \geq 1)$ of this function are simple and are situated on the sum of rays

$$\Gamma_{2s} = \bigcup_{h=0}^{2s-1} \left\{z = r\exp\left(i\frac{\pi h}{s}\right) : 0 \leq r < +\infty\right\}. \tag{10}$$

Besides, we gave an explicit order of numeration of these zeros earlier.

(**(d)** Using notation (4) one can easily write the identity 9.2(6_1) in the form

$$2s\sigma z^s E_{1/2}\left(-\sigma^2 z^{2s}; 1 + \nu\right) = \sum_{h=0}^{2s-1} \beta_s^{-(h+1/2)s} E_s\left(z\zeta_{\sigma,h}^{1/s}; \nu\right) \qquad (s \geq 1). \tag{11}$$

12.3 Cauchy type problems and boundary value problems containing the operators $\mathbb{L}_s$ and $\mathbb{L}_s^*$.

In this section we use the operators $D^{-\alpha}, D^\alpha$ and $D_{\zeta_0}^{-\alpha}, D_{\zeta_0}^\alpha$ $(0 \le \alpha < +\infty, \zeta_0 \in G^\infty)$ defined on the Riemann surface G^∞ in the beginning of Section 11.3. Besides, here, as in Section 11.3, we do some calculations on the base of the simple formulas 11.3(4)-(5) which are true for arbitrary points $\zeta_\sigma = \sigma e^{i\varphi} \in G^\infty$ $(0 < \sigma < +\infty, -\infty < \varphi < +\infty)$. Henceforth we assume $\sigma \in (0, +\infty)$ to be a preassigned number.

(a) First we introduce some notations. Assuming $s \ge 1$ is an integer, $0 < \mu, \mu^* < +\infty$, and that λ, λ^* are arbitrary complex numbers, we put

$$\mathcal{Y}_{s,\mu}(\zeta; \lambda) \equiv E_s\left(\lambda \zeta^{1/s}; \mu\right) \zeta^{\mu-1}, \qquad \zeta \in (0, \zeta_\sigma), \tag{1}$$

$$\mathcal{Z}_{s,\mu^*}(\zeta; \lambda^*) \equiv E_s\left(\lambda^*(\zeta_\sigma - \zeta)^{1/s}; \mu^*\right)(\zeta_\sigma - \zeta)^{\mu^*-1}, \qquad \zeta \in (0, \zeta_\sigma) \tag{2}$$

and observe that these functions are defined in the domain $G_\sigma^\infty = \{\zeta \in G^\infty : |\zeta| < \sigma\}$, since $\zeta_\sigma \in G^\infty$ is not a fixed point.

(b) Now assume that the parameters μ and μ^* satisfy the conditions

$$\begin{aligned}
\mu = \mu^* = 1 \quad &\text{when} \quad s = 1, \\
1/2 < \mu, \mu^* < 1/2 + 1/s \quad &\text{when} \quad s \ge 2.
\end{aligned} \tag{3}$$

Then, as in Section 11.3, denote by $AC[0, \zeta_\sigma]$ (where $\zeta_\sigma \in G^\infty$ is an arbitrary point) the class of functions absolutely continuous in $[0, \zeta_\sigma]$. Further, the functions $y(\zeta)$ and $z(\zeta)$ are said to be of classes $AC_\mu[0, \zeta_\sigma]$ and $AC_{\mu^*}^*[0, \zeta_\sigma]$ respectively, if they satisfy the following conditions:

$$\text{(i)} \quad y(\zeta) \in L_1(0, \zeta_\sigma) \text{ and } z(\zeta) \in L_1(0, \zeta_\sigma), \tag{4}$$

$$\text{(ii)} \quad \begin{aligned} L_0 y(\zeta) &\equiv D^{-(1-\mu)} y(\zeta) \in AC[0, \zeta_\sigma], \\ L_0^* z(\zeta) &\equiv D_{\zeta_\sigma}^{-(1-\mu^*)} z(\zeta) \in AC[0, \zeta_\sigma]. \end{aligned} \tag{5}$$

Now introduce the following operators respectively in $AC_\mu[0, \zeta_\sigma]$ and $AC_{\mu^*}^*[0, \zeta_\sigma]$:

$$\mathbb{L}_s y(\zeta) \equiv D^{-(\mu-1/s)}\left\{\frac{d}{d\zeta} L_0 y(\zeta)\right\} = D^{-(\mu-1/s)} D^\mu y(\zeta), \quad \zeta \in (0, \zeta_\sigma), \tag{6}$$

$$\mathbb{L}_s^* z(\zeta) \equiv D_{\zeta_\sigma}^{-(\mu^*-1/s)}\left\{\frac{d}{d(\zeta_\sigma - \zeta)} L_0^* z(\zeta)\right\} = D_{\zeta_\sigma}^{-(\mu^*-1/s)} D_{\zeta_\sigma}^{\mu^*} z(\zeta), \zeta \in (0, \zeta_\sigma). \tag{7}$$

Observe that the functions $\mathbb{L}_s y(\zeta)$ and $\mathbb{L}_s^* z(\zeta)$ are of $L_1(0, \zeta_\sigma)$, since they are fractional integrals of functions summable on $(0, \zeta_\sigma)$. Note also that, if $s = 1$ and consequently $\mu = \mu^* = 1$, then the classes $AC_\mu[0, \zeta_\sigma]$ and $AC_{\mu^*}^*[0, \zeta_\sigma]$ coincide with $AC[0, \zeta_\sigma]$, and the operators $\mathbb{L}_s$ and $\mathbb{L}_s^*$ take the following simple forms:

$$\mathbb{L}_1 y(\zeta) = \frac{d}{d\zeta} y(\zeta), \quad \mathbb{L}_1^* z(\zeta) = \frac{d}{d(\zeta_\sigma - \zeta)} z(\zeta) = -\frac{d}{d\zeta} z(\zeta). \tag{8}$$

Finally, it follows from (5) that the quantities

$$m_0(y) \equiv L_0 y(\zeta)|_{\zeta=0} \quad \text{and} \quad m_0^*(z) \equiv L_0^* z(\zeta)|_{\zeta=\zeta_\sigma} \tag{9}$$

are finite for any integer $s \geq 1$, provided $y(\zeta) \in AC_\mu[0, \zeta_\sigma]$ and $z(\zeta) \in AC_{\mu^*}^*[0, \zeta_\sigma]$.

(c) Henceforth, unless otherwise stated, we assume that $s \geq 1$ is an integer, and the parameters μ and μ^* satisfy the condition (3). The following two lemmas are similar to Lemmas 11.3-1 and 11.3-2.

Lemma 12.3-1. Let $y(\zeta) \in AC_\mu[0, \zeta_\sigma]$ and $z(\zeta) \in AC_{\mu^*}^*[0, \zeta_\sigma]$, where $\zeta_\sigma \in G^\infty$ is a fixed point. Then the following representations are true almost everywhere in the interval $(0, \zeta_\sigma) \subset G^\infty$:

$$1^\circ. \; \mathbb{L}_s y(\zeta) = D^{1/s} y(\zeta) - m_0(y) \frac{\zeta^{\mu-1-1/s}}{\Gamma(\mu - 1/s)} \quad (s \geq 1) \tag{10}$$

$$2^\circ. \; \mathbb{L}_s^* z(\zeta) = D_{\zeta_\sigma}^{1/s} z(\zeta) - m_0^*(z) \frac{(\zeta_\sigma - \zeta)^{\mu^*-1-1/s}}{\Gamma(\mu^* - 1/s)} \quad (s \geq 1). \tag{11}$$

Proof. When $s = 1$, these representations pass to formulas (8). Thus, it remains to prove representations (10) and (11) for the case $s \geq 2$. In this case (3) obviously implies $1/2 < \mu, \mu^* < 1$, $\mu - 1/s > 0$ and $\mu^* - 1/s > 0$. Therefore, using (6) we obtain

$$D^{-1/s} \mathbb{L}_s y(\zeta) = D^{-1/s} D^{-(\mu-1/s)} D^\mu y(\zeta) = D^{-\mu} D^\mu y(\zeta). \tag{12}$$

Further, as in the proof of Lemma 11.3-1, we obtain that

$$D^{-\mu} D^\mu y(\zeta) = y(\zeta) - m_0(y) \frac{\zeta^{\mu-1}}{\Gamma(\mu)} \tag{13}$$

almost everywhere in $(0, \zeta_\sigma)$. Finally, applying the operator $D^{1/s}$ to both sides of (12), and using (13), we easily arrive at representation (10). Representation (11) is proved in a similar way.

Lemma 12.3-2.. 1°. *The Cauchy type problem*

$$\mathbb{L}_s y(\zeta) - \lambda y(\zeta) = 0, \qquad \zeta \in (0, \zeta_\sigma),$$
$$L_\sigma y(\zeta)|_{\zeta=0} = 1 \tag{14}$$

has the unique solution

$$\mathcal{Y}_{s,\mu}(\zeta;\lambda) = E_s\left(\lambda\zeta^{1/s};\mu\right)\zeta^{\mu-1} \in L_2(0,\zeta_\sigma) \tag{15}$$

in the class $AC_\mu[0,\zeta_\sigma]$.

$2°$. *The Cauchy type problem*

$$\mathbb{L}_s^* z(\zeta) - \lambda^* z(\zeta) = 0, \qquad \zeta \in (0,\zeta_\sigma),$$
$$L_\sigma^* z(\zeta)|_{\zeta=\zeta_\sigma} = 1 \tag{16}$$

has the unique solution

$$\mathcal{Z}_{s,\mu^*}(\zeta;\lambda^*) = E_s\left(\lambda^*(\zeta_\sigma - \zeta)^{1/s};\mu^*\right)(\zeta_\sigma - \zeta)^{\mu^*-1} \in L_2(0,\zeta_\sigma) \tag{17}$$

in the class $AC_{\mu^*}^*[0,\zeta_\sigma]$.

Proof. If $s = 1$ and $\mu = \mu^* = 1$, then, according to (8), the Cauchy type problems (14) and (16) become respectively

$$\frac{d}{d\zeta}y(\zeta) - \lambda y(\zeta) = 0, \qquad \zeta \in (0,\zeta_\sigma),$$
$$y(\zeta)|_{\zeta=0} = 1 \tag{14'}$$

and

$$-\frac{d}{d\zeta}z(\zeta) - \lambda^* z(\zeta) = 0, \qquad \zeta \in (0,\zeta_\sigma),$$
$$z(\zeta)|_{\zeta=\zeta_\sigma} = 1. \tag{16'}$$

Besides, one can easily verify that the functions

$$\mathcal{Y}_{s,\mu}(\zeta;\lambda) \equiv \mathcal{Y}_{1,1}(\zeta;\lambda) = \exp\{\lambda\zeta\}, \qquad \zeta \in (0,\zeta_\sigma), \tag{15'}$$

$$\mathcal{Z}_{s,\mu^*}(\zeta;\lambda^*) = \mathcal{Z}_{1,1}(\zeta;\lambda^*) = \exp\{\lambda^*(\zeta_\sigma - \zeta)\}, \qquad \zeta \in (0,\zeta_\sigma) \tag{17'}$$

are indeed the unique solutions of these problems in the class $AC[0,\zeta_\sigma]$. If $s \geq 2$, then the proof is the same as that of Lemma 11.3-2. The only difference is that $s + 1/2$ is replaced by s.

Theorem 12.3-1. *The following formulas are true for the solutions* $\mathcal{Y}_{s,\mu}(\zeta;\lambda)$ *and* $\mathcal{Z}_{s,\mu^*}(\zeta;\lambda^*)$ *of Cauchy type problems (14) and (16):*

$$\int_0^{\zeta_\sigma} \mathcal{Y}_{s,\mu}(\zeta;\lambda)\mathcal{Z}_{s,\mu^*}(\zeta;\lambda^*)d\zeta$$
$$= \frac{E_s\left(\lambda\zeta_\sigma^{1/s};\mu+\mu^*-1/s\right) - E_s\left(\lambda^*\zeta_\sigma^{1/s};\mu+\mu^*-1/s\right)}{\lambda - \lambda^*}\zeta_\sigma^{\mu+\mu^*-1/s-1}. \tag{18}$$

Proof. If $s = 1$ and $\mu = \mu^* = 1$, then using (15′) and (17′) we obtain

$$\int_0^{\zeta_\sigma} \mathcal{Y}_{1,1}(\zeta; \lambda) \mathcal{Z}_{1,1}(\zeta; \lambda^*) d\zeta = \frac{\exp(\lambda \zeta_\sigma) - \exp(\lambda^* \zeta_\sigma)}{\lambda - \lambda^*}, \tag{18′}$$

which obviously coincides with (18). So, it remains to prove (18) for $s \geq 2$. In the proof we use the operations of ordinary and fractional integration by parts.

First note that, by Lemma $12.3 - 2(1°)$ and by definition (6) of the operator $\mathbb{L}_s$,

$$
\begin{aligned}
\lambda \int_0^{\zeta_\sigma} \mathcal{Y}_{s,\mu}(\zeta; \lambda) \mathcal{Z}_{s,\mu^*}(\zeta; \lambda^*) d\zeta &= \int_0^{\zeta_\sigma} \left[\mathbb{L}_s \mathcal{Y}_{s,\mu}(\zeta; \lambda) \right] \mathcal{Z}_{s,\mu^*}(\zeta; \lambda^*) d\zeta \\
&= \int_0^{\zeta_\sigma} \left\{ D^{-(\mu - 1/s)} \left[\frac{d}{d\zeta} L_0 \mathcal{Y}_{s,\mu}(\zeta; \lambda) \right] \right\} \mathcal{Z}_{s,\mu^*}(\zeta; \lambda^*) d\zeta \\
&= \int_0^{\zeta_\sigma} \left[\frac{d}{d\zeta} L_0 \mathcal{Y}_{s,\mu}(\zeta; \lambda) \right] D_{\zeta_\sigma}^{-(\mu - 1/s)} \mathcal{Z}_{s,\mu^*}(\zeta; \lambda^*) d\zeta \\
&= \int_0^{\zeta_\sigma} \left[\frac{d}{d\zeta} L_0 \mathcal{Y}_{s,\mu}(\zeta; \lambda) \right] \mathcal{Z}_{s,\mu+\mu^* - 1/s}(\zeta; \lambda^*) d\zeta.
\end{aligned}
\tag{19}
$$

Next, observe that for any $\zeta \in (0, \zeta_\sigma)$

$$\mathcal{Z}_{s,\mu+\mu^* - 1/s}(\zeta; \lambda^*) \equiv \frac{(\zeta_\sigma - \zeta)^{\mu+\mu^* - 1/s - 1}}{\Gamma(\mu + \mu^* - 1/s)} + \lambda^* \mathcal{Z}_{s,\mu+\mu^*}(\zeta; \lambda^*). \tag{20}$$

Consequently,

$$\lambda \int_0^{\zeta_\sigma} \mathcal{Y}_{s,\mu}(\zeta; \lambda) \mathcal{Z}_{s,\mu^*}(\zeta; \lambda^*) d\zeta = I_1 + I_2, \tag{21}$$

where

$$I_1 = \frac{1}{\Gamma(\mu + \mu^* - 1/s)} \int_0^{\zeta_\sigma} (\zeta_\sigma - \zeta)^{\mu+\mu^* - 1/s - 1} \frac{d}{d\zeta} L_0 \mathcal{Y}_{s,\mu}(\zeta; \lambda) d\zeta, \tag{22}$$

$$I_2 = \lambda^* \int_0^{\zeta_\sigma} \left[\frac{d}{d\zeta} L_0 \mathcal{Y}_{s,\mu}(\zeta; \lambda) \right] \mathcal{Z}_{s,\mu+\mu^*}(\zeta; \lambda^*) d\zeta. \tag{23}$$

To calculate I_1, we use the simple relation

$$
\begin{aligned}
\frac{d}{d\zeta} L_0 \mathcal{Y}_{s,\mu}(\zeta; \lambda) &= \frac{d}{d\zeta} D^{-(1-\mu)} \mathcal{Y}_{s,\mu}(\zeta; \lambda) \\
&= \frac{d}{d\zeta} \mathcal{Y}_{s,1}(\zeta; \lambda) = \lambda \mathcal{Y}_{s,1/s}(\zeta; \lambda).
\end{aligned}
\tag{24}
$$

We obtain

$$
\begin{aligned}
I_1 &= \frac{\lambda}{\Gamma(\mu + \mu^* - 1/s)} \int_0^{\zeta_\sigma} (\zeta_\sigma - \zeta)^{\mu+\mu^*-1/s-1} \mathcal{Y}_{s,1/s}(\zeta;\lambda) d\zeta \\
&= \lambda D^{-(\mu+\mu^*-1/s)} \mathcal{Y}_{s,1/s}(\zeta;\lambda)\Big|_{\zeta=\zeta_\sigma} \\
&= \lambda \mathcal{Y}_{s,\mu+\mu^*}(\zeta;\lambda)|_{\zeta=\zeta_\sigma} = \lambda E_s\left(\lambda \zeta_\sigma^{1/s}; \mu+\mu^*\right) \zeta_\sigma^{\mu+\mu^*-1}.
\end{aligned}
\tag{25}
$$

As to I_2, integration by parts gives

$$
\begin{aligned}
I_2 &= \lambda^* \left[L_0 \mathcal{Y}_{s,\mu}(\zeta;\lambda)\right] \mathcal{Z}_{s,\mu+\mu^*}(\zeta;\lambda^*)\big|_0^{\zeta_\sigma} \\
&\quad + \lambda^* \int_0^{\zeta_\sigma} \left[L_0 \mathcal{Y}_{s,\mu}(\zeta;\lambda)\right] \frac{d}{d(\zeta_\sigma - \zeta)} \mathcal{Z}_{s,\mu+\mu^*}(\zeta;\lambda^*) d\zeta \\
&= \lambda^* \mathcal{Y}_{s,1}(\zeta;\lambda) \mathcal{Z}_{s,\mu+\mu^*}(\zeta;\lambda^*)\big|_0^{\zeta_\sigma} \\
&\quad + \lambda^* \int_0^{\zeta_\sigma} \left[D^{-(1-\mu)} \mathcal{Y}_{s,\mu}(\zeta;\lambda)\right] \mathcal{Z}_{s,\mu+\mu^*-1}(\zeta;\lambda^*) d\zeta \\
&= -\lambda^* \mathcal{Z}_{s,\mu+\mu^*}(\zeta;\lambda^*)|_{\zeta=0} + \lambda^* \int_0^{\zeta_\sigma} \mathcal{Y}_{s,\mu}(\zeta;\lambda) \mathcal{Z}_{s,\mu^*}(\zeta;\lambda^*) d\zeta.
\end{aligned}
\tag{26}
$$

Finally, combining formulas $(21), (25)$ and (26), we obtain

$$
\begin{aligned}
(\lambda - \lambda^*) &\int_0^{\zeta_\sigma} \mathcal{Y}_{s,\mu}(\zeta;\lambda) \mathcal{Z}_{s,\mu^*}(\zeta;\lambda^*) d\zeta \\
&= \lambda E_s\left(\lambda \zeta_\sigma^{1/s}; \mu+\mu^*\right) \zeta_\sigma^{\mu+\mu^*-1} - \lambda^* E_s\left(\lambda^* \zeta_\sigma^{1/s}; \mu+\mu^*\right) \zeta_\sigma^{\mu+\mu^*-1}.
\end{aligned}
\tag{27}
$$

It remains to observe that the last formula coincides with (18).

Remark. Identity (18) can be directly derived also from formula 1.2(10), if we assume $\rho = s, \alpha = \mu$ and $\beta = \mu^*$.

(d) Now we shall formulate two boundary value problems assuming again that $s \geq 1$ is any integer, and that the parameters μ and μ^* satisfy conditions (3). We assume also $\nu \in [0, 2)$.

Problem I_s *consists in finding those values of λ for which the solution $\mathcal{Y}_{s,\mu}(\zeta;\lambda)$ of the Cauchy type problem (14) satisfies the additional boundary condition*

$$
\sum_{h=0}^{2s-1} \beta_s^{-\nu(h+1/2)s} D^{-(\nu-\mu)} \mathcal{Y}_{s,\mu}(\zeta;\lambda)\bigg|_{\zeta=\zeta_{\sigma,h}} = 0
\tag{29}
$$

at the endpoints $\zeta_{\sigma,h}$ $(0 \leq h \leq 2s - 1)$ of the sum of segments $\gamma_{2s}(\sigma)$.

Problem I_s^* consists in finding those values of λ^* for which the solution $\mathcal{Z}_{s,\mu^*}(\zeta;\lambda^*)$ of the Cauchy type problem (16) satisfies the additional boundary condition

$$\sum_{h=0}^{2s-1} \beta_s^{-\nu(h+1/2)s} D_{\zeta_{\sigma,h}}^{-(\nu-\mu^*)} \mathcal{Z}_{s,\mu^*}(\zeta;\lambda^*)\Bigg|_{\zeta=0} = 0 \tag{30}$$

at the common endpoint $\zeta = 0$ of the segments of $\gamma_{2s}(\sigma)$.

In this chapter we use the results of Chapters 8 and 9 which assume that the parameter ν varies only in one interval, namely in $\Delta_s = (\mu + 1/2 - 1/s, \mu + 1/2)$. This is the reason why we consider only one pair of boundary value problems. Further, we have already mentioned that, when $s = 1$ and $\mu = \mu^* = 1$, the Cauchy type problems (14) and (16) take simpler forms (see (14$'$) and (16$'$)). If we assume, in addition, $\nu = 1$, then the boundary value problems $I_s = I_1$ and $I_s^* = I_1^*$ also take simpler forms. Indeed, when $\nu = 1$, boundary conditions (29) and (30) become

$$\mathcal{Y}_{1,1}(\zeta;\lambda)|_{\zeta=i\sigma} = \mathcal{Y}_{1,1}(\zeta;\lambda)|_{\zeta=-i\sigma}, \tag{29$'$}$$

$$\mathcal{Z}_{1,1}(\zeta;\lambda^*)|_{\zeta=0\in[0,i\sigma]} = \mathcal{Z}_{1,1}(\zeta;\lambda^*)|_{\zeta=0\in[0,-i\sigma]}, \tag{30$'$}$$

where the functions $\mathcal{Y}_{1,1}(\zeta;\lambda)$ and $\mathcal{Z}_{1,1}(\zeta;\lambda^*)$ are determined by formulas (15$'$) and (17$'$). Finally, we further assume the notions of *eigenvalues* and *eigenfunctions* of the boundary value problems I_s and I_s^* ($s \geq 1$) to be defined in the same way as those of the boundary value problems of Section 11.3.

(e) Denoting the sums (29) and (30) respectively by $R_s(\lambda)$ and $R_s^*(\lambda^*)$, we prove the following lemma.

Lemma 12.3-3. *The following identities are true for any $s \geq 1$:*

$$R_s(\lambda) = 2s\sigma^\nu \lambda^s E_{1/2}\left(-\sigma^2\lambda^{2s}; 1+\nu\right),$$
$$R_s^*(\lambda^*) = 2s\sigma^\nu (\lambda^*)^s E_{1/2}\left(-\sigma^2(\lambda^*)^{2s}; 1+\nu\right). \tag{31}$$

Proof. We shall prove only the first of these identities, since the proof of the second one is similar. As

$$D^{-(\nu-\mu)}\mathcal{Y}_{s,\mu}(\zeta;\lambda)\Bigg|_{\zeta=\zeta_{\sigma,h}} = \mathcal{Y}_{s,\nu}(\zeta;\lambda)|_{\zeta=\zeta_{\sigma,h}}$$
$$= E_s\left(\lambda\zeta_{\sigma,h}^{1/s}; \nu\right)\zeta_{\sigma,h}^{\nu-1} \qquad (0 \leq h < 2s-1), \tag{32}$$

identities 12.2(11) imply

$$R_s(\lambda) = \sum_{h=0}^{2s-1} \beta_s^{-\nu(h+1/2)s} E_s\left(\lambda\zeta_{\sigma,h}^{1/s}; \nu\right)\sigma^{\nu-1}\beta_s^{(\nu-1)(h+1/2)s}$$

$$= \sigma^{\nu-1}\sum_{h=0}^{2s-1} \beta_s^{-(h+1/2)s} E_s\left(\lambda\zeta_{\sigma,h}^{1/s}; \nu\right) = 2s\sigma^\nu \lambda^s E_{1/2}\left(-\sigma^2\lambda^{2s}; 1+\nu\right).$$

The following theorem is an immediate consequence of the preceding lemma and of the formulations of the boundary value problems I_s and I_s^* given above.

Theorem 12.3-2. *If $s \geq 1$ is any integer, μ and μ^* satisfy the conditions*

$$\mu = \mu^* = 1 \quad \text{when } s = 1,$$
$$1/2 < \mu, \mu^* < 1/2 + 1/s \quad \text{when} \quad s \geq 2$$

and $\nu \in [0,2)$, then the sets of eigenvalues of both boundary value problems I_s and I_s^* coincide with the sequence $\{\mu_n\}_0^\infty$ ($\mu_0 = 0$) of zeros of the entire function

$$\lambda^s \mathcal{E}_{s,\sigma}(\lambda;\nu) \equiv \lambda^s E_{1/2}\left(-\sigma^2\lambda^{2s}; 1+\nu\right).$$

Thus, all these eigenvalues are simple with the exception of $\mu_0 = 0$, which is of order s.

Remark. If $\nu = 1$, then the sets of eigenvalues of both boundary value problems I_s and I_s^* coincide with the sequence $\{\mu_n\}_0^\infty$ ($\mu_0 = 0$) of zeros of the function $\sin(\sigma\lambda^s)$. This sequence can be expressed in the following explicit form:

$$\{\mu_n\}_1^\infty = \bigcup_{h=0}^{2s-1} \left\{ \left(\frac{\pi k}{\sigma}\right)^{\frac{1}{s}} e^{i\frac{\pi}{s}h} \right\}_{k=1}^\infty. \tag{33}$$

(f) It is useful to give the explicit representations of eigenfunctions and adjoint functions of both considered boundary value problems. To this end it is first necessary to introduce some general notations.

Let $s \geq 1$ be an integer, $0 < \mu, \mu^* < +\infty$ and also let $\nu \in [0,2)$. As in Section 12.2(c), we denote by $\{\mu_n\}_1^\infty$ the sequence of zeros of the entire function

$$\mathcal{E}_{s,\sigma}(z;\nu) \equiv E_{1/2}\left(-\sigma^2 z^{2s}; 1+\nu\right).$$

Next we introduce in $(0, \zeta_\sigma)$ the functions

$$\mathcal{Y}_{\nu,n}(\zeta) \equiv \left.\frac{\partial^{-n}}{\partial\lambda^{-n}}\mathcal{Y}_{s,\mu}(\zeta;\lambda)\right|_{\lambda=0} = \frac{\Gamma(1-n)}{\Gamma(\mu - n/s)}\zeta^{-n/s+\mu-1} \quad (n \leq 0), \tag{34_1}$$

$$\mathcal{Y}_{\nu,n}(\zeta) \equiv \mathcal{Y}_{s,\mu}(\zeta;\mu_n) = E_s\left(\mu_n\zeta^{1/s};\mu\right)\zeta^{\mu-1} \quad (n \geq 1), \tag{34_2}$$

$$\mathcal{Z}_{\nu,n}(\zeta) \equiv \left.\frac{\partial^{s-1+n}}{\partial\lambda^{s-1+n}}\mathcal{Z}_{s,\mu^*}(\zeta;\lambda)\right|_{\lambda=0}$$
$$= \frac{\Gamma(s+n)}{\Gamma\left(\mu^* + \frac{s-1+n}{s}\right)}(\zeta_\sigma - \zeta)^{\frac{s-1+n}{s}+\mu^*-1} \quad (-(s-1) \leq n \leq 0), \tag{35_1}$$

$$\mathcal{Z}_{\nu,n}(\zeta) \equiv \mathcal{Z}_{s,\mu^*}(\zeta;\mu_n) = E_s\left(\mu_n(\zeta_\sigma - \zeta)^{1/s};\mu^*\right)(\zeta_\sigma - \zeta)^{\mu^*-1} \quad (n \geq 1). \tag{35_2}$$

These formulas obviously define the introduced functions in the whole domain $G_\sigma^\infty = \{\zeta \in G^\infty : |\zeta| < \sigma\}$. Using these functions and Theorem 12.3-2 we easily obtain the following assertion.

Theorem 12.3-3. *If $s \geq 1$ is an integer, μ and μ^* satisfy the conditions*

$$\mu = \mu^* = 1 \quad \text{when} \quad s = 1,$$
$$1/2 < \mu, \mu^* < 1/2 + 1/s \quad \text{when} \quad s \geq 2$$

and $\nu \in [0,2)$, then the sequence of functions

$$\{\mathcal{Y}_{\nu,n}(\zeta)\}_{-(s-1)}^{+\infty} \tag{36}$$

is a system of eigenfunctions (when $n \geq 0$) and adjoint functions (when $s \geq 2$ and $-(s-1) \leq n \leq -1$) of the boundary value problem I_s. On the other hand, the sequence of functions

$$\{\mathcal{Z}_{\nu,n}(\zeta)\}_{-(s-1)}^{+\infty} \tag{37}$$

is a system of eigenfunctions (when $n > 0$ and $n = -(s-1)$) and adjoint functions (when $s \geq 2$ and $-(s-1) < n \leq 0$) of the boundary value problem I_s^*.

Remark. If $s = 1$ (and consequently $\mu = \mu^* = 1$) and $\nu = 1$, then all the eigenvalues of the boundary value problems $I_s = I_1$ and $I_s^* = I_1^*$ are simple, and they coincide with the sequence of zeros $\{\pi k/\sigma\}_{-\infty}^{+\infty}$ of the function $\sin(\sigma\lambda)$. Moreover, in this case the systems of eigenfunctions of the boundary value problems I_1 and I_1^* can be expressed respectively in the forms

$$\left\{\exp\left(\frac{\pi k \zeta}{\sigma}\right)\right\}_{-\infty}^{+\infty} \quad \text{and} \quad \left\{\exp\left(\frac{\pi k(\zeta_\sigma - \zeta)}{\sigma}\right)\right\}_{-\infty}^{+\infty}. \tag{38}$$

12.4 Expansions in $L_2\{\gamma_{2s}(\sigma)\}$ in terms of Riesz bases

It is first necessary to give a summary of some notations and results of Chapter 9.

(a) For any integer $s \geq 1$ and for any $\omega \in (-1,1)$

$$\mu = \frac{s + \omega + 1}{2s} \in (1/2, 1/2 + 1/s). \tag{1}$$

Besides, we assumed

$$\kappa_0 = \frac{1 + \omega - s}{s} = 2(\mu - 1). \tag{2}$$

Further, the condition $\nu \in [0,2)$ was assumed to be satisfied, provided the parameter ν was present, and $\{\mu_n\}_1^\infty$ was assumed to be the sequence of zeros of the entire function

$$\mathcal{E}_{s,\sigma}(z;\nu) \equiv E_{1/2}\left(-\sigma^2 z^{2s}; 1 + \nu\right), \tag{3}$$

and the interval $\Delta_s = (\mu + 1/2 - 1/s, \mu + 1/2) \subset [0, 2)$ of variation of ν was considered.

(b) When $\nu \in [0, 2)$, the sequence of vector functions

$$\{\kappa_n(\tau)\}_{-\infty}^{+\infty} \equiv \left\{ \{\kappa_{n,h}(\tau)\}_{h=0}^{2s-1} \right\}_{-\infty}^{+\infty}, \qquad \tau \in (0, \sigma) \tag{4}$$

was defined by the formulas

$$\kappa_{n,h}(\tau) = \beta_s^{-(h+1/2)n} \frac{\Gamma(1-n)}{\Gamma(\mu - n/s)} \tau^{-n/s+\mu-1} \qquad \begin{array}{l}(0 \le h \le 2s - 1, \\ -\infty < n \le 0),\end{array} \tag{5_1}$$

$$\kappa_{n,h}(\tau) = E_{1/2}\left(\mu_n \tau^{1/s} \beta_s^{h+1/2}; \mu\right) \tau^{\mu-1} \qquad \begin{array}{l}(0 \le h \le 2s - 1, \\ 1 \le n < +\infty).\end{array} \tag{5_2}$$

When $\nu \in \Delta_s$, the parameter

$$\eta_{2,s} = \nu + \frac{1 - s - \omega}{2s} = \nu + \frac{1}{s} - \mu \tag{6}$$

was considered, and by use of it the sequence of vector functions

$$\{\Omega_m(\tau)\}_{-(s-1)}^{+\infty} \equiv \left\{ \{\Omega_{m,h}(\tau)\}_{h=0}^{2s-1} \right\}_{-(s-1)}^{+\infty} \tag{7}$$

was defined on $(0, \sigma)$ in the following way:

$$\Omega_{m,h}(\tau) = \beta_s^{(h+1/2)m} \frac{\sigma^{-\nu}\Gamma(1+\nu)}{2s\Gamma(1-m)} \frac{(\sigma - \tau)^{\frac{s-1+m}{s}+\eta_{2,s}-1}}{\Gamma\left(\frac{s-1+m}{s} + \eta_{2,s}\right)} \tag{8_1}$$

when $0 \le h \le 2s - 1$ and $-(s - 1) \le m \le 0$, and

$$\Omega_{m,h}(\tau) = \frac{\beta_s^{-(h+1/2)(s-1)} \sigma^{-\nu}}{2s\mu_m^s \mathcal{E}'_{s,\sigma}(\mu_m; \nu)} \tag{8_2}$$
$$\times E_s\left(\mu_m(\sigma - \tau)^{1/s}\beta_s^{h+1/2}; \eta_{2,s}\right)(\sigma - \tau)^{\eta_{2,s}-1},$$

when $0 \le h \le 2s - 1$ and $1 \le m < +\infty$. Also, a passage was done from system (7) to the system

$$\{\omega_m(\tau)\}_{-(s-1)}^{+\infty} \equiv \left\{ \{\omega_{m,h}(\tau)\}_{h=0}^{2s-1} \right\}_{-(s-1)}^{+\infty}, \qquad \tau \in (0, \sigma), \tag{9}$$

where

$$\omega_{m,h}(\tau) \equiv \overline{\Omega_{m,h}(\tau)} \qquad (0 \le h \le 2s - 1, -(s-1) \le m < +\infty). \tag{10}$$

(c) We shall also use the following results of Chapter 9.

1°. If $\nu \in \Delta_s$, then the systems of vector functions

$$\{\kappa_n(\tau)\}_{-(s-1)}^{+\infty} \quad \text{and} \quad \{\omega_m(\tau)\}_{-(s-1)}^{+\infty} \tag{11}$$

are biorthogonal in the space $L_2^{2s}(0,\sigma)$ of $2s$-dimensional vector functions, i.e.,

$$\{\kappa_n; \omega_m\} = \sum_{h=0}^{2s-1} \int_0^\sigma \kappa_{n,h}(\tau)\overline{\omega_{m,h}\tau})d\tau = \delta_{n,m} \qquad (-(s-1) \le n, m < +\infty). \tag{12}$$

2°. If $\nu \in \Delta_s$, then each of the biorthogonal systems of vector functions

$$\left\{(1+|n|)^{\mu-1}\kappa_n(\tau)\right\}_{-(s-1)}^{+\infty} \quad \text{and} \quad \left\{(1+|m|)^{1-\mu}\omega_m(\tau)\right\}_{-(s-1)}^{+\infty} \tag{13}$$

is a Riesz basis of $L_2^{2s}(0,\sigma)$.

(d) Now we introduce on the sum of segments $\gamma_{2s}(\sigma)$ (see formulas 12.2(5)-(6))
the Hilbert space $L_2\{\gamma_{2s}(\sigma)\}$ of measurable functions $\Phi(\zeta)$ having finite norms

$$\|\Phi\| \equiv \left\{\int_{\gamma_{2s}(\sigma)} |\Phi(\zeta)|^2 |d\zeta|\right\}^{1/2} = \left\{\sum_{h=0}^{2s-1} \int_0^{\zeta_{\sigma,h}} |\Phi(\zeta)|^2 |d\zeta|\right\}^{1/2}$$
$$= \left\{\sum_{h=0}^{2s-1} \int_0^\sigma \left|\Phi\left(\tau\beta_s^{(h+1/2)s}\right)\right|^2 d\tau\right\}^{1/2}. \tag{14}$$

Observe that the inner product of any two functions $\Phi(\zeta)$ and $\Psi(\zeta)$ of $L_2\{\gamma_{2s}(\sigma)\}$
is naturally the integral

$$[\Phi; \Psi] \equiv \int_{\gamma_{2s}(\sigma)} \Phi(\zeta)\overline{\Psi(\zeta)}|d\zeta| \tag{15}$$

and obviously

$$\|\Phi\| = [\Phi; \Phi]^{1/2}. \tag{16}$$

(e) Assuming that $s \ge 1$ is an integer, $0 < \mu, \mu^* < +\infty$ and $\nu \in [0,2)$, we
introduced, by formulas $(34_1) - (34_2)$ and $(35_1) - (35_2)$ of Section 12.3(f), the
systems of functions

$$\{\mathcal{Y}_{\nu,n}(\zeta)\}_{-\infty}^{+\infty} \quad \text{and} \quad \{\mathcal{Z}_{\nu,n}(\zeta)\}_{-(s-1)}^{+\infty}. \tag{17}$$

It appears that the values of functions of these systems on $\gamma_{2s}(\sigma)$ are closely
connected with the vector functions of the systems

$$\{\kappa_n(\tau)\}_{-\infty}^{+\infty} \quad \text{and} \quad \{\Omega_m(\tau)\}_{-(s-1)}^{+\infty} \tag{18}$$

which were considered above. Indeed, one can easily verify the validity of the following formulas.

(i) Let $1/2 < \mu < 1/2 + 1/s$ and let $\nu \in [0, 2)$. Then for any $\tau \in (0, \sigma)$ and h, n $(0 \le h \le 2s - 1, -\infty < n < +\infty)$

$$\mathcal{Y}_{\nu,n}(\zeta)|_{\zeta \in [0,\zeta_{\sigma,h}]} = \mathcal{Y}_{\nu,n}\left(\tau \beta_s^{(h+1/2)s}\right) = \beta_s^{(\mu-1)(h+1/2)s} \kappa_{n,h}(\tau). \tag{19}$$

(ii) Let $1/2 < \mu < 1/2 + 1/s$, $\nu \in \Delta_s$, and also let μ^* be determined by the relation $\mu + \mu^* = \nu + 1/s$. Then $1/2 < \mu^* < 1/2 + 1/s$ and, in addition $\eta_{2,s} = \mu^*$, as follows from (6). Therefore, for any $\tau \in (0, \sigma)$ and h, m $(0 \le h \le 2s - 1, -(s-1) \le m < +\infty)$

$$\mathcal{Z}_{\nu,m}(\zeta)|_{\zeta \in [0,\zeta_{\sigma,h}]} = \mathcal{Z}_{\nu,m}\left(\tau \beta_s^{(h+1/2)s}\right)$$
$$= \beta_s^{(\mu^*-1)(h+1/2)s} \beta_s^{(h+1/2)(s-1)} C_{s,\nu,m} \Omega_{m,h}(\tau), \tag{20}$$

where

$$C_{s,\nu,m} = \begin{cases} 2s\sigma^\nu \Gamma(s+m)\Gamma(1-m)\Gamma^{-1}(1+\nu) & \text{when } -(s-1) \le m \le 0, \\ 2s\sigma^\nu \mu_m^s \mathcal{E}'_{s,\sigma}(\mu_m; \nu) & \text{when } 1 \le m < +\infty. \end{cases} \tag{21}$$

We introduce a new system of functions

$$\left\{\tilde{\mathcal{Z}}_{\nu,m}(\zeta)\right\}_{-(s-1)}^{+\infty} \tag{22}$$

setting

$$\tilde{\mathcal{Z}}_{\nu,m}(\zeta) = \left(\frac{\zeta}{|\zeta|}\right)^{1-\nu} \overline{\frac{\mathcal{Z}_{\nu,m}(\zeta)}{C_{s,\nu,m}}} \quad (-(s-1) \le m < +\infty). \tag{23}$$

Then obviously

$$\tilde{\mathcal{Z}}_{\nu,m}(\zeta) = \beta_s^{(1-\nu)(h+1/2)s} \overline{\frac{\mathcal{Z}_{\nu,m}(\zeta)}{C_{s,\nu,m}}} \quad (-(s-1) \le m < +\infty) \tag{24}$$

for any $\zeta \in [0, \zeta_{\sigma,h}]$ $(0 \le h \le 2s - 1)$. Besides, the following lemma is true for the introduced system.

Lemma 12.4-1. *If $1/2 < \mu < 1/2 + 1/s$, $\nu \in \Delta_s$ and $\mu^* = \nu + 1/s - \mu$, then the systems of functions*

$$\{\mathcal{Y}_{\nu,n}(\zeta)\}_{-(s-1)}^{+\infty} \quad \text{and} \quad \left\{\tilde{\mathcal{Z}}_{\nu,m}(\zeta)\right\}_{-(s-1)}^{+\infty} \tag{25}$$

are biorthogonal in $L_2\{\gamma_{2s}(\sigma)\}$, i.e.,

$$[\mathcal{Y}_{\nu,n}; \tilde{\mathcal{Z}}_{\nu,m}] = \delta_{n,m} \quad (-(s-1) \le n, m < +\infty). \tag{26}$$

Proof. Indeed, if $-(s-1) \le n, m < +\infty$, then we have

$$[\mathcal{Y}_{\nu,n}; \tilde{\mathcal{Z}}_{\nu,m}] = \sum_{h=0}^{2s-1} \int_0^{\zeta_{\sigma,h}} \mathcal{Y}_{\nu,n}(\zeta)\overline{\tilde{\mathcal{Z}}_{\nu,m}(\zeta)}|d\zeta|$$

$$= \frac{1}{C_{s,\nu,m}} \sum_{h=0}^{2s-1} \beta_s^{(1-\nu)(h+1/2)s} \int_0^{\zeta_{\sigma,h}} \mathcal{Y}_{\nu,n}(\zeta)\mathcal{Z}_{\nu,m}(\zeta)|d\zeta|. \tag{27}$$

If we use formulas $(19), (20)$ and the simple relation

$$(1-\nu)\left(h+\frac{1}{2}\right)s+(\mu-1)\left(h+\frac{1}{2}\right)s+(\mu^*-1)\left(h+\frac{1}{2}\right)s+\left(h+\frac{1}{2}\right)(s-1)=0, \tag{28}$$

then from (27) it will follow that

$$[\mathcal{Y}_{\nu,n}; \tilde{\mathcal{Z}}_{\nu,m}] = \sum_{h=0}^{2s-1} \int_0^{\sigma} \kappa_{n,h}(\tau)\Omega_{m,h}(\tau)d\tau$$

$$= \sum_{h=0}^{2s-1} \int_0^{\sigma} \kappa_{n,h}(\tau)\overline{\omega_{m,h}(\tau)}d\tau = \{\kappa_n; \omega_m\} = \delta_{n,m}.$$

(f) Now we are ready to prove one of the main theorems of this section.

Theorem 12.4-1. *Let $1/2 < \mu < 1/2+1/s$, $\nu \in \Delta_s$ and let $\mu^* = \nu + 1/s - \mu$. Then any function $\Phi(\zeta)$ of the space $L_2\{\gamma_{2s}(\sigma)\}$ can be expanded in the series*

$$\Phi(\zeta) = \sum_{n=-(s-1)}^{+\infty} [\Phi; \tilde{\mathcal{Z}}_{\nu,n}]\mathcal{Y}_{\nu,n}(\zeta), \qquad \zeta \in \gamma_{2s}(\sigma) \tag{29}$$

which is convergent in the norm of this space. Besides,

$$\|\Phi\|^2 \asymp \sum_{n=-(s-1)}^{+\infty} \left|[\Phi; \tilde{\mathcal{Z}}_{\nu,n}]\right|^2 (1+|n|)^{2(1-\mu)}. \tag{30}$$

Proof. Let $\Phi(\zeta) \in L_2\{\gamma_{2s}(\sigma)\}$ be an arbitrary function. We associate with it the vector function

$$\varphi(\tau) = \{\varphi_h(\tau)\}_{h=0}^{2s-1} \in L_2^{2s}(0,\sigma), \tag{31}$$

where

$$\varphi_h(\tau) \equiv \beta_s^{(1-\mu)(h+1/2)s}\Phi_h(\tau) \tag{32_1}$$

and

$$\Phi_h(\tau) \equiv \Phi(\zeta)|_{\zeta\in[0,\zeta_{\sigma,h}]} = \Phi\left(\tau\beta_s^{(h+1/2)s}\right). \tag{32_2}$$

Obviously

$$\|\Phi\| = \|\varphi\|, \tag{33}$$

i.e., the norm of Φ in $L_2\{\gamma_{2s}(\sigma)\}$ is equal to the norm of φ in $L_2^{2s}(0,\sigma)$. In addition, the equalities

$$[\Phi; \tilde{\mathcal{Z}}_{\nu,m}] = \{\varphi; \omega_m\} \qquad (-(s-1) \le m < +\infty) \tag{34}$$

follow from formulas $(24),(20)$ and from relations (28), $(32_1) - (32_2)$. Further, the systems of vector functions (13) are biorthogonal Riesz bases of $L_2^{2s}(0,\sigma)$. Therefore, the expansions

$$\begin{aligned}
\varphi_h(\tau) &= \sum_{n=-(s-1)}^{+\infty} \{\varphi; (1+|n|)^{1-\mu}\omega_n\} (1+|n|)^{\mu-1}\kappa_{n,h}(\tau) \\
&= \sum_{n=-(s-1)}^{+\infty} \{\varphi; \omega_n\}\kappa_{n,h}(\tau), \qquad \tau \in (0,\sigma), 0 \le h \le 2s-1
\end{aligned} \tag{35}$$

and the two-sided estimate

$$\|\varphi\|^2 \asymp \sum_{n=-(s-1)}^{+\infty} |\{\varphi; \omega_n\}|^2 (1+|n|)^{2(1-\mu)} \tag{36}$$

are true. Using relations $(19), (32_1) - (32_2)$ and (34) we can write (35) in the form of the following expansions in spaces $L_2[0, \zeta_{\sigma,h}]$ $(0 \le h \le 2s - 1)$:

$$\Phi(\zeta)|_{\zeta \in [0,\zeta_{\sigma,h}]} = \sum_{n=-(s-1)}^{+\infty} [\Phi; \tilde{\mathcal{Z}}_{\nu,n}]\mathcal{Y}_{\nu,n}(\zeta)\Bigg|_{\zeta \in [0,\zeta_{\sigma,h}]}. \tag{37}$$

Finally, it remains to observe that these expansions imply (29). As to estimate (30), it follows from (33), (34) and (36).

Remark. If stated in another way, the preceding theorem says that if the parameters μ, μ^* and ν satisfy the suitable conditions, then the systems of functions

$$\left\{(1+|n|)^{\mu-1}\mathcal{Y}_{\nu,n}(\zeta)\right\}_{-(s-1)}^{+\infty} \quad \text{and} \quad \left\{(1+|m|)^{1-\mu}\tilde{\mathcal{Z}}_{\nu,m}(\zeta)\right\}_{-(s-1)}^{+\infty} \tag{38}$$

are biorthogonal Riesz bases of the space $L_2\{\gamma_{2s}(\sigma)\}$.

(g) The following theorem, relating to expansions in terms of eigenfunctions and adjoint functions of the boundary value problem I_s, is a consequence of Theorem 12.4-1.

Theorem 12.4-2. *Let $s \geq 1$ be an integer and let μ satisfy the conditions*

$$\mu = 1 \quad \text{when} \quad s = 1,$$
$$1/2 < \mu < 1/2 + 1/s \quad \text{when} \quad s \geq 2.$$

Further, let $\nu \in \Delta_s$ and let $\mu^* = \nu + 1/s - \mu$. Then any function $\Phi(\zeta)$ of $L_2\{\gamma_{2s}(\sigma)\}$ can be expanded in terms of eigenfunctions and adjoint functions $\{\mathcal{Y}_{\nu,n}(\zeta)\}^{+\infty}_{-(s-1)}$ of the boundary value problem $\mathbf{I}_s$:

$$\Phi(\zeta) = \sum_{n=-(s-1)}^{+\infty} [\Phi; \tilde{\mathcal{Z}}_{\nu,n}] \mathcal{Y}_{\nu,n}(\zeta), \qquad \zeta \in \gamma_{2s}(\sigma). \tag{39}$$

In addition,

$$\|\Phi\|^2 \asymp \sum_{n=-(s-1)}^{+\infty} |[\Phi; \tilde{\mathcal{Z}}_{\nu,n}]|^2 (1 + |n|)^{2(1-\mu)}. \tag{40}$$

Remark. If we take into account the remark to Theorem 12.3-3, then in the case when $s = 1$ and $\nu = 1$ (and consequently $\mu = 1$) we can express one of the assertions of the preceding theorem in the following way: any function $\Phi(\zeta)$ of $L_2\{\gamma_2(\sigma)\}$ (where $\gamma_2(\sigma) = [0, i\sigma] \cup [0, -i\sigma] = [-i\sigma, i\sigma]$) can be expanded in terms of functions $\{\exp(\pi k\zeta/\sigma)\}^{+\infty}_{-\infty}$. Obviously this assertion is equivalent to the well-known theorem of classical analysis relating to the expansions of functions of $L_2(-\sigma, \sigma)$ in terms of the Fourier system

$$\left\{ e^{i\pi kx/\sigma} \right\}^{+\infty}_{-\infty}. \tag{41}$$

(h) As was mentioned in Section 12.2, $w = \zeta^{1/s}$ maps the sum of segments $\gamma_{2s}(\sigma) \subset G^\infty$ onto the sum of segments $\Gamma_{2s}(\sigma)$ situated in the w-plane. The segments of $\Gamma_{2s}(\sigma)$ are all of length $\sigma^{1/s}$ and have a common endpoint at $w = 0$. In addition, these segments are at angles $\pi(h + 1/2)/s$ $(0 \leq h \leq 2s - 1)$, so they form pairwise equal angles of opening π/s. Further, we assume that $\omega \in (-1, 1)$ and $\mu = (1 + s + \omega)/2s$, as always. We denote by $L_{2,\omega}\{\Gamma_{2s}(\sigma)\}$ the space of functions $\Psi(w)$ measurable on $\Gamma_{2s}(\sigma)$ and having finite norms

$$\|\Psi\|_\omega \equiv \left\{ \int_{\Gamma_{2s}(\sigma)} |\Psi(w)|^2 |w|^\omega |dw| \right\}^{1/2}. \tag{42}$$

Then evidently

$$[\Psi_1; \Psi_2]_\omega \equiv \int_{\Gamma_{2s}(\sigma)} \Psi_1(w) \overline{\Psi_2(w)} |w|^\omega |dw| \tag{43}$$

is the inner product of arbitrary two functions $\Psi_1, \Psi_2 \in L_{2,\omega}\{\Gamma_{2s}(\sigma)\}$. Now observe that

$$\Psi(w) \equiv \sqrt{s}\,\Phi\left(w^s\right) w^{s(1-\mu)} \tag{44}$$

is a transformation of a function $\Phi(\zeta)$ defined in $\gamma_{2s}(\sigma)$ into a function $\Psi(w)$ defined in $\Gamma_{2s}(\sigma)$. A simple calculation shows that (44) represents an isometric isomorphism between the Hilbert spaces $L_2\{\gamma_{2s}(\sigma)\}$ and $L_{2,\omega}\{\Gamma_{2s}(\sigma)\}$. Therefore, using this transformation we can easily move from Theorem 12.4-1 to the following theorem on expansions of functions of $L_{2,\omega}\{\Gamma_{2s}(\sigma)\}$ in terms of definite systems of entire functions.

Theorem 12.4-3. *Let $s \geq 1$ be an integer, $-1 < \omega < 1$ and let the parameter $\mu \in (1/2, 1/2 + 1/s)$ be defined by the equality*

$$\mu = \frac{1 + s + \omega}{2s}.$$

Further, let $\nu \in \Delta_s$ and $\mu^* = \nu + 1/s - \mu$. Then any function $\Psi(w)$ of $L_{2,\omega}\{\Gamma_{2s}(\sigma)\}$ can be expanded in the series

$$\Psi(w) = \sum_{n=-(s-1)}^{0} C_n \frac{\Gamma(1-n)}{\Gamma(\mu - n/s)} w^{-n} + \sum_{n=1}^{\infty} C_n E_s(\mu_n w; \mu), \quad w \in \Gamma_{2s}(\sigma) \tag{45}$$

which converges in the norm of $L_{2,\omega}\{\Gamma_{2s}(\sigma)\}$, and

$$\|\Psi\|_\omega^2 \asymp \sum_{n=-(s-1)}^{+\infty} |C_n|^2 (1 + |n|)^{2(1-\mu)}. \tag{46}$$

In addition, the following formulas are true for the coefficients of expansion (45):

$$C_n = \left[\Psi(w); s\tilde{\mathcal{Z}}_{\nu,n}\left(w^s\right) w^{s(1-\mu)}\right]_\omega, \qquad -(s-1) \leq n < +\infty.$$

12.5 Notes

The results of this chapter were established in somewhat different forms in the papers of S.G. Raphaelian [4, §3] and M.M. Djrbashian [7,8] considering the simplest cases $s = 1$ and $s = 2$.

Bibliography

AGARWAL, R.P.
 [1] A propos d'une note de M.Pierre Humbert. C.R. Acad. Sci. Paris **236**, N 21, 2031–2032 (1953).

AKOPIAN, S.A.
 [1] Two constants theorem for functions in class H_p (Russian). Izv. Akad. Nauk Armjanskoy SSR, Matematika **2**, N 2, 123–127 (1967).

BATEMAN, H.
 [1] Higher transcendental functions. In: Bateman Manuscript Project, vol. **3**. New York-Toronto-London 1955, 292 p.p.

BHUL, A.
 [1] Séries analytiques. Sommabilité. In Mem. Sci. Math., Fasc. 7. Gauthier-Villars, Paris 1925, 53 p.p.

CLEOTA, G.F.–H.K.HUGENS
 [1] Asymptotic developments of certain integral functions. Duke Math. J. **9**, 791–802 (1942).

DJRBASHIAN, M.M.
 [1] On the integral representation and uniqueness of some classes of entire functions (Russian). Dokl. Akad. Nauk SSSR **85**, N 1, 29–32 (1952); Mat. Sb. **33(75)**, N 3, 485–530 (1953).
 [2] On the integral representation of functions continuous on some rays (generalization of Fourier integral) (Russian). Izv. Akad. Nauk SSSR, Ser. Mat. **18**, 427–448 (1954).
 [3] On asymptotic behavior of Mittag-Leffler type function (Russian). Dokl. Akad. Nauk Armjanskoy SSR **19**, N 3, 65–72 (1954).
 [4] On a new integral transform and its application in the theory of entire functions (Russian). Dokl. Akad. Nauk SSSR **95**, N 6, 1133–1136 (1954); Izv. Akad. Nauk SSSR, Ser. Mat. **19**, 133–180 (1955).
 [5] Integral transforms and representations of functions in the complex domain (Russian). Nauka, Moscow 1966, 672 p.p.
 [6] Boundary value problem for a Sturm-Lieouville type differential operator of fractional order (Russian). Izv. Akad. Nauk Armjanskoy SSR, Matematika **5**, N 2, 71–96 (1970).
 [7] Interpolation and spectral expansions associated with differential operators of fractional order (Russian). Izv. Akad. Nauk Armjanskoy SSR, Matematika **19**, N 2, 81–181 (1984); English transl. in Sov. J. of Contemp. Math. Anal. **19**, N 2, 1–116 (1984).
 [8] Interpolation and spectral expansions associated with differential operators of fractional order. Colloquia Mathematica Societatis János Bolyai 49, Alfred Haar Memorial Conference, Budapest, 1985, 303–322.
 [9] On a boundary value problem in the complex domain (Russian). Izv. Akad. Nauk Armjanskoy SSR, Matematika **22**, N 6, 523–542 (1987); English transl. in Sov. J. of Contemp. Math. Anal. **22**, N 6 (1987).
 [10] On the basisness of systems of eigenfunctions generated by one boundary value problem in the complex domain (Russian). Izv. Akad. Nauk Armjanskoy SSR, Matematika **23**, N 5, 409–421 (1988); English transl. in Sov. J. of Contemp. Math. Anal. **23**, N 5, 1–14 (1988).
 [11] Differential operators of fractional order and boundary value problems in the complex domain. Operator Theory: Advances and Applications **41**, 153–172 (1989).

DJRBASHIAN, M.M.–A.E.AVETISIAN
 [1] Integral representations of some classes of functions analytic in corner domain (Russian). Dokl. Akad. Nauk SSSR **120**, N 3, 457–460 (1958); Sibirsk. Mat. Ž. **1**, N 3, 383–426 (1960).

DJRBASHIAN, M.M.–R.A.BAGIAN
 [1] On integral representations and measures associated with Mittag-Leffler type functions (Russian). Izv. Akad. Nauk Armjanskoy SSR, Matematika **10**, N 6, 483–508 (1975); Dokl. Akad. Nauk SSSR **223**, N 6, 1297–1300 (1975).

DJRBASHIAN, M.M.–A.B.NERSESIAN

[1] On the construction of some special biorthogonal systems (Russian). Izv. Akad. Nauk Armjanskoy SSR, Ser. Phys.-Mat. Nauk **12**, N 5, 17–42 (1959).

[2] Expansions with respect to special biorthogonal systems and boundary value problems for differential equations of fractional order (Russian). Dokl. Akad. Nauk SSSR **132**, N 4, 747–750.

[3] Expansions with respect to some biorthogonal systems and boundary value problems for differential equations of fractional order (Russian). Trudi Moskov. Mat. Obshch. **10**, 89–179 (1961).

[4] Fractional derivatives and the Cauchy type problem for differential equations of fractional order (Russian). Izv. Akad. Nauk Armjanskoy SSR, Matematika **3**, N 1, 3–29 (1968).

DJRBASHIAN, M.M.–S.G.RAPHAELIAN

[1] On entire functions of exponential type in weighted classes L_p (Russian). Dokl. Akad. Nauk Armjanskoy SSR **73**, N 1, 29–36 (1981).

[2] Interpolation theorems and expansions with respect to Fourier type systems (Russian). Dokl. Akad. Nauk SSSR **285**, N 4, 782–787 (1985).

[3] Interpolation theorems and expansions with respect to Fourier type systems. J. Approx. Theory **50**, N 4, 297–325 (1987).

[4] Interpolation expansions in classes of entire functions and Riesz bases generated by them (Russian). Izv. Akad. Nauk Armjanskoy SSR, Matematika **22**, N 1, 23–63 (1987); English transl. in Sov. J. Contemp. Math. Anal. **22**, N 1, 20–61 (1987).

[5] Special boundary value problems generating Fourier type systems (Russian). Izv. Akad. Nauk Armjanskoy SSR, Matematika **23**, N 2, 109–122 (1988); English transl. in Sov. J. Contemp. Math. Anal. **23**, N 2, 1–16 (1988).

DJRBASHIAN, M.M.–B.A.SAAKIAN

[1] Classes of formulas and Taylor-Maclaurin type expansions associated with differential operators of fractional order (Russian). Izv. Akad. Nauk Armjanskoy SSR, Ser. Mat. **39**, N 1, 69–122 (1975).

GARNETT, J.B.

[1] Bounded analytic functions. Academic Press, New York 1981, 467 p.p.

GOHBERG, I.–M.G.KREIN

[1] Introduction to the theory of linear nonselfadjoint operators in a Hilbert space (Russian). Nauka, Moscow 1965, 448 p.p.; English transl., Amer. Math. Soc., Providence 1969, 378 p.p.

GRIGORIAN, Sh.A.

[1] On a property of functions in H^p ($0 < p < +\infty$) over a half-plane (Russian). Izv. Akad. Nauk Armjanskoy SSR, Matematika **12**, N 5, 335–340 (1977).

HARDY, G.H.

[1] Divergent series. Clarendon Press, Oxford 1949, 396 p.p.

HILLE, E.–J.TAMARKIN

[1] On the theory of linear integral equations. Annals of Math. (2nd ser.) **31**, N 3, 479–528 (1930).

HOFFMAN, K.

[1] Banach spaces of analytic functions. Prentice-Hall, Englewood Cliffs, N.J. 1962.

HUMBERT, P.

[1] Quelques resultats relatifs a la function de Mittag-Leffler. C.R. Acad. Sci. Paris **236**, 1467–1468 (1953).

HUMBERT, P.–R.P.AGARWAL

[1] Sur la function de Mittag-Leffler et quelques — unes de ses généralisations. Bull. Sci. Math. (2) **77**, 180–185 (1953).

KACZMARZ, S.–H.STEINHAUS
[1]   Theory of orthogonal series (Russian). Fizmatgiz, Moscow 1958, 507 p.p.

LEVIN, B.Ya.
[1]   Distribution of zeros of entire functions (Russian). Gostekhizdat, Moscow 1956, 632 p.p.;
English transl. in Transl. Math. Monographs, v. 5, Amer. Math. Soc., Providence, Rhode
Island 1980.

MARTIROSIAN, V.M.
[1]   Closure and basisness of some biorthogonal systems and solution of the multiple interpolation problem in angular domains (Russian). Izv. Akad. Nauk Armjanskoy SSR, Matematika
13, N 5–6, 490–531 (1978).

MITTAG-LEFFLER, G.M.
[1]   Sur la nouvelle fonction.... C.R. Acad. Sci. Paris (2) **137**, 554–558 (1903).
[2]   Sopra la functione $E_\alpha(x)$. R. Acad. dei Lincei, Rendiconti (5) **13**, 3–5 (1904).
[3]   Sur la représentation analytique d'une branche uniforme d'une function monogène (cin-
quième note). Acta Math. **29**, 101–182 (1905).

NATANSON, I.P.
[1]   Theory of functions of real variable (Russian), Gostekhizdat, Moscow 1957, 552 p.p.; Ger-
man transl., Academie-Verlag, 1961, 590 p.p.

PALEY, R.E.A.C.–N.WIENER
[1]   Fourier transforms in the complex domain. Amer. Math. Soc. Colloq. Publ. **19**, Amer.
Math. Soc., New York 1934, 184 p.p.

POLLARD, H.
[1]   The representation of e^{-z^λ} as a Laplace integral. Bull. Amer. Math. Soc. **52**, 908–916
(1946).
[2]   The completely monotonic character of the Mittag-Leffler function $E_\alpha(-x)$. Bull. Amer.
Math. Soc. **54**, 1115–1116 (1948).

POLYA, G.
[1]   Über die Nullstellen gewisser ganzer Funktionen. Math. Zeitsch. **2**, 352–383 (1918).

RAPHAELIAN, S.G.
[1]   On a class of entire functions of finite growth (Russian). Dokl. Akad. Nauk Armjanskoy
SSR **70**, 65–71, (1980).
[2]   On basisness of some systems of entire functions (Russian). Dokl. Akad. Nauk Armjanskoy
SSR **70**, 198–204 (1980).
[3]   Interpolation and basisness in weighted classes of entire functions of exponential type
(Russian). Izv. Akad. Nauk Armjanskoy SSR, Matematika **18**, N 3, 167–186 (1983); English
transl. in Sov. J. Contemp. Math. Anal. **18**, N 3, 1–21 (1983).
[4]   Basisness of some biorthogonal systems in $L_2(-\sigma,\sigma)$ with weights (Russian). Izv. Akad.
Nauk Armjanskoy SSR, Matematika **19**, N 3, 207–218 (1984); English transl. in Sov. J.
Contemp. Math. Anal. **19**, N 3, 21–32 (1984).

SAMKO, S.G.–A.A.KILBAS–O.I.MARICHEV
[1]   Integrals and derivatives of fractional order and some of their applications (Russian). Nauka
i Tekhnika, Minsk 1987, 688 p.p.

SEDLETSKI, A.M.
[1]   Equivalent definition of spaces H^p over a half-plane and some applications (Russian). Mat.
Sb. **96**(138), N 1, 75–82 (1975).

SMIRNOV, V.I.
[1]   Course of higher mathematics (Russian), v. 4, Gostekhizdat, Moscow 1957, 812 p.p.

STEIN, E.M.
[1]   Note on singular integrals. Proc. Amer. Math. Soc. **8**, 250–254 (1957).

TITCHMARSH, E.C.
[1]   Introduction to the theory of Fourier integrals. Univ. Press, Oxford, (1937,1948), 395 p.p.

WIMAN, A.
[1]   Über den Fundamentalsatz in der Theorie der Functionen $E_\alpha(x)$. Acta Math. **29**, 191–201 (1905).
[2]   Über die Nullstellen der Functionen $E_\alpha(x)$. Acta Math. **29**, 217–234 (1905).

WINTER, A.
[1]   Stratifications of Cauchy's "stable" transcendents and of Mittag-Leffler entire functions. Amer. J. Math. **80**, N 1, 111–124 (1958).
[2]   On Heaviside and Mittag-Leffler's generalizations of the exponential function, the symmetric stable distributions of Cauchy-Lévy, and a property of the Γ-function. J. de Math. Pures et Appliquées **38**, 165–182 (1959).

Titles previously published in the series

OPERATOR THEORY: ADVANCES AND APPLICATIONS
BIRKHÄUSER VERLAG

27. **A. Bultheel:** Laurent Series and their Pade Approximation, 1987, (3-7643-1940-2)
28. **H. Helson, C.M. Pearcy, F.-H. Vasilescu, D. Voiculescu, Gr. Arsene** (Eds.): Special Classes of Linear Operators and Other Topics, 1988, (3-7643-1970-4)
29. **I. Gohberg** (Ed.): Topics in Operator Theory and Interpolation, 1988, (3-7634-1960-7)
30. **Yu.I. Lyubich:** Introduction to the Theory of Banach Representations of Groups, 1988, (3-7643-2207-1)
31. **E.M. Polishchuk:** Continual Means and Boundary Value Problems in Function Spaces, 1988, (3-7643-2217-9)
32. **I. Gohberg** (Ed.): Topics in Operator Theory. Constantin Apostol Memorial Issue, 1988, (3-7643-2232-2)
33. **I. Gohberg** (Ed.): Topics in Interplation Theory of Rational Matrix-Valued Functions, 1988, (3-7643-2233-0)
34. **I. Gohberg** (Ed.): Orthogonal Matrix-Valued Polynomials and Applications, 1988, (3-7643-2242-X)
35. **I. Gohberg, J.W. Helton, L. Rodman** (Eds.): Contributions to Operator Theory and its Applications, 1988, (3-7643-2221-7)
36. **G.R. Belitskii, Yu.I. Lyubich:** Matrix Norms and their Applications, 1988, (3-7643-2220-9)
37. **K. Schmüdgen:** Unbounded Operator Algebras and Representation Theory, 1990, (3-7643-2321-3)
38. **L. Rodman:** An Introduction to Operator Polynomials, 1989, (3-7643-2324-8)
39. **M. Martin, M. Putinar:** Lectures on Hyponormal Operators, 1989, (3-7643-2329-9)
40. **H. Dym, S. Goldberg, P. Lancaster, M.A. Kaashoek** (Eds.): The Gohberg Anniversary Collection, Volume I, 1989, (3-7643-2307-8)
41. **H. Dym, S. Goldberg, P. Lancaster, M.A. Kaashoek** (Eds.): The Gohberg Anniversary Collection, Volume II, 1989, (3-7643-2308-6)
42. **N.K. Nikolskii** (Ed.): Toeplitz Operators and Spectral Function Theory, 1989, (3-7643-2344-2)
43. **H. Helson, B. Sz.-Nagy, F.-H. Vasilescu, Gr. Arsene** (Eds.): Linear Operators in Function Spaces, 1990, (3-7643-2343-4)
44. **C. Foias, A. Frazho:** The Commutant Lifting Approach to Interpolation Problems, 1990, (3-7643-2461-9)

45. **J.A. Ball, I. Gohberg, L. Rodman:** Interpolation of Rational Matrix Functions, 1990, (3-7643-2476-7)

46. **P. Exner, H. Neidhardt** (Eds.): Order, Disorder and Chaos in Quantum Systems, 1990, (3-7643-2492-9)

47. **I. Gohberg** (Ed.): Extension and Interpolation of Linear Operators and Matrix Functions, 1990, (3-7643-2530-5)

48. **L. de Branges, I. Gohberg, J. Rovnyak** (Eds.): Topics in Operator Theory. Ernst D. Hellinger Memorial Volume, 1990, (3-7643-2532-1)

49. **I. Gohberg, S. Goldberg, M.A. Kaashoek:** Classes of Linear Operators, Volume I, 1990, (3-7643-2531-3)

50. **H. Bart, I. Gohberg, M.A. Kaashoek** (Eds.): Topics in Matrix and Operator Theory, 1991, (3-7643-2570-4)

51. **W. Greenberg, J. Polewczak** (Eds.): Modern Mathematical Methods in Transport Theory, 1991, (3-7643-2571-2)

52. **S. Prössdorf, B. Silbermann:** Numerical Analysis for Integral and Related Operator Equations, 1991, (3-7643-2620-4)

53. **I. Gohberg, N. Krupnik:** One-Dimensional Linear Singular Integral Equations, Volume I, Introduction, 1992, (3-7643-2584-4)

54. **I. Gohberg, N. Krupnik:** One-Dimensional Linear Singular Integral Equations, Volume II, General Theory and Applications, 1992, (3-7643-2796-0)

55. **R.R. Akhmerov, M.I. Kamenskii, A.S. Potapov, A.E. Rodkina, B.N. Sadovskii:** Measures of Noncompactness and Condensing Operators, 1992, (3-7643-2716-2)

56. **I. Gohberg** (Ed.): Time-Variant Systems and Interpolation, 1992, (3-7643-2738-3)

57. **M. Demuth, B. Gramsch, B.W. Schulze** (Eds.): Operator Calculus and Spectral Theory, 1992, (3-7643-2792-8)

58. **I. Gohberg** (Ed.): Continuous and Discrete Fourier Transforms, Extension Problems and Wiener-Hopf Equations, 1992, (ISBN 3-7643-2809-6)

59. **T. Ando, I. Gohberg** (Eds.): Operator Theory and Complex Analysis, 1992, (3-7643-2824-X)

60. **P.A. Kuchment:** Floquet Theory for Partial Differential Equations, 1993, (3-7643-2901-7)

61. **A. Gheondea, D. Timotin, F.-H. Vasilescu** (Eds.): Operator Extensions, Interpolation of Functions and Related Topics, 1993, (3-7643-2902-5)

62. **T. Furuta, I. Gohberg, T. Nakazi** (Eds.): Contributions to Operator Theory and its Applications. The Tsuyoshi Ando Anniversary Volume, 1993, (3-7643-2928-9)

63. **I. Gohberg, S. Goldberg, M.A. Kaashoek:** Classes of Linear Operators, Volume 2, 1993, (3-7643-2944-0)

64. **I. Gohberg** (Ed.): New Aspects in Interpolation and Completion Theories, 1993, (3-7643-2948-3)

65. **M.M. Djrbashian:** Harmonic Analysis and Boundary Value Problems in the Complex Domain, 1993, (3-7643-2855-X)